AF615053

CRITICAL MASS

Publishers' Note

When we, Aurora Publishers, contacted the author, Jacque Srouji, to research and write a book on the positive and negative aspects of nuclear energy, we were totally unprepared for the furor that arose as a consequence of her investigations. Although we, as surely all people, have certain biases, we were committed to producing a work as objective as humanly possible. Therefore Mrs. Srouji was given a free hand and no editorial restrictions were placed on the final manuscript.

One aspect that struck us as rather odd was the adamant opposition from certain segments of our society who object to this or to any work—no matter how objective it may be—that is not anti-nuclear. While it is extremely difficult to pin-point and prove motivation, we believe that they have used and will use any means to discredit this work and the author.

After some lengthy negotiations with investigators of Congressman Dingell's subcommittee on Energy and Environment, Mrs. Srouji was persuaded to appear and testify regarding the somewhat mysterious circumstances surrounding the death of one Karen Silkwood, a technician at the Kerr-McGee nuclear plant in Crescent, Oklahoma (details and supportive documents are all in Chapter 13). Immediately following Mrs. Srouji's appearance, a storm broke and a tidal wave of half-truths and recriminations were hurled at her. She was simultaneously accused of being an FBI agent, FBI informant, KGB agent, Russian spy, counterspy, apologist for the nuclear power agencies and lastly, but not least, the instrument by which some devious element was attempting to sidetrack the Dingell Committee from its avowed purpose of seeking out truth and light.

Immediately after appearing before the Committee she was fired from her position as a copy editor for a local newspaper, the Nashville *Tennessean,* suffered a miscarriage, was sued by a Nashville bookie for alleged violation of his Civil Rights, was sued by a Nashville welfare couple for alleged violation of their Civil Rights, and lastly sued by the heirs of Karen Silkwood for alleged violation of Ms. Silkwood's rights. The latter suit seems particularly bizarre since Karen Silkwood's death preceded Mrs. Srouji's investigation by one full year. And if this is not enough she is presently the subject of an FBI investigation for the alleged possession of FBI documents.

What has all this to do with the efficacy of nuclear energy and whether or not it is the best and safest source of our future power needs? We frankly cannot answer, but we have addressed ourselves, we believe, to this vital question in detail and trust the reader finds sufficient evidence, information and, in some instances, an opinion, to make an intelligent analysis of our present and future energy system.

—Dominic de Lorenzo

CRITICAL MASS

Nuclear Power, the Alternative to Energy Famine

by Jacque Srouji

Aurora Publishers Incorporated
Nashville / London

Library of Congress Cataloging in Publication Data

Srouji, Jacque, 1944-
Critical mass.

1. Energy policy — United States. 2. Power resources — United States. 3. Atomic power-plants — United States. I. Title.
HD9502.U52S7 333.7 76-55841

ISBN 87695-188-4

AURORA PUBLISHERS, INCORPORATED
NASHVILLE, TENNESSEE 37203
LIBRARY OF CONGRESS CATALOG NUMBER 76-55841
STANDARD BOOK NUMBER: 87695-188-4
MANUFACTURED IN THE UNITED STATES
Designed by Nina Chin-Fung Lin

To my husband, *Suheil*
and our children,
John, Joseph, and
Yousra Claire

Contents

Acknowledgments 11

Introduction 14

1 Odyssey 20

2 Odyssey Unveiled 27

3 Critic Psychology 41

4 Frank Zarb: Power Man 61

5 Alternative Energy Sources 71

6 Why a Reactor Won't Go "Poof" in the Night 91

7 Backup Systems to Back Up the Backup Systems 97

8 The Most Toxic Substance Known to Man? 127

9 Garbage-Dumping on a Grand Scale 147

10 Beach Balls by the Sea 167

11 ABC's of LMFB's 203

12 Getting It All Together: Fusion 221

13 "Silkwood, Karen Gay: Former Kerr-McGee Employee" 261

14 Kernenergie 361

15 My Friend, the Russian 379

Epilogue 397

Acknowledgments

It is perhaps the best part of a book to say "thank you" to those who have provided both personal and intellectual assistance on such a project. Specifically, I would like to acknowledge technical advice from a terrific group at Oak Ridge: Dr. William R. Bibb, chief, Research and Technical Branch, Division of Research and Technical Support, ERDA; Floyd L. Culler, deputy director, Oak Ridge National Laboratory; Dr. Chester R. Richmond, associate director for Biomedical and Environmental Sciences, Oak Ridge Lab; and Dr. John F. Clarke, director, Thermonuclear Division, Oak Ridge.

At Los Alamos Scientific Laboratory in New Mexico, Dr. Charles Fenstermacher, group leader for the Electric Discharge Gas Laser Group, took time from a busy schedule for an interview and extensive tour.

The San Onofre Nuclear Plant, midway between San Diego and Los Angeles, produced willing assistance from Plant Superintendent Hans Ottoson and Dr. Alek Ronald Strachan, senior marine biologist for Southern California Edision. Dr. Strachan was also thoughtful enough to loan valuable data on a thermal effects study. This material included research prepared by Environmental Quality Analysts, Inc., Marine Biological Consultants Inc., plus six additional independent consultants. Their reports represented nine years of data collected off the waters around the San Onofre nuclear facility.

And what can I say to the people of the Tennessee Valley Authority? In my first magazine article I blasted and blistered them with virtually groundless accusations. Then, rather than waving fists and angry words in my direction, they merely tossed out documented data and caused me to do a very legal U-turn. So, a red-faced "thanks" to Dr. Robert L. Davidson, director of the Nuclear Engineering Branch and chairman of TVA's Nuclear Safety Review Board; Harry J. Green, superintendent of Browns Ferry Nuclear Plant in northern Alabama; and Nat B. Hughes, director of TVA's Power Supply Planning Branch.

In addition, I would like to acknowledge the support and interest of TVA Board Chairman Aubrey Wagner, who willingly shared his forty years of expertise in this field of energy.

You might say that public relations people are paid to do just that, so why thank them? I disagree. A good PR man or woman knows enough to do the unasked. Their instinct and experience tells them what material a writer can actually use. You don't have to spell it out to these people, and often you aren't quite sure yourself what direction to take. So, God bless "good" PR people like Milton H. Transchel, supervisor, Public Programs for Southern California Edison; Charles McBride, director of information for TVA's Power Branch in Chattanooga; Wayne Range, assistant to the manager for public information at ERDA's Oak Ridge operations; and Edward Aebischer, associate director for public relations for the Nuclear Division, Union Carbide. These men made my work easier without influencing its direction in any way.

A very special thanks to several anonymous (at their request) individuals in Oklahoma, Texas and New Mexico, who provided valuable and, until now, unpublished data on the controversial Kerr-McGee case involving the death of a plutonium worker named Karen Gay Silkwood. One was a Justice Department official; two were top Kerr-McGee executives anxious to set the record straight but equally anxious that their names not be used; and several were FBI agents and independent security investigators, including one on the Oklahoma State Highway Patrol. Again, my thanks for trusting me wih the documented facts to unravel this strange case.

I would also like to express my appreciation to a man that I would ordinarily have never met. He is Dr. Sergey F. Zaitsev, a nuclear physicist assigned to the Embassy of the Union of Soviet Socialist Republics, Washington, D.C. Dr. Zaitsev accepted me—a stranger—as an American writer and graciously offered his help and particular expertise to my research on the technical aspects of nuclear power, especially in the Soviet Union.

There are four other individuals who provided not technical expertise, but steady encouragement and a firm faith that somehow my course would be completed. Without the support of William R. Steltemeier Jr., Father Charles Giacosa, and Sister Mary Alma, this book would have been very difficult, if not impossible, to write. Thank you very, very much.

A special thanks to my English guide and critic, Professor WM, who taught a profound truth: "If you flutter with the owls, you can't soar with the Eagles."

Any acknowledgment would be incomplete without mention of Dominic de Lorenzo, my publisher, whose patience, wit, and honesty makes my first book a very special one.

There is one other person to whom I would like to acknowledge a lasting debt. Were it not for his trust, it is doubtful that I would be here at this particular point in time. He is the late Charles Moss, former executive editor of the *Nashville Banner,* who looked at an exuberant eighteen-year-old girl many, many years ago and decided that "Jacque" just might make a good cub reporter. Charlie took a chance and hired me and proceeded to teach me what he knew about people, integrity, and the world of writing. I will always be grateful for the trust and friendship of that great newspaperman.

In conclusion, if I have omitted anyone who deserves mention, my apologies. Actually, there is not enough space to list everyone who willingly shared their own special expertise. Most people, with only one or two exceptions, were interested and eager to see *Critical Mass* become a reality. Thank you.

Introduction

You might say that my expertise in writing this book has been a lack of expertise. True, my credentials as a journalist are strong—some fifteen years of writing and editing the news and magazine articles plus assorted awards and honors. I have my share of eagle feathers, and I've earned them by digging and then digging some more. The facts, it seems, don't often rise quickly to the surface. Sometimes, we have to pull and tug and even get scratched. Sometimes, we may even have to say we goofed and goofed bad. In newspaper jargon, this is known as a retraction and it may come in various forms. Only, sometimes it doesn't come at all . . .

A year ago, my understanding of nuclear power was virtually nil. I knew it had something to do with the atom, but terms like *fission* and *fusion* could be new breakfast cereals for all I knew. Like most Americans of modest means and average intelligence, I generally left such "heavy" thinking to the professionals—men and women with impressive degrees and doctorates who looked after such matters of the higher realm. In my mind, such people were almost infallible and I trusted their judgment. This wall of credibility started cracking, however, as more and more questions were raised about nuclear power. Like many Americans, I began to reason that the Establishment (whatever that means) was trying to rip-off John Q. Citizen and on a radioactive scale.

You see, we seem to be at a point in time where there is a need for a visible symbol to vent our frustration and anger . . . a scapegoat at which to toss not pebbles, but sharp and jagged boulders. Very simply, the nuclear power industry and 'Government' in general, are tailor-made targets. And, there's a prime rule of play: Even if the carcass falls, continue to kick and jab and then stomp some more for the old red, white and blue . . . Reason is gone and blind fury prevails.

Nor am I without original sin. My ego was hooked initially when it seemed that here was a chance for one journalist to take up the peoples' fight and exercise freedom of the press in its most noble form. The "Light Brigade"—led by yours truly—rallied for a grand and glorious charge . . . straight to a mud hole.

This writer and the magazine which I represented decided to take on the United States Government and the so-called nuclear power conspiracy. We were not afraid. We would overcome this evil that had descended on mankind. The will of the people would prevail.

One point I learned immediately: anti-nuke copy makes great reading and is a writer's dream to shape and color. There are so many scary possibilities, ranging from the ominous China Syndrone Theory to terrorist bombings. Nor is it really necessary to substantiate the tidbits that are tossed out like pablum to a waiting public. For example, I was actually convinced that a reactor could blow at any minute, and that plutonium was ready and waiting to be dispersed over the countryside. Here loomed Armageddon and I determined that it should not happen. Now, let me inject a further point of honesty: it is a very pleasant thing to be a crusader, and I thoroughly enjoyed the vantage point it afforded in terms of homage by the masses you seek to represent. You represent the American press—the magic pen that will carry their thoughts and anger to the highest point. Herein lies your power and it is an awesome one.

So, in the fall of 1974, when Larry Bogart (national director of the Citizens Energy Council which is based in New Jersey) inquired about the possibility of a magazine article on the dangers of nuclear power, I yelped like an excited puppy. It was "go" all the way, and I was quickly supplied with pamphlets, statistics and dire warnings on the hazards of radiation.

Shortly thereafter, the first article appeared in *Nashville!* magazine with a modernized *American Gothic* showing the traditional farmer and his wife. Only in this adaptation, the farmer was half-eaten with radiation, and in the background loomed an ominous-looking nuclear plant surrounded by acres of desolate or dying farmland. Most copies were sold out once they reached the newsstand. As I said, it makes hot news.

Needless to say, the anti-nukes were delighted, for their pupil had performed well, and I stood in center stage to receive—humbly of course—their applause. Civic groups asked me to address their noon luncheons, so I went on the roast beef and green beans circuit for several weeks. It was fun and my telephone was constantly ringing as other concerned citizens inquired about this or that.

Soon, however, my early satisfaction began to flicker and dim. I had an uneasy feeling that something wasn't right. You might say that my reporter's instinct began to twich at a strange odor that was beginning to creep under the doorsill. Fanaticism is no virtue and more and more people in the anti-nuke camp were beginning to look glassy-eyed and foaming at the mouth. Some I could excuse but most began to bother me, especially when they had no answers for questions I began to toss in their direction. It was a congratulatory note from Bogart that really activated the burr in my conscience. Then I decided to make from my pedestal a rapid descent back to earth. It turned out to be quite a fall from grace.

In early interviews with Bogart, he had repeatedly stressed an excessive number of power plants slated for Tennessee. In his mind—as he stated on a taped interview—there were *seventeen* docketed for my state. I grimaced to think that the lush, green countryside would be choked by such a pack of "monsters." I decided to do some checking, and after a thorough examination of Atomic Energy Commission and Tennessee Valley Authority records (I demanded to see the original copies), I learned that Bogart's figures were far from being accurate. For the record, Tennessee is docketed for *three main plants* and *five units*. In addition, the world's largest breeder reactor is slated for Oak Ridge at Clinch River. The practicality of this type plant is that it "breeds" more fuel than it consumes, which is a nice ratio in an energy conscious world.

So, when I questioned a critic about this discrepancy in numbers, she seemed perturbed and merely replied: "Larry has so much on his mind. He must have made a mistake." That wasn't much of an answer. Another critic insisted that Tennessee would receive seventeen and claimed to have inside information that no one else knew about. She lacked documentation that would verify her statement, however. Many of the comments hinted of paranoia, and I began to realize something just wasn't right and that, to be fair, a second article was in order.

So, early one morning, notebook and tape recorder in hand, I dropped by to see Turney Stevens, publisher of *Nashville!* magazine. Any doubts I had were dispelled as Turney came forth instantly as a conscientious journalist who shared a concern for truth. He listened thoughtfully to the evidence and then agreed that we did indeed have a responsibility to our reading public. We then set about preparing a second article on nuclear power which would provide an in-depth appraisal of the other side's viewpoint: namely industry and government. It was certainly not to our advantage to do this, because we had gained some powerful leverage as a crusader of the people. Now, it seemed that their white knight had stopped for a few precious minutes to test the wind and decide whether that windmill really needed jostling.

What I'm saying is that the anti-nukes were not pleased, and my earlier laurel wreath withered overnight. Regardless, we proceeded. It was to be a valuable lesson on what happens when you go against the popular grain. Most people, I have since learned, demand freedom of expression for themselves and their turf, but find it difficult to grant it to others. The "we are right" syndrome is well-entrenched in our society and leaves little tolerance for dissent. Indeed, it seems that the confusion in the public mind in interpreting *what might happen in the most unfavorable combination of circumstances* with what actually *will happen,* has been one of the major sources of difficulty in discussing the issues of nuclear energy.

Admittedly, there was some difficulty in researching the pro-article since government and utilities maintain a policy of refusing to engage in a shouting match with their critics. In fact, sometimes their profile is so low that you miss it altogether. I learned that you don't present an agency with a list of objections to nuclear power and get an immediate Xerox reply. There is a belief, bordering on official policy, that the public and its journalists, given adequate material, are quite capable of making up their own minds. I truly respect such openness and lack of hysteria, but I have some honest doubts as to the length most people will go to decipher truth from fiction. Americans have a naïveté about them that is both beautifully innocent and beautifully deadly. We also hate to say that we goofed . . .

The government agencies and utilities did provide ample literature and make top officials available for consultation—often on short notice. There was no visible attempt made at any time to cover up the facts; all responses seemed quite candid and honest. Most important, it was possible to *document* all statements. Surprisingly enough, such a policy of public disclosure has earned industry not an expected accolade of praise, but very often a red eye which is beginning to turn an ugly blue.

Finally, on January 1, 1975, Part Two of my series on nuclear power greeted the new year. It was an article that soundly refuted—almost point-by-point—all that was contained in the original article. This time, however, there were no bands playing nor congratulatory letters. Only silence. Regardless, as a writer I was satisfied that my professional code of ethics had been restored. I could live with myself again, and in the final tally that's important, although it was nice to be a hero of the people. . . .

Exactly one month later the silence was broken by a telephone call that caught me mixing a batch of Gerber cereal for the baby. Her two-year-old brother, Joseph, seized the ringing distraction as an opportunity to rush in and pinch his sister. She responded with a loud scream which Joseph answered with an equally loud scream . . . and so it went as I wiped my hands and lifted the receiver. I silently prayed that it wouldn't be a business call, because the background noises were clearly not the professional clicks of IBMs.

It turned out to be a business call and one of the most important ones that I have ever received. My caller was Dominic de Lorenzo, publisher of Aurora Publishers, Inc. Very simply, he had read both magazine articles, was especially impressed with the data contained in the second, and would like me to write a book on nuclear power. . . .

The words dropped like steel beads on a china plate. There I stood —the baby was making choking sounds interspersed with wails as she grabbed for the phone cord, while down below, Joseph was busy smearing cereal on the tile and fretting "Mama . . . Mama." And there's a book publisher on the line wanting to know if I'm interested in preparing a manuscript on nuclear power. As far as I was concerned, the "Hallelujah" Chorus was singing on all sides. My answer was "yes" and it wasn't a cool, detached one. I was genuinely elated, because this afforded me the opportunity to expand on a subject that was really beginning to interest me. It was also to be one of the greatest challenges I have ever faced. . . .

My eyes are the eyes of an average observer. My expertise has been a lack of expertise because I stopped and paused where the seasoned professional would probably never look. Another vital point: Never at any time was any pressure applied or suggestion made that this book should be slanted either pro or con. The conclusions and work are my own, acquired through visits and contacts with major installations throughout the world.

As a wife and mother, I share the concerns of other citizens who want to see a land in which our children, indeed all children, can grow and prosper in peace and health. I have also found that this is pretty rhetoric that covers much ground . . . for if we are to truly chisel such a brave new world of tomorrow, then it is going to be necessary to do away with our "sticks and stones" mentality and really open the windows. We are going to have to think and think hard, and sometimes this is painful. There is no way that we can continue to utilize seventeenth century methods in dealing with the technology of the twentieth century. Our penny firecracker is just about ready to go "bang-bang" unless we get a longer fuse and handle it with care. After all, who wants to get burnt? And for only a penny. . . .

Dr. William R. Bibb, chief of the Research and Development Branch at the Oak Ridge operations of the Energy Research and Development Administration.

Chapter 1

ODYSSEY

A very spoiled schefflera basked her leaves in the late afternoon sun while a group of the nation's top scientists gathered at its spindly base and talked of mice and men. Here were the Zen, the monks of the infinite realm who held within their sanctuary of numbers the power to make war or peace or even make love if the formula was right . . .

Today was no ordinary day. Today the conclave had met to discuss energy, and the talk wove in and out like the staccato click of an IBM plotting its own program. A scratching sound from the right, and an Oak Ridge physicist tapped quietly for attention.

Gradually, like the hands of a massive grandfather clock, heads and eyes focused toward Dr. William R. Bibb, chief of research and development of the Division of Research & Technical Support for the Oak Ridge operations of the Energy Research and Development Administration. ERDA, fledging newcomer who was spawned with the Nuclear Regulatory Commission during those final, spasmic hours of the Atomic Energy Commission. The old AEC, a relic from the past, gasping and struggling for air—a point of leverage—had finally split to avoid any interest conflict in its supervision and construction of nuclear plants.

And thus rested ERDA, a slice of youth, a slice of promise on the sacrificial altar of mankind. Not yet a year old and already beginning to turn gray. Poor ERDA. . . .

Doctor Bibb stood before his colleagues, some glaring in their scarlet robes and others appearing like thin, gray lines on bleached paper: the greater and the lesser stood waiting. Some knew this man from Washington where he had served for two years as a technical assistant to former AEC Commissioner James T. Ramey. Others remembered when he was assigned as the U.S. representative to the International Atomic Energy Agency Panels in London, Paris, Vienna and Moscow from 1965 to 1970. His credentials were approved and respected. All was in order.

"Give us imagery," the Assembly cried to their heavens. "Give us imagery."

Suddenly, a hollow voice boomed through the Great Hall, silencing this mental anguish of a thousand, tormented minds. "So, let him speak," it decreed. "So, let him speak."

Quiet fell like a blink in time, and the Oak Ridge scientist came forth like a tow-headed country boy with a freshly caught string of fish, and did begin to speak, and his words bounced and scattered through the Great Hall.

"The United States is like the ocean liner *Titanic,*" he began slowly, with precision. Those in the back rustled slightly and strained forward. The schefflera stiffened and quivered her leaves in anticipation . . .

"We, the People of América, have been ruptured by a great iceburg and the first three decks are flooded, but the passengers are still wining and dining on the main level.

"The water," he said, almost as an afterthought, "is rising rapidly. We must save the ship. But can it be done?"

Silence. Silence. Silence.

The point was made and declaration given. William Bibb had chosen his swiftest steed and rode vigorously toward that North Church Tower which still looms on the horizon. Torn by the wind, ridiculed by some, it was still his hope that somehow a great lantern of reason would once again outshine the shrieks of darkness which threatened to plunder and destroy all that is sacred and true. The still unanswered question was whether the ship was willing to be saved. It was to this matter that the conclave attempted to address itself.

Humming, stirring, noises from right and left . . . like the drone of a monstrous hive in which the bees suddenly find themselves torn by a force that cuts their very existence . . . their manner of being . . . their manner of thinking. What to do? They shake and toss and turn but there is no instant answer. It is left for them to decide their fate. They are coming of age in the twentieth century and there is much pain . . .

A revered old leader from the back rose and moved slowly forward, touching his beard thoughtfully as he glanced kindly at those around him. Smiling, he said with the calmness of his years: "Now, now. We are getting worked up over nothing. This can be resolved. What we need to do is tighten our belts a bit and cut down on electrical consumption. Maybe, use our electric toothbrushes less and even do away with those pesky water picks. Never liked them anyway. You know, just do without. Dear children, we'll make it. We always have. Have faith and be at peace."

Slow murmurings of agreement rose and fell. It was good to hear such words. They wanted to hear these words. . . .

Another rose and moved forward, his paunch rolling like a giant blob of jelly. Puffing, he quietly took his cue from the previous speaker. Now was not the time to dissent from what was obviously the popular norm. No, not now. Later, perhaps, when the power was his and his alone.

"We, the People of America, are in the midst of a gigantic hoax," he began, punctuating his remark, almost hissing, to gain the greatest momentum. Eyebrows rose. There was an air of expectation. It could go either way. The schefflera strained . . .

"The oil companies are against us, squeezing the lifeline of the average citizen, and the Government has not been acting in our best interests. Someone is making a massive profit off this entire matter and it is not the people of this great country. Come, let us attack the enemy at hand. Let us draw anew on the great legacy of our proud past."

Wisdom. Such wisdom. There were nods of agreement and suggestions by more than one that this learned speaker might perhaps be a candidate for First Chairman. The man's words trailed off into the void they created, and he returned, exhausted, to his pallet. Thought was effort and of this, he knew well. . . .

Suddenly, a bald-headed economist stood boldly in the midst of his colleagues, most of whom were unaware that he had the ability to speak above a mouse-like quiver. "The last speaker's point is well made," he declared. "My latest findings show that the price of natural gas is in the process of being de-regulated because the gas companies have been deliberately holding back for a year or more. I suspect collusion."

For added impact, he said it again, with more force, more determination than he had ever given forth: "I charge collusion!"

Heavy rumbling began in the back and moved swiftly to the front like a giant tidal wave that swept all in its path. "Ahah. Ahah." This was the chant in unison. This was the litany of the people. Here was participatory democracy in its finest hour.

Cameras clicked and popped as the pale-faced economist found himself the center of attention in a very pleasant way. Here was an orgasm of delight. Ignored for years, he had finally been discovered and his worth recorded for all to see and marvel. He was needed.

"Wait a minute," shrieked a voice near the front. "W-A-I-T JUST A MINUTE. THINK. This is democracy and that's fine, but too often we operate on the basis of a short-term policy while ignoring the future.

"What if Dr. Bibb is right? What if this country is going to hell and what if we end up doing the same thing? What if we deplete our uranium and coal supplies? Then what is left? "P-l-e-a-s-e, P-l-e-a-s-e, think carefully about this matter. The future must be painted accurately. Is there no reason left among you?. . ."

"I can answer the ravings of that maniac," shouted a mustached militiaman on the far right. "If we deplete our energy resources, then there are always the tar sands, which can easily be tapped."

"But those are in Canada," quivered a small voice on the left. It was a nameless face among a strangely silent crowd. . . .

"So, we conquer Canada," the surly militiaman answered without so much as a flicker of his bushy eyebrows. With that he slowly rubbed an ivory-handled .45, then moved rapidly to a better vantage point. He sensed that here was a mass in search of firm, decisive leadership, and he was ready to be called.

Somewhere, in the far distance, just beyond the Great Doors of the Great Hall, a band began to play, and its strange melody held the people in its grip. It was a carnival atmosphere, and near the front of the Assembly, another face jerked to its feet and screamed: "Everybody! Everybody! Everybody listen! Let's have one hell of a joy ride for the next fifteen years. We can do it. We can do it," the man glistened, shaking his hands wildly. "Let's pump all the gas and oil. To hell with the environment. Let's devastate the place and have one good time because all we are really doing now is just postponing the inevitable.

And what the hell is our right to future generations? Why should we worry about them? It's their mess to take care of."

"Yeah. Yeah," drooled an oilman from the back. "Hell, don't cut off my electricity."

"Yes! Yes!" shouted the great mass of people as they moved forward and stomped and kicked until the words of any and all doomsday prophets were silenced forever. Enough of such realism, they nodded and wiped their hands and minds in satisfaction.

This was history as The People decreed that it should be written. There's was a joy unto oblivion, for the Great Hall soon vibrated and shook as if in the grip of a giant tremor. The stately schefflera whithered and died. The sun ceased to shine. And, somewhere among the twisted and charred ruins of what could have been, there still murmurs the ghost of a long-dead Assembly: Dr. Bibb was right . . . Dr. Bibb was right . . . Dr. Bibb was. . . .

But, there are none to hear that such was so.

Chapter 2

ODYSSEY UNVEILED

A fanciful beginning, you might say. Yes, that's true, but the words point directly to some immediate problems facing this Great Hall of life in the twentieth century. And just for the official record, Oak Ridge's National Laboratory does have a Dr. Bibb, and he did make that statement to me during an informal session with Dr. Herman Postma, who directs the entire operation at Oak Ridge.

The doctor's thoughts and credentials, then, are very much in order, and he does have a very solid vantage point. Also, like most of us, Dr. Bibb has a wife and children and a sizeable stake in their future.

Now, what about our stake in the future? Is it in order? You might say that the ball—or atom—rests with *us* now. That's right, that great body of people who normally sit in the stands and cheer rather than getting down on the field and into the game. We have to decide whether we want to take the ball and run, squat on it, maybe think a spell longer, or just scream "help" and expect some delivering hand to pull us up and over.

In answer to the latter, I can still remember the words of an ageless Sunday School teacher who somehow reached my childish mind with what is still a profound thought applicable to the so-called mod adult world. "The Lord helps those who help themselves," she was fond of repeating each Sunday as we sat at our tiny tables and read scripture. I believed that then and I believe it now. So, the precise problem is what we, the conclave, the keepers of the Great Hall, intend to do about our energy situation. Should we seal our doors and row for land or just continue to scramble for deck chairs?

In a real sense, we've been in trouble ever since a couple of hunched-back humanoid types grunted and twisted two pieces of odd-looking flint stone. As luck would have it, between dodging hungry dinosaurs, Mutt and Jeff, as they might have been called, finally hit the big time. A few sparks broke loose and these cave men of yesteryear squinted in awful reverence of their new god: fire.

Here was a discovery of the primal order, and as the older tribesmen rightly deducted, it did have implications. For one thing, fire made an irreplaceable dent in what was man's virgin energy supply. On another level, it initiated the chain called pollution. That thin, gray smoke oozing skyward had the somewhat dubious distinction of making a first scratch on Earth's pure circular layer of atmosphere. And scientists tell us that remnants are still there today.

Looking back on the twentieth century, historians of the future will probably record—if our present pace continues—that Western man discovered and then systematically burned up virtually all the earth's resources of petroleum and natural gas. The picture is not a pretty one and the landscape is beginning to look jumbled. Irreplaceable fossil fuels—specifically oil, coal, and natural gas—are formed through millions and millions of years and are getting harder and harder to come by. Once they are gone, they are gone forever and ever and ever. No more.

Dr. Bernard L. Cohen, director of the Scaife Nuclear Laboratories at the University of Pittsburgh and chairman of the American Physical Society's Division of Nuclear Physics, has pointed out that continued use of fossil fuels should be at least only a temporary solution. Uneasiness is beginning to grow over the enormous amounts of carbon dioxide added to the atmosphere as a result of combustion produced by these fossil fuels. What this does, notes Dr. Cohen, is trap more of the radiation from the sun within the earth's atmosphere, thereby raising this planet's average temperature. Based on this, you can readily deduct that over an extended period of time, the situation could get hot. So, that's something else we need to work on.

The United States, with only 6 percent of the world's population, consumes about 35 percent of the global production of energy and minerals to produce nearly half of the world's goods and services. American farmers number one-tenth of one percent of the world's population but produce one-fifth of its total food supply.

There's a little booklet which ERDA is distributing now that shows a man-figure, obviously upset, holding aloft one tiny candle. No, the caption doesn't read: "It's better to light one candle than curse the darkness." Although this might be worth thinking about as we brace for the days, weeks, months and years ahead. In fact, the Christophers* may have been way ahead of their time. . . .

To get back to ERDA's pamphlet, however. It's titled the "Energy Crisis" and lists four reasons why we've found ourselves in this mess: (1) our total energy consumption has been rising rapidly as population and the rate of per capita demand and living standards have grown; (2) our domestic supplies of natural gas and oil are running out, and dependence on overseas sources could create international political and financial risks; (3) the production of energy is now affected by standards of environmental quality concerning our air, water, land, recreational, and esthetic resources; and (4) we are not developing new sources of energy and new energy production systems fast enough to keep up with the increasing demand.

Now, let's add a few postscripts to this. Between 1950 and 1970, the nation's energy demands grew by 3.6 percent each year, and today every American has the energy equivalent of 178 full-time servants. There has also been a noticeable energy trend regarding the use of electricity: it has been rising at the rate of 7 percent a year, almost twice as fast as total energy use. Projections show that if this trend continues, then the 25 percent of our energy that is going into electric generating plants today will jump to 40 percent by the year 2000.

Figures show that energy use and life styles have risen together—37 percent of the total energy figure goes to homes and cars. Eight percent is allotted trains, airplanes and trucks, 14 percent for commerce, and a big 41 percent for industry. The latter means jobs that feed, clothe and shelter people. Did you know, for example, that if public moratoriums are successful and the nuclear industry is completely axed, it will mean the immediate loss of six million jobs? Right away, six million men and

* The Christophers are a New York-based, Catholic group that has an international motto, "It's better to light one candle than to curse the darkness." Founded by Father Kelley but nondenominational in scope.

women are going to be walking the pavement or applying for welfare. That's a thought to ponder as we continue a brief trek through the world of energy figures.

Wait, is that six million figure rubbing like a burr in your mind? How about this one: Construction of the Hartsville Nuclear Plant in Tennessee will involve a work force of more than 6,500 people during the next eight years (1975–1983). Economy-wise, putting this many people to work is helping on all levels, and Tennessee is just one example.

The President's plan to involve private enterprise in manufacturing nuclear fuels could mean 220,000 new jobs by 1990. This figure comes from Labor Secretary John T. Dunlop in an address to a congressional hearing in late 1975. He testified that these jobs would be created if private industry can build and have in operation by the 1990s at least two new uranium enrichment plants, as envisioned by the President's proposal.

Dunlop was just one of a number of top administration officials who testified before the House Senate Joint Atomic Energy Committee. As such hearings were in session, the Energy Research and Development Administration was waiting approval to enter into contracts with private firms wishing to build the enrichment plants. A second bonus to this plan would empower government to share its nuclear technology with these firms and even sweeten the pot with up to $8 million in government guarantees.

As to the manpower implication of such a move, Dunlop stated that the new jobs which would be created would be largely "high paying" ones requiring highly skilled manpower. He also stressed another vital point: ". . . because of the time it takes to find such manpower and the long lead times involved in building nuclear facilities, it is my hope that we can get going. The time is long overdue for us to get to work in this country and to build some more nuclear enrichment plants."

There is another thorn, one which should prick the conscience of any thoughtful individual. A third of the world's people have annual incomes of less than $200 and average less than 200 kilowatt hours per year. Many of these so-called less fortunate know—through modern communication—they are missing a sizeable chunk of the good life. They want to be like the boy or girl in the ad that smiles out to them. They want dishwashers and other luxury items that many Americans regard as routine fixtures. So, the pressure to get these continues to grow.

Bringing this lower third of the world's people up to just 2,500 kilowatts per year, however, would involve doubling the world's present consumption of energy. Fossil fuels, alone, just can't do the job. And that's a fact.

Oil is not really an answer. World use reached a rate of 55 million barrels a day during the 1973-74 period. This was just before the oil embargo, price increases and recession. Records show that the United States consumed some 18 million barrels a day, in addition to 600 million tons of coal per year and gas at the rate of 26 trillion cubic feet per year. Put this on a world scale and you will see that the ten countries who produce the most electricity together produce three times as much as all of the other world countries combined.

Here's how it stacks up. As far as we know, about one-half of the oil believed to be reachable in the United States had been found by January 1, 1971. Most of the remaining oil is located in such frontier areas as the Alaskan North Slope, the Atlantic Offshore and the deep waters of the Gulf of Mexico.

Granted, the oil and gas finds in Alaska do look promising. One is estimated to have 25 trillion cubic feet of gas. Sound like a lot? This figure is slightly more than the United States used in 1973-74, which means that just to stay even we have to find another oil or gas field of comparable size *each year*. That's called walking a tightrope to eternity.

Another problem is that as these oil and gas supplies run lower and lower, it becomes increasingly difficult and expensive to extract the part which remains. By the year 2010—less than four decades from now—estimates based on projected rates of consumption say we will reach the last 10 percent of the earth's supply of natural gas, which is the cleanest burning of the fossil fuels. A few decades later we will arrive at the last 10 percent of petroleum. Any so-called promising news about low-sulfur oil is not that encouraging. First, it's not all that clean. Second, it must be imported from remote areas like Indonesia, which measures some 8,000 miles away. Oh yes, low-sulfur oil is already in short supply, and the price is skyrocketing due to increased competition from would-be buyers. Dear readers, it seems an unwritten and universal law of profit decrees that when the pinch is on, there's money to be had. Further defined, it spells "rip-off" of the highest and most lucrative order. Still, facts are facts, and we have to live with the real and not the unreal.

Don't despair. Coal can be stretched out a little longer. Year 2300 is the estimated date for reaching the last 10 percent, but coal poses serious air pollution problems. It's one of those situations where you can't live with it, nor can you afford to live without it. Even a superficial glance warns that there's more to that chunky, black stuff than we thought originally. Ask a miner what it's like to pull coal out of the ground. Very few see it as a labor of love. The experts advise that we need to dramatically expand the extraction system. There must be new mines that meet modern health and safety standards with a minimum of impact on the environment. Realistic regulations that require sane reclamation of the land have to be implemented. This creates a necessity for restoring a dependable railway system. It is possible that we may even become dependent on synthetic gases and liquid fuels obtained from coal.

At this point, says Congressman Mike McCormack (Democrat-Washington), the need for a systematic and integrated national energy policy becomes obvious. Here's a man who will tell you directly—minus a lot of fancy language—that the United States will not make it over the energy hump unless some rapid reshuffling is done. Credentials? McCormack is one of two scientists in Congress and in the summer of 1971, was selected to chair the House Task Force on Energy, an appointment unprecedented for a freshman congressman. In 1973, at the beginning of his second term, McCormack was appointed chairman of a new Subcommittee on Energy of the Science and Astronautics Committee. This made him the only second-term congressman in modern history to hold a major subcommittee chairmanship.

He is also the author of legislation establishing the U.S. Comprehensive Geothermal Energy Research, Development and Demonstration Program; of the Solar Heating and Cooling Demonstration Act of 1974; and of legislation creating a new National Institute of Solar Technology and a public data bank on solar energy. He is regarded as the leading Congressional advocate for the Nuclear Fusion Program and, in this capacity, obtained substantial increases in funding for fiscal year 1974-75. McCormack also sponsored legislation which would allow greater participation by individual states in the siting of nuclear power plants and a plan to create a nuclear power park in specific regions of the country.

Now, for a bit of candor. To many people, Government has become almost a non-entity that seemingly has nothing to do with the day-to-day struggle just to survive. It seems to imply fancy-dressed men and their ladies in far-away offices, passing laws that most cannot understand or really influence in any way. Such feelings aren't unique. They are becoming mainstream America.

I had heard the name "McCormack" during various energy seminars but had dismissed him as just another voice on the Hill, and I don't mean that as a compliment. He simply didn't reach me. My mind was blocked and had very neatly catalogued him in a specific slot: "Washington, Politician."

Then came an Energy Technology Conference that took me to the Capitol, and it was there that I met THE MIKE MCCORMICK. The man took one look at my press badge and seemed to hate me instantly. He then proceeded in a very loud and harsh tone to accuse any and all journalists of preferring the sensational rather than the real. Writers, he said, like to distort the facts and are sacrificing the country as a result. Red-faced, I stared. Then he really laid it on the line and his tone never broke: America is in bad shape; energy-wise we need to move and move fast. It's going to take all of us pulling together and not in opposite directions. That means the press, too.

It was all over in a matter of seconds. McCormack, in a puff of non-polluting smoke, was gone . . . leaving me as part of the ashes. His words won't win him the White House or even a popularity contest, because the truth seems to make people uncomfortable. We want to squirm and itch a little and then we try to forget. I left that first session stinging, but honestly convinced that Congressman McCormack is one light that just might set Capitol Hill on fire in a positive way. And, that's a nice surprise for a taxpayer to experience, because such sparks seem few and far between. A quick check later showed that this man has a good pivot point from which to speak: twenty years (1950–1970) as a research scientist at the AEC's Hanford (Washington) Project.

Detour's over . . . back to coal. The ERDA (Energy Research & Development Administration) Fossil Energy Program involves two objectives: (1) creation of synthetic fuels—coal liquefaction, and gasification to high and low BTE gas, and (2) direct utilization of coal for power and combustion. These goals are being researched under a greatly expanded budget—$323 million is the request for fiscal year 1976–77—and industry is playing a major role. Recent major events include progress in operation of pilot plants such as one which is used to product a synthetic natural gas from coal, and plans for additional demonstration plants, plus research initiatives involving universities.

The hydrogenation or HTG process is one of the most advanced coal-gasification technologies available, but it presents problems of another sort. It is the process of treating coal with oil and hydrogen under heat and pressure, then separating the liquid mixture into useful products. Figures show, however, that half the cost of this operation alone—just to close the presently-predicted gap between supply and demand for natural gas—would be a cool $200 billion. Such a gasification operation alone would require 40 percent of all coal mined today and the equivalent of most of the flow of the Potomac River just for processing—not cooling. That is quite a load in any man's mind.

And Congressman McCormack guarantees that even if these projections are off by even a low margin, one can still grasp the implications a project of such magnitude would have. What we are talking about is coal, water, steel, dollars and manpower, not to mention logistics, plus sheer environmental impact.

If you're into the energy trip even a smattering, the term "oil shale" is probably bobbling around. The suggestion is often made that since most of our oil is going down the drain, perhaps we should expand operations in this area. It sounds good but that's about all. Oil shale is found out West, specifically the Green River Formation that underlies the Colorado Plateau just west of the Rockies. It covers some 16,500 square miles of Colorado, Utah and Wyoming. It is neither oil nor shale but a marlstone which is mostly clay and contains a brown to dark gray organic material called kerogen. When heated to about 900° Fahrenheit in a huge vessel called a *retort,* the kerogen reacts to form shale oil, plus some gas that can be recycled to heat additional shale. Shale oil is low in sulfur and, although different in some respects from conventional petroleum, can be refined into most petroleum products.

The Green River oil shales are estimated to represent two trillion barrels of oil, but only a portion of this has the potential of being economically recoverable using present technology. The deposits of current interest are those that assay 25 gallons or more of shale oil per ton of oil shale. What we are talking about here is some 600 billion barrels of in-place oil. Two tons of shale—about the volume of an office desk—will yield more than a barrel of oil. Oil people normally use 42-gallon barrels as a basis for resource estimates.

Several techniques for recovering oil shale have been available for some time, although it has never been demonstrated on a large scale. Until just recently, it could not compete economically with natural crude oil, but the price of crude oil now has risen so radically that many companies are vying for shale oil. For example, a consortium of companies has begun work on facilities near Rifle, Colorado, to demonstrate a process for extracting oil from shale. At another Colorado site, an experimental program featuring a different processing approach is under way. With proper incentives and planning, it is estimated, production of shale oil could be as much as 1.5 million barrels a day by 1985.

Oil shale in the Green River Formation of Colorado covers about 2,600 square miles, that in Utah about 4,700 square miles, and about 9,200 square miles of it lie under Wyoming. It also occurs in at least twenty-seven other states, but none compare with the Green River Formation. The so-called "black" shale of the central United States constitute a large energy resource and occur from Texas to New York and from Alabama through Michigan. There may be anywhere from 200 billion to a trillion barrels of 10 gallons per ton of in-place oil within this area, but it is not economically recoverable at this time.

Initial production on the Colorado Plateau will probably involve mining shale by conventional methods, either room-and-pillar or open-pit mining, and retorting on the surface. Such retorted shale from above-ground operations will probably represent the primary waste disposal problem. It occupies about the same volume as the raw crushed shale fed into the retorts, but the volume is considerably greater than that of the in-place material before mining and crushing. Consequently, even if as much as possible of the solid waste is returned to the mined-out areas, some will require above-ground disposal. The procedure being studied would involve conveying the material to a nearby wasteland arroyo area. In other words, once the oil is extracted, the customer gets more than he expected: namely, some garbage that has to be carted off.

There's one other matter which is probably the most important limiting factor: large amounts of water will be needed by the communities that will grow in the vicinity of oil shale operations, and significant quantities of water are essential for the operation of the mines and extraction plants. Right. Shale occurs where water is in short supply. It's like that picture of the four fish, each larger than the other, following one after another, with the caption: "There's no such thing as a free lunch." Or, as a bearded dude from Frisco quipped: "Mother Nature is a mean mother."

Regardless, according to Dr. Murray W. Rosenthal, associate director for Advanced Energy Systems at Oak Ridge National Laboratory, oil shale is an obviously abundant resource, and we are going to explore its feasibility. But at the present, you might say that the logistics are on the same par as those who equate solar power with putting a few sun beams in a test tube. Presto! Let there be light! No, not today or tomorrow, but maybe the day after. As Congressman McCormack told some of his fellow lawmakers, the time required—even with a crash program—between the successful laboratory demonstration of a concept involving the conversion of an energy source to a useful form and its actual implementation, can vary from ten to thirty years—usually the latter. There is absolutely no way that a tidal wave of federal funds could make solar, or geothermal for that matter, a significant energy resource for this nation before the year 1990.

No one is saying that an aggressive and well-funded program is not vital for energy research and development. It is equally vital, however, that we realize such benefits are long range and we must live with the here and now to get to that then and there.

This question of relying on imports—while essential for the present—could cut our throat. It involves tremendous political and economic complications, the magnitude of which barely surfaced with recent events in the Middle East. Nor is there any indication that America is experiencing an upsurge of friendly allies. One presidential hopeful described the Arab oil embargo to a group of professionals in Washington as a "bunch of dirty little men playing dirty little tricks." While he may be right, the coin has two sides. Other factions say that those sheiks unknowingly did the United States a favor by stopping the show. For one thing, it provided an interesting lesson on the logistics of interdependence. We got torpedoed in the rear and Uncle Sam is still busy doing some patchwork. Granted, this purse-tightening did have the traumatic effect of letting John Q. Citizen know that the United States of America is not an island that exists unto itself. Memories are short, however. Once the gas lines returned to near-normal, more than one member of the intelligentsia laped into their earlier state of general apathy. It was all a hoax, they might reason.

Another more positive point to be noted is that twice within the past hundred years the United States has managed to switch from one major energy source to another without collapsing. Until the middle of the last century, wood was our major source of energy, and it remained so until 1885, when coal took over. This trend continued until just after World War II, when natural gas and oil jointly kicked coal back to third base.

In summary, we can all agree that Americans enjoy the most comfortable and affluent society in the history of this planet. It takes electrical power, however, to keep the great cogs going, and the need has been doubling and doubling and then doubling some more.

In Southern California, for example, the demand for electrical power has been doubling every ten years, and of this percentage less than 25 percent goes into any kind of residential use. The bulk goes into factories, business establishments, agriculture, government services and the like—uses which provide jobs and payrolls that keep our economy afloat. Chop this percentile down even further and you find that all but a fraction of 1 percent goes for the heavy duty jobs of heating, cleaning, cooking and cooling. Somewhere in that 1 percent are all those so-called convenience gadgets like knives, can openers, shavers, and so forth. So, your agonizing decision whether or not to use an electric toothbrush or a "pesky" water pick won't even make a ripple in the big pond.

There is also one ironic fact for environmentalists to weigh very carefully. If the opponents of nuclear power are successful, they will defeat their own purpose for one simple reason: *It is going to take a lot more—not less—electricity to clean up our environment.* That's a law that bears repeating, because just about every important environmental project requires an enormous amount of electric power. So, you end up with a situation like the tiger who goes 'round and 'round, biting his own tail and wondering why he's in pain. For example:

- Re-cycling: Whether it's plastics, paper or aluminum cans, recycling plants require large volumes of electric power. Ditto that for the plants that compress old car bodies into neat little blocks of scrap.
- Air Pollution Control Equipment: Literally thousands of business and industrial operations see this as a basic need, but equipment for such cleanup requires pumps, motors and fans, which are all run by electricity. For example, just one air pollution device em-

ployed by Bethlehem Steel requires as much electricity as utilized by 1,700 average homes.

- Sewage Treatment Plants: Again, thousands of these are needed to clean up waterways, but just one secondary sewage plant, serving a population of 90,000, requires about 500,000 kilowatt hours a year.
- Mass Rapid Transit: This could take many smog-producing automobiles off urban areas and help conserve fossil fuels, but again, such systems run on millions of kilowatt hours.

The remaining question, then, is just how much do nuclear power plants contribute to the total picture of air pollution? Using the Southern California coastal area as an example and including all air pollutants such as carbon monoxide and hydrocarbons, the answer comes out to something less than 1 percent of the total. General Electric, in fact, has been running an advertisement that reads: "Light a match, and you put more smoke in the air than a nuclear power plant." And, a few quick paragraphs below this headline, there's a short section that deals with the question of radioactivity: "Nuclear plants typically add an average of less than 5 millirems per years to the background levels of their sites. The nearest neighbor to a typical nuclear plant would be exposed to about the same extra radiation that he'd receive during a round trip crosscountry on a jet."

Hovering over the West Coast, a deadly haze known as smog stings the eyes and burns the lungs. The chemical reaction which produces smog starts with two man-made pollutants, hydrocarbons and oxides of nitrogen, which interact in the presence of sunlight and oxygen to form an unpleasant cloud. Power plants contribute almost no hydrocarbons, but they do emit significant amounts of oxides of nitrogen, or NOx. Realizing this, since the early 1950s Southern California Edison has been engaged in reducing its emission of NOx. As a result of Edison's research, emissions along the Southern California coastal area have dropped from 11 percent to 6 percent during the past few years. This can be further appreciated when realizing that the automobile contributes a major share of oxides of nitrogen—73 percent.

The total picture is strikingly clear. Man's exploration of energy sources is laid out just far enough to reach the next rung of a ladder that stretches toward infinity. Fossil fuels have lasted until we could discover the potential of uranium. Uranium reserves, too, are limited, but these are sufficient to carry man into the age of the "breeder reactor," which is an advanced nuclear fission system in which more fuel is created than is consumed. In some circles, it has been irreverently referred to as an over-sexed hen, but it still keeps on producing energy, and right now, that's a vital part of the pecking order.

Chapter 3

CRITIC PSYCHOLOGY

Dear Sir:
I know right now that I am against the entire fission program. There's no sense in going over the report—point by point.
I am against the whole insane idea. I am highly aware and alerted to the fact that my well-being, even my life, is being threatened
So, whether there are five volumes or 100 volumes doesn't make no difference.

Yours very truly,
An American citizen

This letter, legibly written on plain white paper that crinkled to the touch, was received by Thomas A. Nemzek, director, Division of Reactor Research and Development, for the Energy Research and Development Administration (ERDA).

Sitting in his Washington office, he looked at the script for a few thoughtful minutes and then sighed: "This writer's tone reflects genuine emotion and, in a real sense, fear. But it also illustrates an important personal concern of mine. It seems to me that in the emotions generated over legitimate concerns about the impact of nuclear power, many citizens have simply lost their perspective. They act as if we live in a risk-free society, which is just not true."

". . . which is just not true. They act as if we live in a risk-free society, which is just not true . . . which is just not true. They act . . ."

Nemzek's words are a familiar litany that acts like a kind of sonar, bouncing out and then back without seriously penetrating an object. Its presence is detected and noted, but the body continues on its path, unless forcibly halted. And force is not the American way.

Proponents of nuclear power are asking the general public to compare alternate solutions and balance benefits against "probable" chances and consequences. They argue that the people are forgetting benefits and focusing instead—in an almost obsessive way—on the risk factor. Granted, this may be viewed as a natural reaction in view of the "Nagasaki Syndrone," which compels Americans to see nuclear energy in one perspective: The Bomb. Points Two Three, and Four are then reached accordingly. Such a reaction, however, makes a sensible choice—a rational decision-making process based on appreciable variables—damn near impossible. . . .

Safety, of course, is and should be a major concern. As a successor of the old Atomic Energy Commission, ERDA has maintained such a policy. During its sixteen-year period, the nuclear utility industry has accumulated over 180 operational power reactors, with a 100 percent safety record, which means *no* radiation-induced injuries. The National Safety Council tags this figure as "nothing short of extraordinary."

"Our basic strategy," explained Nemzek, "is to understand safety hazards and design the reactor in a manner to minimize the probability of their potential occurrence. We have duplicate and independent shutdown systems. These include a monitoring protection system, which senses abnormalities and automatically shuts down a reactor in the unlikely event of a break or leak in the main cooling system."

One of the most important ways of identifying issues and answering public concerns is through a mountain of paper known as the Environmental Impact Statement. This paper (and it generally goes into thousands and thousands of pages) is required by the National Environmental Policy Act of 1969, better known as NEPA. The implementation of NEPA—as interpreted by a 1973 Court decision—requires the documentation of the environmental effects of ERDA's entire program.

For example, the Clinch River Liquid Metal Fast Breeder Reactor (see Chapter 11) has provided the first Environmental Impact Statement, which attempts to address, in great detail, all the possible, future implications of a large-scale source of electrical energy *that is still under development.* You might compare such a massive effort to trying to describe, in the early 1920s, the total environmental effect of the automobile or airplane in the 1970s. By mid-1975 a proposed final EIS had been issued on the Clinch River LMFBR. *It involved a seven-volume statement consisting of more than 5,000 pages and representing more than fifty man-years of effort by the government and its research labs.* That's called in-depth analysis.

"You can also term it a difficult task of forecasting," according to James P. Darling, who heads the Power Supply Planning Branch for the Tennessee Valley Authority's Division of Power Resource Planning in Chattanooga. In addition to the Clinch River Project near Oak Ridge, it has also been Darling's responsibility to direct impact statements on the Hartsville Nuclear Plant near Nashville.

As he explained procedures, a well-used bottle of Excedrin stared out from a shelf where it sat tucked between a sheaf of curling papers and several bottles of water samples bearing a Clinch River tag. Darling smiled a tired Friday-afternoon smile and then looked out at the interstate traffic that was beginning to build up. He thought for a couple of long minutes, then shook a head of blond hair. "We'll make it," he grinned. "We'll make it." And he probably will.

As nuclear power has grown, so has its opposition. A few of the people are scientists representing a variety of disciplines related and unrelated to nuclear power technology. Many are members of legitimate environmentalist groups, and a fewer number are opportunists who seem to surface like water spiders wherever there is controversy or dissent. An even greater number are simply unaffiliated citizens who associate nuclear energy with Nagasaki-Hiroshima and are genuinely frightened. As a rule, these people will seek and often accept the advice of a neighbor, member of the clergy, or business associate rather than a qualified professional. Probably more major decisions take place in beauty shops and barber chairs than in all the executive conference rooms combined.

Most people in the nuclear industry don't know how to combat hostile publicity because they're used to working with formulas and are ill-equipped to survive in the political environment. And the coin has another side. Those people with political savvy and media prowess often don't understand the technical side.

The situation becomes so polarized that the pro-nuclear forces think anyone against nuclear power development is condemning all Western civilization. Meanwhile, the anti-nuclear people believe unquestionably that power plant development will result in killing babies and producing mutations. The intensity reaches such a level that reason is blinded and both sides lose ground.

It has reached the point that more than a few are wondering just what type psychology makes a critic a critic, and a believer a believer. Are most sincerely in search of a "Golden Grail" that will save mankind from a disaster of some dire sort? Or does the answer lie in the power that simply comes from being recognized and needed? After all, we all want to be loved, don't we? But, as one scientist noted, "dragging the country to her knees is a hell of a way to go about it."

Proponents, on the other hand, are wandering around like sheep in search of a master. Some are so badly entangled in the briars that they stand a good chance of losing all or part of their wool. Granted, there are utilities, like the Tennessee Valley Authority or Commonwealth Edison, that use nuclear power, vendors who earnestly sell it, and government scientists who gather in their Zenish conclaves to marvel over it. Indeed, the list may stretch on and on. It ends at the point where organization begins. Proponents of nuclear power are just plain lacking in this vital ingredient called organization and they have a lot of PR homework to pursue.

Let me cite one quick example. The 1975 Energy Technology Conference in Washington, D.C., drew 850 participants, many of whom were top level officials from throughout the world. In addition, there was an Energy Capabilities Display. Here was a chance for a genuinely sincere layman or professional to come and get the facts, most of which were presented neither pointedly pro or con. Perhaps most important, press passes were available to journalists, who were thus given an excellent chance to spend three days studying the basics of energy technology. It was sponsored by *Research/Development, Power Engineering, Pollution Engineering* magazines, and Government Institutes, Inc. publishers devoted to energy and the environmental field. And they did a terrific job, keeping the three-day conference right on schedule. In fact, it was so successful that a third Energy Technology Conference was held in March, 1976, in Washington.

Yet, other than myself, there were virtually no reporters representing the non-technical press. None whatsoever. Yes, there were plenty technical writers, but very few of our Aunt Pollys or Uncle Joes read magazines like *Pollution Engineering* or *Scientific American,* which are geared toward the professional community. In short, the word just didn't get out.

Prior to attending the seminar, I asked the publisher of a Southern afternoon daily if he was planning to have the conference covered.

"Who's sponsoring it?" he boomed. When I replied, he expressed concern that first, the conference was a CIA trick to lure participants and pick their brains; and second, "I doubt there's anything coming out of that which we could use."

With that the matter was closed, because this publisher was on his way to a very important golf game. Shocked, I hung up the phone and thought about what a ridiculous situation it really was. Here was an issue dividing the nation, an issue which newspapers are sensationalizing over and beyond normal, and one publisher doesn't think it's worthwhile to cover a conference that would dissect the question pro and con.

In contrast, the publisher of Nashville's *Tennessean,* a major Southern daily, expressed a genuine interest in the conference and requested additional information. As a rule, the newspaper doesn't have the reputation of being overly friendly to the nuclear industry, and yet, its publisher showed far more interest in both sides of the issue than his conservative counterpart. So, I guess it goes to show that in this business you never know which side is going to be up.

There is still another aspect. Most answers to the anti-nukes, as they are affectionately called by the other side, are so shrouded in scientific gobbily-gook that the uninitiated find it impossible to decipher. As a result, the average layman has developed a very dangerous habit of merely tuning out anyone classed as a proponent of nuclear power. After all, *critics* are just plain folks like you and me, and they don't use such fancy language. Besides they speak from the heart. Unfortunately, that's what most do—speak from the heart and not the mind.

In the past, the public has not been served by those who exaggerate the risk of nuclear energy, nor by those who claim there is none. Here we have a thin gray line called reality that should not be breached by either side. This is not to say that legitimate questions have not been raised that deserve some equally legitimate answers. Much of the outcry seems to fall within these specific areas:

- Radioactive Waste Disposal
- Sabotage and Diversion

- Radiation from Nuclear Plants
- Thermal Discharges
- Breeder Reactors
- Overall Safety of Power Plant Systems

Critics come in all shapes and sizes. Ralph Nader, for example, derived his early power by slicing the jugular of General Motors and putting the consumer out there in front. It equipped him with a notoriety that has diminished little since he moved into the larger arena of nuclear power. Adding to his aura of mystery is an almost ascetic-type life that lends itself well to the theory that there indeed is a modern-day Sir Galahad. This youngish-looking Lebanese has been often quoted as saying that if the public knew what the facts are, and if they had to choose between nuclear reactors and candles, they would choose candles.

Potential accidents are a "bug" with Nader. In fact, a journal of the American Health Physics Society quoted him as saying: ". . . the risk of accident is at a point of catastrophic consequence unparalleled in the history of mankind." (Hint: The man was referring to nuclear accidents, and his fury was genuine, despite the fact that not a single civilian has ever been killed in the standard operation of a nuclear power plant. Furthermore, the "dangerous" breakdowns alluded to are, for the most part, related to *non-nuclear* aspects of the operation and actually occur less often than similar breakdowns in conventional power plants.)

Just what is the Nader alternative to nuclear power? It is the use of solar energy—a kind of Rube Goldberg dream which he argues is just around the corner. Granted, solar sounds good but no serious scientist sees it as anything but a remote possibility. For example, it would require a solar energy plant some twenty square miles in area just to produce the energy obtained today from 9,000 tons of coal (figures from the American Health Physics Society). And the cost of a solar electric system as of 1975-76 would be in the $40,000 to $80,000 per-kilowatt range. It's something to think about the next time a 100-watt bulb is casually flipped on and off.

As the early rumblings of nuclear criticology first began sweeping the country, Nader gathered the forces to Washington for a national meeting. The conference was successful from one aspect: It produced virtually every critic in the country (except David Comey and Arthur Tamplin, who were both lecturing in Europe).

There were some 750 persons in attendance, representing 165 groups from thirty-nine states, plus Britain, Japan and France. The registration charge was a modest $10 (except for some industry types who were billed $100). Cost for the endeavor was partially supported by the J.M. Kaplan Fund, Inc. of New York. Basically, the meeting served as a freeze frame picture of the organized nuclear opposition. Several points emerged:

- The anti-nuclear forces had finally crystalized under their own distinctive banner and thus blossomed into a full-fledged political movement. Interestingly enough, some of the most popular speakers were not traditional nuclear opponents but rather leaders in the antiwar and civil rights movement.
- Few at Nader's meeting voiced interest over distinctive issues pertaining to nuclear power (i.e., the operation of an Emergency Core Cooling System or other back-up mechanisms), but rather the assembly agreed that nuclear power was an enemy of the people and, as such, must be stopped. Their emphasis was on tactics and approaches.

Nader offered a glimpse of the other side by allowing a former AEC Commissioner and Dr. Ralph Lapp to speak. A typical question and answer session went like this:

AUDIENCE: "How many atomic explosions in our cities would you accept before deciding that nuclear power is not safe?—no complexities, just a number."

Both men faced the huge conference and urged the critics to look before leaping into an untenable position. Former AEC official William Doub warned that "you lose whatever creditability you still have if you continue to follow four trends," trends which he identified as: a failure to be specific, repeating discredited statements, harping on conditions of the past after they have changed, and implying an infallability.

Undaunted, Nader responded by calling for no quarter in the fight to halt nuclear power. He stressed the use of avenues that he had found effective: the news media—and if reporters are unresponsive, "talk to their editors"; Congress—reach members through state grassroots organizations and directly in Washington. The audience nodded.

Not long ago, the scientific attaché to the British Embassy related, not too happily, to Dr. Herman Postma, director of Oak Ridge Laboratory, how "Nader-type" groups are beginning to make a beachhead in

Great Britain. "He said it was very strange," Postma would later muse, "because they all have American accents."

Funny to a degree. The international aspects are changing very fast because such groups that are opposing nuclear power—whether right or wrong—are rapidly moving beyond the United States and on to the shores of France, Britain, and elsewhere.

Another interesting postscript to this cult of Naderese is that his sister, I was told, is employed—ironically enough—at Oak Ridge nuclear labs. So, while it may be "all in the family," they apparently occupy opposite seats at the table. . . .

Then there's Larry Bogart, an otherwise nice chap who could be somebody's gentle old grandfather rather than the messiah of mankind. He could be viewed as a modern-day Moses attempting to lead the new Children of Israel out of the desert, toward a bleary promised land that looms further and further on the horizon.

Now retired, Bogart was formerly the assistant to a board member of Allied Chemical Corporation. Most of his adult life has been spent taking notes and orders from someone else and probably catching plenty of static when things went wrong. Now he sits in the executive chair as the "voluntary" national director of the Citizens Energy Council, which is headquartered in New Jersey.

It all started in 1966 when he decided to retire and oppose the construction of a nuclear plant near his home in Allendale, New Jersey. This crusade was to his liking and provided a purpose to life that was nothing short of exhilarating. He has been fighting the nuclear industry since then in his capacity as coordinator of the forty-state coalition of anti-nuclear groups (a job that requires him to jet-hop across these United States and sometimes overseas).

Nor is the path he has chosen not beset with obstacles. Bogart detailed how, on a wintry evening in 1971, he was driving his Volkswagen on the New Jersey Turnpike, heading for a meeting where he was the scheduled speaker. Another car, he vows, zoomed alongside and forced him off the road. His VW bounced off a guardrail, skidded wildly for a few long seconds, but somehow managed to regain its balance. Bogart says he was shaken, but otherwise unhurt.

Bogart maintains that on two other occasions, his car has been tampered with. Once, he said, the accelerator rod was even sawed through. Visitors to his office, he further warns, risk being photographed by sinister forces lurking outside. He declined to identify whom he believed such evil entities represented. . . .

"Environmentalists tend to be naive sorts," Bogart once told a small gathering in Nashville. "They think that if they prove nuclear power is dangerous, then it will automatically follow that change will come. They don't realize that big money is controlling the business."

I can still remember the fall of 1974 when Bogart approached me to inquire about the possibility of a magazine article on the dangers of nuclear power. It was at a meeting of the Metropolitan Nashville Board of Health, and a group of critics had gathered to ask Dr. Joseph Bistowish, director, to investigate possible radiation contamination in milk. The Board agreed to look into the matter, and I secretly wondered if perhaps they were politely putting Bogart and his crew out-of-sight and out-of-mind. I later learned that a conscientious Bistowish had personally ordered the study and received a report that showed radiation levels registered well within the normal range.

At that time, Bogart found this writer to be a willing pupil, eager to learn about the dangers of radiation. I was provided with pamphlets, statistics and dire warnings about infant mortality by the famous or infamous, depending on your vantage point, Dr. E.J. Sternglass of Pittsburgh. Much, much later I learned a long list of reputable professionals had stoutly discredited this man. Like many laymen, I had made the mistake of associating an impressive title with instant credibility. Now I know that this just isn't necessarily so.

For example, Dr. Sternglass's statement that the operation of nuclear power plants and reprocessing facilities has increased infant mortality in the areas where these facilities are operating makes good copy and especially great headlines. Not only has the good doctor been unable to prove his claims of increased infant mortality, but over a four-year period his methods of analysis and treatment of the data have been found to be erroneous by numerous statisticians, epidemiologists, and public health officials. His claims have also been repudiated by the Health Physics Society in a statement signed by all fourteen living past-presidents; by the American Academy of Pediatrics; the Public Health Departments of California, New York, Michigan, Pennsylvania and Illinois; and by the Environmental Protection Agency. In fact, the Health Physics Society noted in its statement that "without exception, these agencies and scientists have concluded that Dr. Sternglass's arguments are not substantiated by the data he presents."

Bogart, however, didn't mention this.

Another factor in my initial leanings toward the opposition was the presence of Mrs. Jeannine Honicker as state representative of Southerners for Safe Power, an environmentalist group that has zeroed in on nuclear power. Both of us are fiery Pisceans and we had discovered our mutual interests and temperament during my employment as a copy editor for the Nashville *Tennessean*. Her husband, Dolph, or "Bunny" as colleagues call him, was then and continues to be news editor of the *Tennessean*. There was one thing I learned very early about this man: if Bunny's for you, then he's for you, but if he's agin' you, brother, is he agin' you. Regardless, I accepted that attitude and his clear-cut Mencken style, and we continue to be friends. During the last few years, Honicker and his typewriter have managed to deal the nuclear industry more than one hefty blow through the power of the press.

The Honicker story is mentioned because it is a tragedy that has, for the most part, a happy ending. Their daughter, Linda, was diagnosed as having leukemia. Months of tests and pain and, finally, hospitalization in Seattle. Brave, tiny, Linda—despite the medical odds—surprised the world and pulled through. Her story was carried in *Reader's Digest* and is an inspiration to read, but the ordeal left its mark on the family. Their rationale goes something like this: Linda had leukemia. Leukemia is caused by radiation. We believe radiation comes from nuclear plants. No one else's child must go through the suffering that Linda did. We will not let it happen.

Understandably, in the Honickers' mind, A + B definitely equals a giant "C" for correct. Their fight against nuclear energy has become a consuming and convincing force, and I, like many others, felt sympathy for them and sought to share their concern.

Unlike some critics of nuclear power, Honicker does not come on like a wild and foaming fanatic. Rather, he reacts as a father who almost had a cub ripped from his protecting arms. In addition, he feels more than a little concern for the world out there. Newspaper people, you see, have a tradition of being very much like the sabra cactus of Israel: tough and thorny on the outside and very soft on the inside. I've seen his expressive, blue eyes water over a wire photo of a Vietnamese child, cut and bleeding, or the horror of a bombed refugee camp. Often, he would deliberately place such blood-and-guts pieces on the front page; then a few days later, without fail, the publisher would receive an irate letter from someone, complaining that the *Tennessean* should remember it is a family newspaper. Honicker, never fazed, would only smile and do it again. Each night I watch him carefully sorting and deciding and I think that somehow he hopes to shock the public into an awareness of the horror that goes on around us. I think that he realizes he will probably never succeed in making more than a tiny splash, if even that much of a ripple.

There was one particular night in the news room that is still vivid in my memory. It was press time and in a matter of minutes, fresh copies of the morning edition would be rolling up and out. Most of the staff were sitting around, either sipping some black liquid resembling coffee, cursing, or just enjoying a quiet, uninterrupted smoke. In all respects, it was a typical city room—dirty, littered and beautiful.

"It's a two-way street," Honicker sighed, after a long silence. "It's a two-way street. Radiation hurt Linda and, at the same time, radiation saved her life. How do you figure that out?"

The newspaper arrived just then, breaking our thought and making further conversation impossible. I have yet to answer this paradox and I'm not sure exactly how to go about it. Nuclear power, contrary to both viewpoints, is not a cut and dried issue where the rules are clearly printed for any and all to see. Occasionally, you run into those faded areas where the print is difficult to read. That's when the going gets a bit tricky. And in the "war" between the nukes and anti-nukes, personalities often blur the fine print. But if we understand what motivates individuals, it is easier to explain or accept their behavior. Even, in some special cases, respect their position for what it is and how it came about. Honest differences of opinion should never divide a people into warring camps. Never.

Researching the pro position, admittedly one encounters difficulty in sifting information from an avalanche of printed matter which government and utilities agencies make readily available. As I mentioned before, you don't present a list of objections to nuclear power and get an immediate Xerox reply. The belief that, given adequate material, the public and its journalists are quite capable of making up their own minds is admirable. But I have my doubts about the length to which most people will go to decipher truth from fiction.

For this reporter, the government and utilities did provide ample literature, plus make top officials available for consultation—often on short notice. There was no visible attempt made at any time to cover up anything, and all responses seemed quite candid and honest. Most important, it was possible to *document* all statements. But for the average stay-at-home citizen who relies on magazines and newspapers for the facts, an article which reports a malfunction in a nuclear plant will attract the eye much more quickly than the follow-up article that explains the actual danger—or lack of danger—involved.

As TVA Board Chairman Aubrey Wagner termed it:

> Critics of nuclear power have warped the significance of these reported abnormal occurrences out of perspective. Part of the safety consciousness in the United States is reflected in reporting procedures which are in effect for nuclear plants where even the most minute malfunctions are reported . . . Such abnormal occurrences reported during plant testing and startup are a natural result of weeding out potentially faulty components, of making construction and systems adjustments.

Wagner, based at TVA headquarters in Knoxville, paused for a while to look toward the hazy rim of the faraway Smoky Mountains. His background includes some forty-five years in the power industry, yet critics almost instantly dismiss what he says. "What does he know?" they say. Plenty, I found out. So much so, that power experts from throughout the world travel to east Tennessee just to tap Mr Wagner's expertise.

The nuclear plants now being built, he explained, are subject to stringent regulations of the Nuclear Regulatory Commission (NRC) and ERDA and are constructed under a "super cautious" philosophy. He noted that as a part of such safety consciousness, it takes from nine to ten years just to finish a nuclear plant. Then there are reporting procedures which require that even the most minute malfunctions be reported and explained.

Consequently, abnormal occurrences reported during testing and startup are a natural result of weeding out potentially faulty components, of making construction and system adjustments. Newspaper headlines, however, often do not tell the story in this particular light. It was George Orwell, I believe, who pointed out that it is possible to distort language so that words take on the reverse of their actual meaning. If words are taken from the language, it can also mean the removal of the concepts they epitomize from a peoples' consciousness.

For example, in the fall of 1974, copy began to move across the country when it was discovered that filters to detect radioactive particles had been left out at the Browns Ferry Nuclear Plant in northern Alabama. True, very true, but the way it appeared in type produced an exaggerated sinister impression. Point One: workmen had neglected to put in filters, but this slight was discovered during routine testing procedures specifically designed to uncover flaws and put things right. Point Two: the safety system caught the error, and Browns Ferry is now filtering okay. There was never any danger, because the unit where the filters were missing was not yet operational, but this aspect never received the kind of play that the original news articles did. Funny, about that.

On another aspect, Wagner noted how a report by the AEC showed that the Browns Ferry Unit One had reported sixty-five abnormal occurrences during 1973, which came out to be the largest number for any unit in the country. "Now, these figures must be viewed in their proper perspective. To say that this unit had experienced more such incidents than other units during a particular time period attests to TVA's diligence in identifying, correcting and reporting these occurrences.

"None of these occurrences," Wagner continued, "was placed in a 'Most Serious' category by the AEC. The health and safety of the public or plant employees was not endangered by these occurrences. Nuclear power critics have warped the significance of these out of perspective. However, the NRC requires and TVA will continue to identify and report even the minor events, because by doing so we can better insure public confidence in the safety and reliability of nuclear generating units."

In fact, newcomers to the world of nuclear power could really get the "willy-nillies" if they came across a copy of "Nuclear Safety," which is prepared for the NRC by the Nuclear Safety Information Center at Oak Ridge National Laboratory. Issues feature a concise listing—usually running several pages—of reactor occurrences and causes. This includes a listing of the reactor and facility, tabulated data, and at what phase the problem occurred. In addition, the docket number and report dates are given, so interested readers may obtain more detailed copies from the Public Document Room, 1717 H Street, Washington, D.C. Here again is a very open policy of disclosure that is unique to the United States.

When we get into this thing of "Critic Psychology," there is one so-called public hearing that still rings in my mind. It was a set-up by opponents, and the sacrificial lamb was Dr. Robert H. Davidson, chairman of the TVA Nuclear Safety Review Board. The speaker had driven through a steady downpour from Chattanooga to Nashville (about 130 miles) to reach the appointed meeting at the appointed time. His gait faltered just for a second when he walked through the door and discovered a sea of familiar faces—all anti-nukes. I watched as they proceeded to roast, barbecue and then slice the TVA representative into very tiny chunks. Floor questions became diatribes that stretched on and on without really allowing him time to reply.

Nearly two hours later, Dr. Davidson was still patiently listening to questions and somehow smiling. He refused to engage in a shouting contest. Astounded, I watched. It was nothing short of fascination to wonder at just what point the cool exterior would break and he would go on a killing rampage. Probably has a sack of plutonium in the car, I mused.

Finally, Bob Davidson, a member of the American Nuclear Society and former research engineer with Atomics International, counted coup (Indian term for unarmed valor against armed enemy) in a most royal way. It went like this:

CRITIC: "Would radioactivity be released from a reactor if it suffered a direct hit by a full-megaton nuclear warhead?"

Well, the man from TVA stared for a full minute at the questioner, a middle-aged woman who was all seriousness. Outside lightning zigzagged across the sky and rain pelted the windows. A heavy wind pushed and tugged at the trees. For a split second, the whole thing took on an air of the unreal. After two hours of explanation on all aspects of nuclear safety, this comes as a sort of finale. . . .

DR. DAVIDSON: "The plain answer is yes, but at that point, it wouldn't make much difference, would it?"

CRITIC: "You still haven't answered my question!"

DR. DAVIDSON: "Ma'am, if a full-megaton nuclear warhead strikes a nuclear plant—dead center—the least of our worries will be whether the reactor will shut down."

Critic psychology, then, might well be defined as a very tight circle of intimates who talk and listen only to themselves. You see, for many of their questions, there are no answerable answers. . . .

And, on this question of critic psychology, during the month of April and May, 1976, I served on active duty with the United States Navy in the Office of the Chief of Information at the Pentagon in Washington. My job was to evaluate Project Seafarer, which is a system designed to provide the last remaining communications link between an

America under siege by a nuclear first strike force and a deeply submerged submarine fleet capable of hurling retaliation against the enemy. I was to study the public relations program in regard to the technical and biomedical aspects to determine if the critics had an argument. "We want you to be factual and honest," Admiral Cooney told me. "Tell us, if we have done anything wrong." As usual I took my boss at his word.

Basically, Project Seafarer is to operate under a principal utilizing a land-based transmitter sending ELF or Extremely Low Frequency Communications. This allows a sub to stay deep and undetected by sonar. In other words, the enemy may attack the United States but he will die in the process. In my military handbook, that is known as aggressor deterrent.

The problem was where to locate the transmitter. Critics and environmentalists had begun coming out of the woodwork. The fruit flies were going wild. Elections produced a horde of anti-navy statements. Wisconsin and Texas, the two states initially selected by the government, said: "NO." Michigan was still pending by June, 1976, but the opposition seemed strong. Everything despite Environmental Impact Statements which showed no adverse effects associated with the system. The navy, I learned, had spent an estimated $82 million on research and development since 1958. I pored through the reports and studied papers. There were simply no biomedical effects harmful to plants or man or even animals.

As one navy doctor told me: "If ELF (extremely low frequency communications) is wrong then we might as well turn off everything electrical in the house."

And, the intellectual quality of the outcries against Project Seafarer and nuclear power may be measured by the outburst of one fervent critic in the Upper Peninsular area:

"Philosophically, it is wrong," the man wrote the navy. "Even if it works, it is wrong. I don't think that we need it. It is a waste of tax payers' money and it would be a shame to plunge it into our beautiful wilderness."

On one point at least such critics are certainly right. Without a trigger for our nuclear submarines and sufficient power to meet America's energy needs until the year 2000, there will be plenty of wilderness and very few—if any—beautiful people.

It seems to me that blocking nuclear power or a missile defense system, is not on the same par as boycotting lettuce, zero population growth and women's rights. I eventually made an unofficial recommendation to the navy as to who would make an ideal spokesman capable of speaking to the people and actually being heard. It would have to be a cyborg built to the following specifications:

The mind of an Einstein, the wit of Bob Hope, the wisdom of a Moses and possibly, the body of Racquel Welch. . . .

James P. Darling, director, Power Supply Planning Branch, Division of Power Resource Planning, Tennessee Valley Authority, Chattanooga.

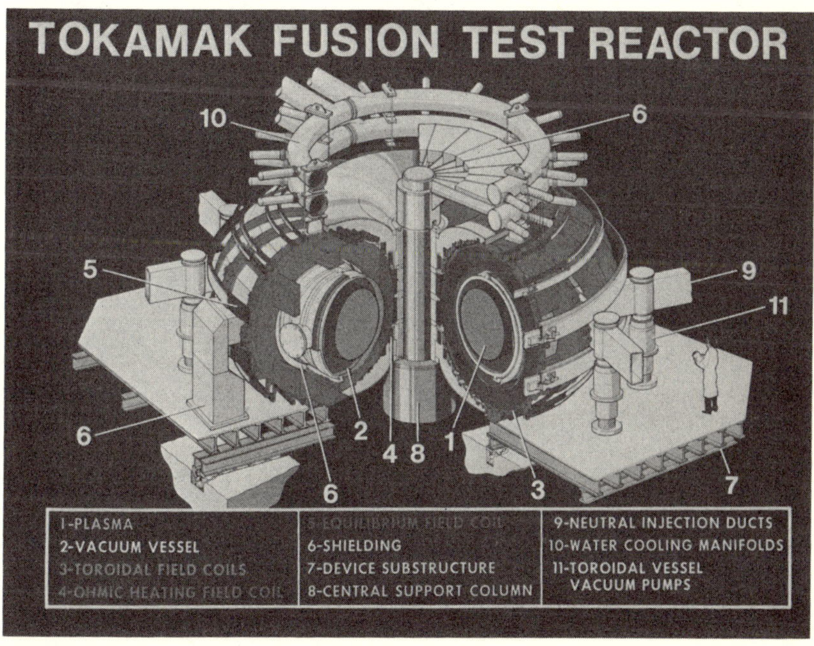

The Browns Ferry Nuclear Plant in northern Alabama, scene of a major fire in 1975 estimated at $100 million in damages and caused by a workman testing for air leaks with a . . . candle.

Sections of the Browns Ferry Nuclear plant in northern Alabama became operational again in mid-1976.

Dr. Robert H. Davidson, chief of the Nuclear Engineering Branch, Division of Power Resource Planning for the Tennessee Valley Authority, Chattanooga, Tennessee.

Chapter 4

FRANK ZARB: POWER MAN

Cool, calm and collected . . . He's been called "Zarb the Czar," plus numerous other less printable names. Regardless, Federal Energy Administrator Frank Zarb comes on like a seasoned broncobuster, staying clear of the dirt and not at all concerned with the gnats flying overhead.

As the third successor to a somewhat shaky throne, Zarb played a major role in ERDA's creation, in addition to serving as executive director of the Energy Resources Council. This means that he advises the President on matters like domestic and foreign policies affecting the production, conservation, use, control, distribution and allocation of energy. Needless to say, the man has more than one iron in the fire. . . .

Prior to Zarb's appointment at the FEA, he was associate director of the Office of Management and Budget for Natural Resources, Energy and Science. One of his jobs there was keeping check on budgeting for all federal energy programs. In addition, he chaired a White House steering committee which worked toward passage of legislation that eventually led to the organizational transition of what was to become ERDA.

During the height of the 1974 energy crisis, then-Administrator William Simon asked Zarb to join the Federal Energy Office—forerunner of FEA—as Acting Assistant Administrator for Operations and Compliance. He was charged with coordinating the preparation of mandatory allocation regulations, and setting up FEO's compliance and enforcement programs. One of his major achievements was organizing and manning FEA's regional offices in ten major cities, and clearing up a heavy backlog of applications for fuel allocations. After returning to OMB, Zarb completed several assignments for Administrator Simon, including playing a major role in settlement of the dispute with truckers over fuel prices and availability.

Accurate biographical listings do more than just occupy space. They establish that a so-called expert does have a solid vantage point from which to speak. As far as Zarb is concerned, perhaps John S. Foster Jr., vice president of T.R.W., a research firm, put it in simplest terms when he told a Washington management conference: ". . . you should listen to Frank Zarb." Here, then, is a briefing with the Power Man:

QUESTION: Coal is obviously important to our future development of energy resources. This reaches all the way from our present ability to burn coal conventionally, to mining it, transporting it, and onward into advance technology. Is there a comprehensive government plan that facilitates all this?

ZARB: ". . . the Clean Air Act prevents some burning coal in specific areas. As to transportation, we need some real thinking and programming in areas where improvements are needed. As to the actual perfection of an overall plan, we are looking toward ERDA to help put that into perspective sometime this year [1975]. We have in our ten-year program some key objectives for coal, and we are tackling it with the attitude that these objectives are rather certain and can be met with a minimum amount of activity . . . but it's going to take some people to get off the dime and face up to the fact that coal is important and that energy is important."

QUESTION: Do you expect a federal order requiring public utilities to switch to coal burning away from natural gas-oil burning?

ZARB: "Yes, we've already had an exciting reaction to our first adventure in these waters, and we expect before it's all over that the whole issue should become exciting. So, I would definitely stick around and watch.

NOTE: Zarb was referring to the initial letters sent out by the FEA in May, 1975, to nine utilities representing some twenty-five facilities in the West. The message was a request for voluntary compliance, and since then other letters have followed. Reactions continue to be varied, suggesting that the federal government may eventually find it necessary to apply a little clout.

QUESTION: There has been a lot of conversation regarding the safety of nuclear power, but not a great deal about any economic impact in regard to *not* going forward with full development of nuclear energy. What is the proper perspective on this?

ZARB: "The base of this question is one level of abstraction. When you say that there are those in the country who would request a moratorium on nuclear power, there are also those in the country who would still debate the fact that we need to develop our own energy base and that this energy base cannot be fed, certainly, by imported oil. So, you really must have this parallel discussion in mind.

"On the first question, you know where I stand. There is absolutely no question in my mind that there's been a legitimate argument raised by anyone which demonstrates our nation should not go forward with the development of our own energy technology. With respect to nuclear power, it fits carefully—and we have studies on this—into the future of this country and its energy requirements. The economics are related to our ability to continue to be a major factor in international world trade. Nuclear power is seen as one of the more efficient means of generating power to make products for trade. We also have data on that.

"The other issue concerns where there would be disastrous effects if we withdrew from the world nuclear community. It's very obvious that the rest of the world isn't going to stop because one or two of our states determine that they are going to declare a moratorium on nuclear technology. *If we withdraw, then we withdraw alone.*

"We are now in the process of looking through energy requirements stretching into the mid-eighties. The earliest figures reflect that if we do not get some of these re-schedulings straight and develop our program, then we will have a recession in this country—due to a lack of power—around the 1980s. There's no way you can cancel out 50 percent of your energy capacity and not do some of the other things that also need to be done. It will be rather interesting for some of those who want to look at the resulting unemployment effects in their particular areas of the country. These will provide fuel for thought. We have an awfully lot of numbers and data if you want to wander through these, and we'll be glad to show them to you."

NOTE: Zarb is referring to the FEA's "open-door" policy that gives a citizen access to facts and figures regarding the status of energy.

QUESTION: What would happen to our program for invulnerability if the OPEC (Organization of Petroleum Exporting Countries) should cut its prices or even disintegrate?

ZARB: "First, let's take a look at the actual prospect of the OPEC disintegrating. Here we have a small group of nations who are both politically and economically bound together in a small cartel which none have known before. They have a plan—pretty much architected by people who have probably been to Harvard for the most part. They believe that they have a valuable resource, and if they can make that valuable resource last for an extended period of time, and in the process maximize revenues, then they are doing what their destiny dictates. Probably, they are doing what we would be doing if our roles were reversed. Except for momentary lapses, there are absolutely no indications that the cartel will disintegrate, or that here will be a sharp break in prices—unless done purposely.

"Now, when we were trying to put together options for the President at our last session at Camp David (May, 1975), the last question that we asked ourselves was: 'What will the other fellows do?' And we looked at all the possibilities, reactions, and counter-reactions, and pretty much worked them down to one possibility: Assuming that we developed a national energy policy and Congress sponsored it—and I realize these are all very big assumptions—but assuming this did take place, what would be the reaction on the other side? We determined that one thing they might say was: 'Maybe we ought to feed those fellows for a couple of years on cheap oil, and they'll forget this funny notion of self-sufficiency and independence and go back to their dirty old habits.'

"We calculated that this might be a possibility, and that's what gave way to the price war debate, which simply said that the President of the United States shall be required to insure that domestic prices are protected against that kind of economic attack. Keep in mind that this related to the conventional forms of energy as tied to oil. This reasoning kinda got tied, someway, in our submitted fuel program, where we had another set of notions. Basically, it was the attitude of in-legislation which meant that the President shall have the authority to detect and isolate those technologies which have the potential and the ability to perform in the eighties and nineties . . . and, starting immediately, shall have the authority, through use of guarantees or subsidies, to help those industries get going now. What this means is that, by 1981, whoever is in this group will be standing firm. These are two different issues. We treated them both in a very balanced way, and we're still waiting for Congress to go along with both of them."

QUESTION: Congressman Mike McCormack has indicated that in 1974 the United States produced less energy and imported more. Is there legislation pending which would either encourage or discourage such activity?

ZARB: "I just don't know. I really don't. It's kind of a funny Congress. We've worked with some key members who really understand the magnitude of the problem. However, it didn't start that way. Took awhile. They came a long way from the initial knee-jerk reaction of 'let's go to rationing and roll-back prices' to thinking in terms of the free market system and its marvelous power processes and the needs of the future. Their ability to get congressional momentum on the big level has been marked with controversy, confusion, and compromise of these initial steps. I don't know what it's going to take. I guess it's two things . . . perhaps just continuing debate. We can be making progress one barrel at a time but we *can* make progress. The other thing is that there are crises here and there. This coming winter [1975] we will see some natural gas problems which will have a direct baring on the well-being of the members of Congress, because it will effect their local economy. And I think that each time you have this kind of problem, you are going to get some reaction. In the meantime you are going to fight like hell to get an energy plan in place this year. And if you stand back and look where we've come in the last few months, it's not bad—considering where we started from.

"You know, we started with all kinds of funny ideas as to what we had to do to roll back prices and other things to where now we are actually debating the control of conservation programs. So, we've made some progress, even though we have a long way to go and I'm not prepared to say I'm absolutely certain that Congress will legislate something this year. I am going to say that the President is not going to stand by and let the process wander into ultimate do-nothing results. We are going to use every administrative legal bid at our disposal to continue to focus attention where it needs to be focused.

QUESTION: What about energy and environment? Where will all this come out in the great national debate?

ZARB: "It's kinda funny the way things shape up in this area. We have two schools that are the most vocal and probably represent the smallest amount of Americans. One school says that anything which has anything to do with nuclear energy—bricks, mortar—is going to destroy the environmental well-being of the nation. The other school says that anybody who raises questions about the implications of energy and these developments and sites, etc., is some kind of environmental kook.

"They are both wrong. They are both wrong. It's clear in everything that I've looked at that we can work out both objectives, and work them out satisfactorily. We need a reasoned people with reasoned information agreeing to reasonable goals, and they can be worked out. We should not allow ourselves to be polarized, because polarizing means—really means—hardening the arteries and doing nothing.

"When an environmental issue is raised, we should look at it in its entirety, because it may be valid and perhaps it should be adjusted for. We shouldn't just dismiss it out-of-hand. But, by the same token, environmental objections should be specific in nature, and once it is, it can be dealt with either with technology or through legislation. So, I continue to ask those who would ask for extended delays: 'Tell me just what you would like to accomplish during that moratorium period. Spell it out. What are the things specifically that you would like done?'

"Once we've done this, then we can determine if a moratorium is required and whether it needs to be twelve hours or twelve months. Then you can close in on specific issues and begin to make progress. What we need in this debate, more than anything, are well-balanced, reasoned people who have an appreciation for both objectives and have the ability to argue a balanced program for Congress . . . Those who would seek to make a public argument, I don't think are doing it with justice in mind."

QUESTION: Are Mike McCormack's forecasts correct? If so, do you see any alternatives for the job losses he forecasts, or changes in life style and redistribution of population, etc?

ZARB: "I haven't seen Mike's forecasts that you are referring to . . . Forecasting in Washington is an interesting science. Particularly, if you go far out. If you get into the '79, '80 or '81 period, then boy, you can have the most satisfying experience of your whole life. You can let your creativity just run, and you don't have to worry about standing up and being counted when the time comes.

"The fact is, we (FEA) have some forecasts, and these are based upon today's reality. We are now importing 38 percent of our oil energy. The price of that oil has been $3 billion in 1970 to $25 billion in 1974. In 1977, that bill will be up to $32 billion and we'll be up to 45 percent dependence. Now, if we sell the argument of national security—which I feel very strongly about—if we sell the argument of the embargo and its effect on the economy—which I feel very strongly about. But if we don't share in those two views, then tell me why the cartel won't raise its prices another 10, 20, 30 or even 40 percent in the next three years, in the face of recovering world economy and in an increasingly heavy amount of imports coming from there to here. Every indication suggests that this will occur. What will that do to inflation? What will that do to the American consumer? We will do well to get by the two problems I just described. So, my forecast—based upon today's realities—would indicate that this nation has got to do something and has got to do it *now*. We must stop our growing vulnerability, begin to dig into the base, so that we can recapture some of the independence we have lost in the last ten years. I don't know exactly what Mike's forecasts are, but I wouldn't be surprised if they didn't come out close to that."

QUESTION: Aren't we reaching the point where, in effect, the OPEC is determining energy prices in the United States in terms of higher market costs? What, then, are future implications of this? Also, what is the administration policy likely to be in this matter?

ZARB: "A good question. I'm not sure what a realistic price for world oil is right now. I think that we have to look back when we launched our own independence and we sought refuge in cheap imported oil and papered over all these many problems, such as the national gas problem, etc. Cheap oil really did a lot of damage to us. So, you have to ask the question: 'Would it not have been wise for the cartel in those days to let their prices rise gradually to reflect its increasing value to the world economy?' We have to play the ball where it is. And where we are now is a nation of declining production and increasing imports. We have to do two things: We have to make consumption more efficient, and we have to encourage the development of domestic production.

"The first step we take—the making of consumption more efficient—by raising the value of oil in our economy, so that the homeowner makes a different set of investment consumption decisions in respect to such things as storm windows, or insulation in the attic. The automobile driver, rather than ordering a chromium-plated gunboat, makes a different kind of determination in respect to the wheels that he is going to use; and the plant manager, in turn, makes a different decision in respect to his equipment. That may be over-simplifying the situation, but if you multiply that many times in our economy, then you can get an idea of the effect that it might have. We are going to have to adjust our value concept in using oil.

"I had dinner recently with the energy chief of France, and I said there are people in this country who are doubting our ability to construct this new consumption-investment formula based upon high prices. He said that he had, in Paris now, gas at $2.40 a gallon. We have a long way to go before we reach that problem. But he went on to say that, 'If you have any doubters in this country, then ask them to come to Paris and look at one of our parking lots, and then come back and look at one of yours. And then tell me that there's no real effect over a long period of time with respect to the value of energy in society.' I think that pretty much tells its own story.

"So, yes, I agree that we are going to let prices go to an artificial level. Artificial or not, it is the world price set by the world cartel. It also gives us the opportunity to take back in the form of windfall profits —if that's the best term for what we think is required to encourage the continuing development of our oil industry."

QUESTION: Is industry's voluntary compliance with energy conservation rules generally regarded as successful?

ZARB: "The success has been spotty . . . I think the ability to achieve conservation in industry can be rated a C-minus. Too bad, and I can't understand exactly why—except, perhaps, that it just hasn't taken effect yet, or perhaps the momentum has yet to show results. But it has been spotty.

"I gave an award recently to a chief executive of a multi-national corporation who had achieved a 20 percent per unit of output savings in a one-year period. It saved the corporation $3.5 million. And yet, we find others who are only making a minimum effort. . . .

"Now, if you are talking about mandatory measures, that's another story. I don't think mandatory conservation is a viable approach, and I don't think we will get the results . . . I just don't think that we will."

QUESTION: Please outline some of the key elements of the President's Energy Program?

ZARB: "I'm happy to do so. There are two major segments. One is the specific area of using energy more efficiently. To do that, we propose a two-dollar tariff on imported oil, the excise tax on domestic oil of two dollars to bring it into parity with imported oil, a thirty-seven-cent excise tax on natural gas, etc. In effect, what we did was raise the value of oil throughout the economy an average of ten cents a gallon, and we proposed a little deeper tilt toward gasoline during the first three years. Now, that, plus some other things that we predict, should give us the elasticity toward a momentum to begin using energy more efficiently. We have already begun to see some of it.

"The other part of the program that we propose is federal legislation on building codes—after coming to the conclusion that the free market does not move in favor of conservation in construction of buildings in this country today. Decision-making between builder and buyer are not consistent with what we think is needed. There is also the aspect of tax credits in conservation that would go to home owners, etc.

"As to the area of coal, we ask for a re-balance of the Clean Air Act. We have also asked for Facilities Siting legislation where the federal government does, indeed, partially override the state's activity with respect to siting of energy facilities. Everybody would have a say-so, but the final decision would be made in federal court.

"There is also a target of one million barrels a day by 1985 of synthetic fuels in the economy, and we have asked for the authority to promote or subsidize these various sciences.

"You might say that these are the major areas . . . It's a 200-page bill geared toward either saving or increasing production of each measure involving some nine million barrels. Any change made in any of these measures really has to be substituted for an equal measure of barrels. Otherwise, we are just not going to get there. Also, our program does get us to the point in 1985 where we will be importing about four million barrels a day.

"To start with the conclusion that the cartel nature of things will continue, and all the residual problems that fall out of that will continue, you then reach a second conclusion that we have to do something *domestically*. There are only a few things that might qualify. First, over the next few years we can bring on no additional resources except for some additional coal burning, etc., which means that we've gotta readjust our *use* pattern. Second, we've got to look closely at our shale, off-shore [oil], coal reserves, nuclear power, etc., and reach the point where I think that we can truly be self-sufficient—or at least invulnerable to the kind of pranks and other political actions taken by the oil-producing nations at this time."

Thus speaks Frank Zarb on energy policy and technology in twentieth century America and the world. The forecast is hazy, with a chance of clearing if we can move a few clouds move up and out. Otherwise, as Zarb noted, darkness could fall in time for the first snows of winter. . . .

Chapter 5

ALTERNATIVE ENERGY SOURCES

I have acquired a solid faith in the essential fairness of Americans. It takes leadership to bring out these qualities. It is the task of such leadership to end our illusions and to talk about the world as it really is.

—Congressman Morris Udall
Arizona

Once upon a time there was a great and powerful meeting in Washington, where energy leaders gathered from throughout the world. Government was represented. Industry was represented. Oil men wearing tags like Shell, Exxon, Mobil and Gulf smiled and caucused. Scientists huddled, scribbling furiously on tiny wisps of paper. The technical press listened with their Sonys and Panasonics on "Record."

And there were speakers. Speakers on technology. Speakers on commerce. Speakers who seized the opportunity as they saw it.

Enter then the politicians. Officially, there were two at this conference. The young one was blunt and crisp like the cool winters of his Pacific Northwest. The older man was tanned and distinguished. His greying temples, touched with just a hint of dye, conveyed confidence in who he was and what he was about; steel-blue eyes smiled purposefully.

Then he opened his mouth . . . and out came a sizeable gust of a geothermal energy source known (off the cuff) as hot air. . . .

Sections of the man's speech were shot with truth and optimism. Others, however, indicated that this one presidential hopeful might perhaps need to add a few courses to his homework. First, take into consideration his audience of upper to upper, upper, upper wage earners (excluding journalists like me). Now, sift through his three-prong approach:

- Proposal of a gasoline tax
- Installation and encouragement of energy-saving machinery
- A government energy bank that would provide funds to create alternative energy sources.

As to the gasoline tax, the speaker admitted: ". . . this wins no popularity contest and the polls show it's very unpopular—and I know a lot of governors across the country who have lost their offices advocating such. But I think we ought to have a gasoline tax."

Studies show, he said (not specifying whose), that more than 15 percent of driving is non-essential driving. First, he proposed a three cents per gallon tax on gasoline which could go to an "energy trust fund." Frankly, he added, "I would like to see it going to five cents this year and five cents the next year and another five and another five until you get up to twenty cents a gallon over the next few years. And the vast majority of that could be rebated in an income tax cut across the board." This, he reasoned, would keep people off the roads and cut useless driving.

Now, to leg 2 of the tripod. No argument here: the encouragement and installation of energy-saving materials (i.e., efficient use of building supplies, purchasing smaller cars) makes good sense.

Round the corner, however, there sits point 3, which he sketched as strong support for alternative energy sources through funding available from a national energy bank. "There's a role for government action here, and it's a financial role. We ought to have an energy bank . . . committed just to the development of alternative energy sources. . . ."

Several of those in my section looked puzzled. One was a man from Mobil, another was a New York industrial engineer, another was an Indian physicist from a university in Massachusetts, who was perhaps not too clear on the English translation, and, finally, an elderly geologist from Sacramento.

Verbalized, our thoughts centered on these questions: Would a government energy bank entail still another government agency? Who would "pay" for such an energy bank? Aren't funds already being appropriated for these specific areas through specific agencies already established and chugging away? Why cut off their steam or tighten the valve? There is already a Hydra-type structure from whose tentacles bounce names like ERDA, NRC, EPA, NSF and FEA, to name a few basic ones. These in turn have divisions and sub-divisions that some have hinted smack of bureaucracy and duplications.

This, however, I can live with, and I didn't really question this man's technical expertise—his private and political record is quite impressive—until he yanked my chain by citing geothermal energy's potential as a "tremendous" alternative energy source. I know that this is rhetoric based on the future tense.

So, exactly 22 minutes and 48 seconds later, during the question-and-answer session, I prepared to ask: "Other than the Geysers in California . . . can you tell me specifically what 'tremendous geothermal reserves' you were referring to, and any attendant problems?"

Apparently, since there were only five or six women scattered throughout the large assembly, my hand looked harmless enough, and the senator smiled boyishly and said: "The little lady there in pink."

For a split second, I couldn't help but wonder if he would have addressed a man in the same manner: "The little man there in blue." Oh, well. Just so he didn't say "big lady there in pink."

The question was posed, and I caught the full brunt of those steel eyes. In fact, several blood-shot and bleary stares turned toward me from the speaker's table. One face, belonging to the publisher of a major research journal, watched curiously for a few seconds and then grinned.

A blinding flash of light (self-contained), and I began to get the same type chill a turkey must feel as she watches a farm lady move nearer and nearer, carrying a wooden stick and a bucket of steaming water . . . on the day before Thanksgiving.

Apparently, one does not ask such questions of the gods from on high. At least, I realized that this was not the way to win friends and influence people who may some day become leaders of our land. Still, as Harry Truman once said: "If you can't stand the heat, then stay out of the kitchen." So, I sat on the stool and waited . . . preheated to 400°.

"The lady [I was no longer referred to as "little"] has asked where the tremendous geothermal resources I referred to are located." A pause. Then, briskly: "The geologists I have talked to tell me that west of the Mississippi, we have very substantial reserves, and I know that in my own state . . . we have very substantial geothermal reserves."

My feathers began to fluff just a mite. It seemed clear that this speaker had casually tossed out "geothermal" without being fully aware of its status. Herein lies a genuine problem in dealing with the reality of our energy crisis, especially where the general public is concerned. High people in high places *must* know fully of what they speak because we want to believe and admire them for their expertise. The trip from "reserves" to a fully operational geothermal plant—capable of generating vast amounts of electricity—is far more than just a switch away.

Encouraged and curious about his earlier reference to a "gasoline tax," I ventured: "One more question—"

"You had one. How about giving someone else a chance." It was an order which, in effect, told "Mrs. Davy Crockett" to please go back—on a one-way ticket—to the hills and hollows of Tennessee.

Later, an economist-planner named Peterson nodded in my direction and remarked, "That was a good question. I'm from Mississippi and I've never heard of those areas he mentioned."

Bless you, I murmured.

There are several reasons why I haven't identified the speaker: his political record is impressive, he has a large following, and he stands a good chance of being elected to a major position of leadership in the United States, possibly the world. It is not my intention to deliberately embarrass anyone, nor to cast doubt on a person's ability. Rather, I believe that those of us who constitute the electorate perhaps need to raise our hands more often and question some of the statements that are often tossed out like goldfish flakes. Is there substance, or is it just hot air?

Speaking of hot air, here is a summary of just what the United States *might* expect from its geothermal resources. This is a concept in which natural steam, or hot water trapped in the earth, is utilized to turn a turbine generator. One factor that limits its potential is the actual availability of suitable sites. The only geothermal field now producing power in the United States is the "Geysers," located about ninety miles north of San Francisco, California. Edison recently sank two "dry holes" in an unsuccessful effort to locate high-temperature steam on the shores of Mono Lake in the High Sierra, and is now doing exploratory work on geothermal development in the Imperial Valley and elsewhere.

Despite the popularity of the geothermal idea with many environmental groups, Edison's Mono Lake project aroused a storm of protest in Mono County from concerned groups who circulated petitions terming it "visual pollution" of a scenic area. Geothermal power does present a number of knotty problems that are both technological and environmental. For example, the Imperial Valley fields produce—along with steam—generous amounts of highly corrosive brines which clog and ruin power-generating equipment. If this isn't enough, there are also the problems of ground subsidence, disposing of ill-smelling gases, and the danger of contaminating nearby water supplies.

The only safety problem is really an environmental one. Well blowouts can occur just as in the case of oil wells. The brine water that is blown out could cause severe damage to vegetation. In geothermal plants, unused steam is either vented to the atmosphere or reinjected into the ground after condensation. The plants can also produce ammonia and hydrogen sulfide, each of which has a very unpleasant rotten egg odor.

Also, the steam supply from geothermal fields can actually decrease. In fact, the "Geysers" are decreasing in output from initial values of seven to fifteen megawatts to about one to three megawatts (note: a megawatt is one million watts; one thousand megawatts will supply a city of about one million people). Geothermal plants are small—50 to 100 megawatts—compared to modern fossil-fired and nuclear power plants which produce 1000 megawatts. One major reason for this is that large turbines designed for use with steam obtained from geothermal sources have just not been developed, although officially the interest is there.

Still another consideration is that an adequate number of large installations just cannot be built all at one time with the present technology, since a field's capacity and lifetime must be determined before a significant fraction of a utility's generating capacity is fully committed. Forecasts of potential geothermal electric power plants that could be installed by 1985 have ranged from pessimistic values of less than 2000 megawatts to overly optimistic values as high as 400,000 megawatts. The Task Force for Geothermal Energy in the FEA's Project Independence Blueprint report of November, 1974, has established a 1985 goal of between 20,000 and 30,000 megawatts. Broken down, the latter value would represent an equivalent energy supply of one million barrels of oil per day.

Paul Kruger, scientific adviser in the Division of Geothermal Energy of the Energy Research & Development Administration, said that if this national goal is to be reached, it will require the "accelerated" development of an industry involving three parallel paths of effort:

- The discovery, proving and extraction of geothermal resources to support the production of more than 600 kilowatt hours of electricity over the amortization period of the investment in resource development and power plant construction.
- The technology to convert the resource found in its various natural types and qualities into electrical and thermal power production.
- The removal of institutional constraints to the development of an economically and environmentally acceptable industry.

Using these as guidelines, ERDA, as lead federal agency for geothermal probes, is completing a plan to implement the demonstration of commercially feasible liquid-dominated hydrothermal systems at an accelerated basis and the technical development of advanced systems such as geopressured basins and hot dry rock formations.

Kruger, former director of the RANN Advanced Geothermal Energy Research and Development program at the National Science Foundation, said field operations are slated for full-scale demonstration plants of near-term systems of resource type and conversion technology, plus pilot plants.

Another idea involving water is tidal power. Two immediate problems surface here. Not only is it expensive, but it is also limited to few suitable locations as far as the United States is concerned. One section that has been studied is the Bay of Fundy between Maine and Nova Scotia. Granted, harnessing the power of ocean tides to spin generators is original, but you still can't escape the fact that few places are actually equipped by nature to do the job. To be basically suitable, a beach area must have first, a tidal rise of at least twenty feet, preferably higher, and second, a natural bay to serve as a holding reservoir for the water. If there is no natural bay, then one must be dredged out along the shoreline. This requires an area perhaps five miles deep and ten miles long to be alternately flooded with water as the tide moves in, and then converted to a mud flat as the tide recedes.

Here is an example: To generate electricity for a 1000-megawatt conventional plant (to supply a city of one million), would require a vast array of equipment spread along some fifty miles of beach. And with recreational areas already at a premium, it requires very little imagination to see how such a project would grab the public.

Continuing in a watery vein, hydroelectric power means utilizing the force of falling water to spin power turbines and generators. It offers the decided advantage of clean water and air, but with one familiar drawback: potential sites suitable for development are scarce. In fact the experts say, as far as future needs go, don't count on the United States being able to generate as much power from hydroelectric plants as we are now doing, although actual wattage will increase.

Most of our great rivers are already dammed to capacity, in more ways than one. Other countries, such as Canada, have much greater potential for dam usage as a means of electrical generation. Prior to the industrial and population boom that followed World War II, hydro power was the backbone of California's Edison system, responsible for more than 80 percent of the total power generation. Now, that 80–20 ratio of hydro to steam has been more than reversed. Edison now has thirty-six hydro plants—mostly in the High Sierra—which contribute a total of about 730,000 kilowatts, or about 6.3 percent of Edison's total generating resources of approximately 12.4 million kilowatts.

Nor was early usage of hydroelectric power alien to other sections of the country. There is the Tennessee River watershed, which is a 40,910 square-mile area encompassing sections of *seven states*. When the Tennessee Valley Authority was created in 1933, it was given the direct task of helping this area utilize its natural resources to the utmost. Average income in the Valley was then less than half the national average; farming was the principal means of making a living. There was a critical need for industry which would provide jobs and generate other activities needed to boost a stagnant economy. Taking note of the situation, TVA decided to turn the juice on. A system of dams was built to harness the Tennessee River to provide flood control and a 650-mile waterway for navigation. The water stored by these dams was then used to generate electricity to serve the region's homes, farms, businesses, and industry.

"When I first came to TVA, it didn't have any coal-fired plants," recalled Nat B. Hughes, director of the Division of Power Resource Planning. "Most all our generation was hydroelectric. It was about

1950 when the first large coal-fired plant was built in Johnsonville. This was the beginning of steam-produced electric power for TVA."

Why?

"Why? Well, we had simply used up all the capacity for the river to produce electrical power. The demands were just sky-rocketing . . . just so much more than anybody had ever visualized. After World War II, everything just zoomed right off the top and we were hard put to keep pace," Hughes said.

Even then there was a crisis. Screams of protest yelped long and hard when TVA proposed to build its first coal-fired steam plant just west of Nashville in 1949. Opposition led by the National Association of Electric Companies managed to delay action for a short time—just enough to push construction behind schedule and into the winter months. Congress, however, came through and finally approved the necessary funds. (Today the Johnsonville Steam Plant is very much a reality.) The era of hydroelectric power had, for all practical purposes, come to the end of its curve in history. Steam was in there and puffing.

Nor has it stopped there. In the Tennessee Valley Region, TVA has undertaken the largest commitment by a single utility in the entire nation to nuclear power generation. This commitment, Hughes explained, was taken after a thorough examination of the safety, environmental and economic factors involved. "We want to make sure our region has a dependable supply of electricity during the coming decade."

In total, TVA's nuclear commitment between now and 1984 involves seventeen reactor units at some seven plants to be constructed at a cost of more than $6 million. These plants will have a combined generating capacity of more than 21 million kilowatts of electricity—the equivalent of the capacity of all the dams, steam plants and gas turbines operating on the TVA system as late as June 1974.

These nuclear units, plus planned hydro-pumped storage capacity, will virtually double the TVA electric power system—from some 24 million kilowatts to some 47 million kilowatts—between now and 1984. Almost half the 1984 capacity will be nuclear, a capacity roughly equal to the entire electric generating capacity of the entire United States at the start of World War II. Needless to say, we are growing at giant strides. Hopefully, our shoes will fit and continue to do so without pinching the instep.

Alternative energy sources can perhaps best be described as ranging from the mundane to the exotic. Senator Mike Gravel of Alaska introduced a bill calling for a moratorium on the development of fission power plants and a total phasing-out of civilian nuclear plants by 1980. To fill the void—and it would be a zagged one—Senator Gravel suggested greater use of coal and coal by-products, burning trash to generate electricity, using methane gas given off by algae, and converting animal waste by-products into low-sulfur oil. A further possibility which he proposed involved windmills that would generate electricity to electrolyze water, separating it into hydrogen and oxygen for later burning to generate heat and electricity.

Windpower was first experimented with in the United States during the 1940s and in Europe during the 1950s. For a period, a 1250-kilowatt wind-powered generator was part of the central Vermont Public Service Corporation system. Eventually, however, it suffered an undignified death: its innards rusted due to metal fatigue, and the unit was never reactivated. *Requiescat in pax. . . .*

Studies in the sixties indicated that small-sized plants (5 to 10 megawatts) are technically and financially possible, but large-scale, economically competitive wind power is considered to be most feasible in the Great Plains. Again, however, the technology required to get something going on a grand scale is simply lacking, although funds for scale models are being sought.

As to the conversion of solid waste (i.e. garbage) to fuel, a final report issued by the Environmental Protection Agency (EPA) predicts that this process could develop the equivalent of several hundred thousand barrels of oil a day by 1980. Several successful pilot plants are now operating in cities like St. Louis, Baltimore, Nashville, and San Diego County. The great advantage here is that this substitute for fossil fuels produces low sulfur oxide emission, since solid waste is generally of low sulfur content.

You might consider, too, that the production of one ton of aluminum from natural resources requires more than thirty times as much energy as producing a ton from recycled sources.

"The technology is there," maintains Farris Deep, executive director since 1956 for the Metropolitan Nashville Planning Commission. "All that remains is simply to put it together. Right now, however, solid waste is a step-child. There is actually little or no funding and little or no interest in waste management."

Nashville's $6 million Thermal Heating Plant is regarded as the nation's first to utilize energy conversion from refuse for heating and air conditioning, according to Deep.

Then there's coal gasification, which basically means that you burn coal slowly under pressure and eventually get methane gas. In fact, some $11 billion are going into some plants to do this in North Dakota, but their plans call for getting the coal from strip mines—which does not solve the environmental problem. There might be a better way and this involves burning coal without mining it. The Bureau of Mines states that there's enough coal available for what's referred to as "in-situ gasification" to equal ninety times today's known American gas reserves. At today's rate, that's enough for just about 300 years of use.

In this method, holes are dug 400 to 2000 feet deep—wherever the coal is—and some giant firecracker-types are tossed down to shatter things around. When the coal is broken up, oxygen and water are piped down (purity of the water is not a factor) and the coal is ignited. Under ideal conditions—theoretically—the top of this coal bed will begin to burn, the water will turn to steam, and the whole chemical process will yield methane. What is referred to as natural gas is actually methane with a little propane and butane that make it more efficient. However, methane alone is usable.

Wastes? These are usable or disposable, depending on how much money is expendable, and there's no known danger from them. There is, however, one drawback to this method: a spectacular ecological problem involving land sinkage—perhaps as much as 200 feet. This is a problem that has yet to be solved.

Alternative energy sources such as these—attractive on paper and good for speculation on what might be—have sparked several states and legislators to introduce legislation aimed at nuclear moratoriums. Some 200,000 out of the required 313,000 signatures have been gathered in California (mid-1975 figures) for putting a moratorium referendum on the state ballot in July, 1976. Other states such as Vermont, Missouri, Wisconsin, Minnesota, Iowa, Michigan and Massachusetts are likewise moving in this legislative direction.

In fact, I was stopped by a leafletteer outside the University of California at Berkeley. She smiled, thrust a legal-sized sheet in my hand and asked: "Are you concerned about the safety aspects of nuclear power plants?"

When I replied "yes," she said, "Good, then sign here."

"Is that all?" I pushed, getting curious.

"Yeah," she replied, the smile beginning to fade.

"What do you see as a viable alternative?"

"Huh?" she answered.

"Take me to your leader," I quipped.

"What are you . . . on some kind of trip or something?" she gawked, edging away toward a more willing subject who nodded, signed and moved on toward class—probably in Logic 101.

Me? I maneuvered my Earth shoes toward a stone bench (that turned out to be artificial) and sat back in the warm California sun. Under the rays of that most conspicuous source of heat and energy, I fantasized just how it would be if the people and their moratoriums did kick out nuclear power and turned their efforts toward utilizing the power of the sun.

Interest in solar energy is already soaring among laymen and professionals. In fact, Dr. Lloyd O. Herwig, acting deputy director of the Division of Solar Energy for ERDA, explained to a group of us that the government's solar energy thrusts are on heating and cooling, electric power generation, and the production of clean synthetic fuels. Dr. Herwig said that of the six general approaches to solar energy development and utilization, there are three subprogram areas which could possibly have near-term (within 10 years) impacts if the proposed plans are successful. These are: solar energy for heating and cooling, wind energy conversion, and bioconversion to fuels.

Three other subprogram areas—solar thermal conversion, photovoltaic conversion and ocean thermal conversion—also hold promise of providing significant amounts of electric power by the end of this century.

The federal budget for solar energy technologies in fiscal year 1975 totals about $55 million, with about 70 percent being allocated through ERDA. And, as expected the federal budget in 1976 was substantially larger, with ERDA being responsible for more than 95 percent of the total funds and programs, according to Dr. Herwig.

Defined in its simplest terms, solar energy is electric power generated by the energy of sunlight falling on the earth. It is predicted that the energy in the sunlight falling on about 3 percent of the country's land area—if utilized at about 10 percent efficiency—could meet the total projected U.S. energy requirements in the year 2000. It's that good. There's only one problem: How do you tap it?

The sun, unlike coal, oil, gas and uranium, is a continually renewable source in a never-ending supply. And the sun, unlike most earth-bound real estate, is simply too hot to handle and therefore cannot be bought or sold; its elusive rays are simply free for the taking. Again, the problem is one of collection.

Since 1958, the U.S. space program has powered satellites with electricity produced from solar cells. The first solar-powered system, flown on Vanguard I in March 1958, produced about 0.1 wattage. The largest U.S. solar array, the combination Skylab/Apollo Telescope Mount, generated about 21 kilowatts.

So far as pollution is concerned, using energy from the sun as a fuel source appears as the ideal. Unfortunately, presently available methods are extremely cumbersome because huge land areas are required to collect the sun's rays. Also, the cost of solar energy is presently about 1000 times the cost of conventional power. Indeed, one of the major problems in its development may be scaling it down to a basis competitive with present-day power plants.

The practicality of solar, or any alternative energy source, is directly related to its expense. The question is whether the average American homeowner, reeling under massive electricity and oil bills, can honestly afford the price tag that goes with solar heating. I'm talking about the guy who makes $20,000 and below per year—minus deductions—and is very likely mortgaged to the hilt, with a nice little vine-covered cottage in the suburbs that he just learned will soon be located next to a major interstate cloverleaf system. The wife is probably working and the kids may be enrolled in several different private schools. Chances, are money is a problem. Here are some thoughts on the subject:

Charles J. Michal Jr. of Total Environmental Action (TEA) in New Hampshire, believes that a good many existing homes could be "retrofitted" (to use solar energy) quite economically.

"Sure it takes some ingenuity," adds Bruce N. Anderson, head of TEA, "and it's somewhat more difficult than, say, designing a unit for a new building. But it surely can be done, and on a relatively large percentage of existing buildings." Michal, whose firm designs systems on a "case-by-case basis," put the cost at roughly $6,000 for a solar system that could, he maintains, cut fuel bills in half—but not immediately.

According to Robert Cushman, manager of sales engineering for

Sol-R-Tech of Hartford, Vermont, it would take approximately ten years to recover, by way of lower fuel bills, the excess cost of installing solar heating. This payoff time could be shorter, Cushman added, if Congress passes some type of bill that would give tax credit to homeowners converting to solar energy, and if oil and electricity costs continue to soar. One proposal in Congress would grant a tax credit of up to $12,000 to a taxpayer installing solar heating in his home.

"We assume," said Dr. Ron Larson of the Office of Technology Assessment, a Congressional agency, "a doubling of fuel costs in the next twenty years, excluding inflation."

Home Federal Savings of San Diego became the first bank in the United States to offer private loans for solar heating and cooling. Under this plan, 25 percent of a homeowner's salary is the maximum allowable expenditure for mortgage costs. Initial response has been cautious, but such practices may become readily available in a few years.

Tax incentives are another possibility. Rhode Island became the second state in the country to consider legislation which would offer tax breaks for installing solar or wind energy systems. The measure would provide unlimited sales tax exemptions for any equipment for retrofit or for new construction. Furthermore, it would not add to the assessed valuation of the home or commercial building. Indiana is the only other state with such a provision, and it has a $2,000 limit on the purchase of solar equipment.

There are some other problems to consider. The facts are that some houses would be relatively easy to change while others would not. The prerequisites, according to experts, include:

- An existing building or house must be insulated to electrical heating standards or better, preferably the latter. This would rule out most homes built within the last twenty years, because the quality of workmanship has deteriorated considerably during this time span.
- There must be space, either in the basement or yard, for a 2,000 gallon water tank to store water heated by the solar collector and to retain the heat.*

---

* Note: For example, 20,000 to 30,000 acres of thermal collector area is required for a 1,000 megawatt electric plant at current (1975-76) collection efficiencies, according to ERDA's Division of Solar Energy Research & Development.

- Ideally, you should have a south-facing room, measuring one-third the area of the interior of the house to be heated, at roughly forty-five degrees to the horizontal.
- If the above option is impossible, solar collection panels could be mounted on fences or other structures on the lawn or garden. These panels, whether on the roof or elsewhere, cannot be shaded by tall trees or other buildings. Nor should Junior hit a home run that lands dead center on one of the panels. Can you imagine sending the children out to play and warning them: "First, don't fall in the 2000-gallon water tank and stop throwing rocks and dirt in it; and second, stay away from the solar collection panels, they are too expensive to replace. . . . A child could very well grow up hating the sun.
- If you still want to give it a try, the hot-air heating system is the least expensive to convert, because existing air ducts can be used. Electric or hot water systems are more expensive to replace.

There is one energy conservant model being built to resemble Habitat at Montreal's Expo '67: it is a condominium village with central station solar collectors and storage. Grassy Brook Village, under development in Brookline, Vermont, by Richard D. Blazej, will have a single collector and storage system for each cluster of ten homes. At central equipment sites, fifteen-ton refrigeration units with double-bundle condensers will furnish hot water for heating and chilled water for cooling. Thermal storage will be provided under some of the housing units in a series of steel or concrete water tanks with a combined capacity of 200,000 gallons. Fan coil units in each residence will be capable of receiving warm or chilled water, and each room will be equipped with thermostatic controls. Just in case the system fails one chilly Vermont day, there is a supplementary measure known as one oil-fired boiler, plus small wood-burning heaters in each unit.

Overall director for the program is Robert Shannon of People-Space Company. Supervising design and overseeing the mechanical support system is Fred S. Dubin of Dubin-Mindell-Bloome Associates.

To further conserve energy, units are clustered to create common walls, foam urethane insulation will be used in exterior walls and doors and glass fiber insulation in ceilings, and, as a pleasant extra, roof top gardens will provide earth insulation. Also, casement and fixed windows are to be double-glazed and equipped with insulating shutters. Skylights will be triple-glazed.

The whole thing sounds beautiful . . . and expensive.

Private estimates for solar-heating systems for a "standard" house of 1,200 square feet run between $5,000 and $10,000 in mass production, compared with about $1,000 for conventional heating systems. And, to avoid exorbitant costs of a solar-heating system, an auxiliary unit supplied with gas, oil or electricity must be provided.

The use of solar energy on a massive scale for power generation is another matter. Several proposals have been voiced. One of the most practical, according to Nobel prize winner Dr. H.A. Bethe (*Scientific American,* January 1976, "The Necessity of Fission Power") would be a large field covered by mirrors which reflect sunlight to a central boiler. Such mirrors, he noted, would be driven by a computer-controlled mechanism and the boiler would generate electricity in a conventional manner. ERDA is supporting at least three separate groups who are working on this project. Their best estimates put costs at about $2,500 per installed kilowatt (power averaged over a 24-hour day). This does not include interest or the effects of inflation during construction. On the same basis, nuclear fission reactors cost about $500 per kilowatt, which puts solar at about *five times* as expensive as nuclear power.

Direct conversion of solar energy to electricity in a solar cell is from 2 percent to 10 percent efficient. Solar thermal conversion systems range up to 25 percent efficiency and, as a result, are used only for extremely small generating stations of from 1000 to 2000 kilowatts. (There are more efficient systems in the laboratory stage at least two or three decades away from practical use.) Consider, for example, a 1000-megawatt plant, which would require about ten square miles of land and cost many times as much as plants utilizing conventional methods of generation. Installation costs are likely to range from $300 to $2,000 per kilowatt.

One environmental advantage would be that a 1000-megawatt plant would occupy less land than the strip coal mines needed to supply a coal plant of the same size. Toss the coin another way, however, and you learn that environmental costs would include disposal of 55 percent to 80 percent of the heat rejected by a plant, plus the damage caused by shading the earth—deserts or other sites—below the solar collectors' surfaces.

Just how is solar energy stored, recovered and then converted to electricity? It's a little more complicated than the childhood trick of exposing a magnifier to the sun's rays, although the idea is the same. Technically, sunshine consists of electromagnetic waves that travel through space at the speed of 186,000 miles per second from the sun (which is 93,000,000 miles away).

Sunlight, finally arriving at the edge of the earth's atmosphere, carries energy at a constant rate of 444 British Thermal Units (Btu) per hour for every square foot of area. You can compare this in electrical terms to the equivalent of 130 watts. Taking into account absorption by the atmosphere, hours of darkness, cloud cover, and geographical location, the average amount falling on a square foot of ground in the United States on a yearly basis is about 13 percent of this, or 58 Btu per hour (which means 17 watts in power units). This amount of heat, if absorbed and retained by a little less than a cubic foot of water, would raise the temperature of the water by 1° Fahrenheit during one hour's time.

Averages, however, can be misleading. For example, when it is hot enough in the summer to "fry an egg on the sidewalk," there is seven times as much solar energy reaching the ground compared to a typical cloudy day in winter. And, of course, there is no solar energy at all during the night. This intermittent characteristic of solar energy is one of the challenges that must be met in order to efficiently utilize the sun's energy. Adequate storage providing for such periods of energy demand might be accomplished in a variety of ways, such as electric storage batteries, pumped water storage, or thermal storage, to name several possibilities.

My first encounter with the awesome aspects of solar power came during a stay in the Middle East in 1966-67. It was mid-December when I first arrived, and cold but not disagreeable. Still, hot showers were one luxury I carried over from America, and in this respect, I am a creature of habit. Everything went well the first few days. Plenty of nice, warm water. In fact, I was totally unaware of the roof-top solar collection unit. Like most people, I was acclimated to flicking a switch or turning a knob and presto! out comes electricity in one form or another.

Then it happened. Rain, rain, rain. Cloudy days that meant no sun and nothing for the collector to collect. It was a fact that came gushing home when I nonchalantly turned the shower knob and was blasted with what felt like the Polar Ice Cap. . . .

Solar energy—once you have caught it—can be stored efficiently as heated materials contained in insulated tanks. The heat in these materials can then be used to produce steam for the turbine-generators used in power plants. Another efficient method of storing solar energy is to pump water to reservoirs at higher elevations during the daylight hours and use this water to operate hydroelectric generators at night. Still another method is to produce hydrogen and oxygen from water (by either electrolysis or high temperature decomposition) during daylight hours and to burn the hydrogen when needed to produce electricity. All of these storage and recovery methods, however, require massive and costly installations. And at the present time, all of these methods require a conventional backup heating system which defeats the original purpose.

A Solar Energy Panel resulted from a merger of ideas between the National Science Foundation and NASA (National Aeronautics and Space Administration). Four types of systems were identified and studied as realistic solar prospects:

- Systems mounted on roofs and sides of residences and commercial buildings
- Large generating plants adjacent to industrial complexes, such as aluminum or caustic soda plants
- Large land power plants
- Solar-powered space stations that would convert the solar-array D.C. power to microwaves, beam these back to earth, and then convert the power back to A.C. or D.C. for conventional transmission to cities.

A major technical problem, as outlined by the panel, concerned the cooling portion of the solar-conditioning system. Specifically, an efficient cooling cycle must operate effectively between ambience (pervading atmosphere) and 200° F for integration into the total system. Since the heaviest cooling loads occur when the solar energy is at its maximum, the loads are in phase with the energy source. Then there is also a thermal pollution (increase or decrease in water temperature) aspect to consider, since solar systems would operate at much lower temperatures (an adverse aspect) than a nuclear power plant, coal-fired plant or the breeder series (LMFBRs).

Which way to go? Up and out is a definite possibility. It now appears that the most attractive conditions for solar-conversion are in outer space. What scientists have in mind is a synchronous orbit with a site fully illuminated twenty-four hours a day. The space shuttle program, an economic earth-to-orbit transportation system, is generally regarded as the vehicle to provide such a program of conversion. It would serve as a sort of umbilical cord between a Satellite Solar Power Station (SSPS) and a receiving plant on Mother Earth. An assessment of SSPS feasibility has merited considerable interest from such groups as Arthur D. Little Inc., Grumman Aerospace Corp., Raytheon Company, and Textron Inc., in addition to NASA researchers.

At the Johnson Space Center some thirty miles toward the Gulf from Houston, NASA scientists are hard at work on solar power stations—it's a priority item and a major part of the forthcoming Space Shuttle program.

Solar power, according to NASA Scientist-Astronaut Story Musgrave, has never received so much attention as in the last few years. As Musgrave, a native of Kentucky, told me: "I think the momentum is definitely there. What we would do is use solar panels 'right to microwave' and then 'right to earth'—a clean operation."

In a nutshell, a space satellite capable of generating 5000 Mw of power on earth means the challenge of creating in orbit a 25 million-pound payload and its propellant supplies for station-keeping purposes of about 30,000 pounds per year. In addition, the present cost of silicon solar cells for use in spacecraft—about $175 per watt—would abruptly kill an SSPS. Solar cell arrays, while a sophisticated part of the space program, cost 200–300 times too much for large-scale ground applications. There are, however, new methods of producing single-crystal silicon, mass production assembly techniques for arrays, and a large market which, all working in conjunction, might drive the cost way down. Here again, though, it's one of those nebulous situations known as an "iffer."

From the scientific community, there are also more than a few mixed feelings regarding biological effects of microwave exposure. The power generated by the SSPS in its synchronous orbit must be transmitted to a receiving antenna on the earth's surface—via microwaves—and then rectified. Because of the lack of internationally accepted standards on microwave exposure, the system would have to be designed to accommodate a wide range of frequencies and microwave-power-flux densities. Basically, its microwave system design must produce only low microwave exposure levels to assure safe operations which are internationally acceptable.

As to levels of microwave toleration, Astronaut Musgrave voiced his own opinion that this need not be a tough problem to handle. "It would seem that you could create restricted areas around microwave. For example, the FAA has endless areas that you can't fly in now. Simply, fence off the ground and that's it."

Talk with NASA researchers and you learn that while solar-generated power sounds good on paper, it is still very much in the lab stage. Still, we are beginning to feel a few encouraging warm rays. One angle currently being pursued involves concentrating the microwave by a signal beamed from earth. In essence, there is a multiplicity of antennas synchronized by a earth-bound antenna to the solar power station orbiting in space. This then translates to a "lock" that directs microwaves to the right place, thus preventing their being "scattered" like pieces of grain in all directions. This produces a feedback loop that assures the concentrated force of a bullseye each time in a designated area.

Finally, absence of a land energy source is one clear advantage to the SSPS: there is no environmental problem tied to mining, transporting, and refining the natural resources of earth.

Taking all factors in consideration, the NSF/NASA Solar Energy Panel agreed that by the year 2000, solar power could provide:

- At least 20 percent of the total U.S. energy needs—an amount nearly equivalent to the total energy consumed in the United States during 1970.
- At least 35 percent of the U.S. cooling and heating needs for buildings from coast-to-coast and border-to-border.
- From renewable organic materials, at least 30 percent of the U.S. gaseous fuel and 10 percent of the liquid fuel needs.
- Less U.S. dependence—by at least 20 percent—on foreign energy sources.

What this involves, in terms of dollars and cents, is research geared around an initial investment of about $2.25 billion dollars over a fifteen- to twenty-year period. The U.S. House of Representatives, for example, passed the Solar Heating and Cooling Demonstration Act of 1974. Authored by Congressman McCormack, it provides an immediate $50 million for a five-year program which would demonstrate the commercial feasibility of using solar energy to heat and air-condition residences and other buildings. But, as even Representative McCormack noted, ". . . it is important to keep the short-term potential of solar energy clearly in perspective. It would require a stupendous effort to provide solar heating and cooling in just 5 percent of the buildings in the United States by the year 1990. Yet, this would represent only 1 percent of the total energy consumption of this country. Clearly then, solar energy will have no impact on relieving our current energy shortages."

There is no dispute that solar power may be utilized extensively in the next century, but the question facing citizens of *this century* is how best to keep the home fires burning while we ready our house of tomorrow.

In a report issued in mid-1975, the Energy Research and Development Administration through its administrator, Robert C. Seamans Jr., outlined plans for the United States to become energy independent. Based upon an analysis of scenarios, the status of the candidate technologies, and the extent of available resources, the following points were identified as "near-term" priorities:

- To preserve and expand domestic energy systems: coal, light water reactors (the highest nuclear priority), and gas and oil both from new sources and from improved recovery techniques.
- To increase the efficiency of energy used in all sectors of the economy and to extract more usable energy from waste materials.

The next stage is mid-term and covers from 1985–2000 and involves the following priorities:

- To accelerate the development of new processes for production of synthetic fuels from coal and for extraction of oil from shale.
- To increase the use of under-used fuel forms, such as geothermal energy, solar energy for heating and cooling, and extraction of more usable energy from waste heat. None of these technologies has a major long-term impact, but each can be quite useful in relieving mid-term shortages.

From these steps, ERDA then projects into the year 2000 its long-term priorities: To pursue vigorously those candidate technologies which will permit the use of essentially *inexhaustible resources:* nuclear breeders; fusion; solar energy from a variety of technological options, including wind power, thermal and photovoltaic approaches; and use of ocean thermal gradients.

A final footnote for one more possibility.

A $3.5 million contract for a three-year program to demonstrate the economic feasibility of generating electricity from high-temperature, fast-moving gas was awarded in late 1975 to the Air Force Systems Command's Arnold Engineering Development Center, Tullahoma, Tennessee. The Office of Coal Research in the Department of Interior awarded the contract to show that the technique, which has been employed at AEDC for aerospace testing applications, can generate power at costs that would be commercially attractive.

In theory, magnetohydrodynamics, or MHD, has been around since the 1950s and in limited use here and abroad since the 1960s. Its operation depends on the fact that hot, fast-moving gases (like the exhaust of a jet engine or rocket motor) become ionized, or electrically charged. If this stream of ionized gas is passed through a strong magnetic field, an electric current is generated.

Application of the principal has been restricted to uses where the need for huge bursts of electricity outweigh the costs involved. Such a need existed at AEDC in the early 1960s, and a MHD generator was constructed there to power a wind tunnel then being proposed for the Center.

Robert Dietz, AEDC Director of Technology, explained that due to changing military aviation requirements, the wind tunnel was not built, but the MHD generator did become operational. During its two- to three-minute running time, it produces 16 to 18 kilowatts of electricity. "The major modifications now required to prepare the generator for the research are nearing completion," Dietz added.

The Interior Department contract was awarded following completion of a year-long study, also under OCR sponsorship, by personnel of ARO Inc., the Center's operating contractor. Published in August, 1975, the study noted that there is a "significant gap" between the demonstrated efficiency of MHD generators and the level of efficiency needed to make the technique an economic source of commercial power, but that "a specific program to verify that a MHD generator can achieve the required performance has not yet been activated."

The report further noted that by using the existing $4.5 million installation at AEDC, the demonstration could be done on a relatively large scale and with a minimum expenditure of new funds. Cost of the program covered under the contract included both necessary modifications to the AEDC equipment and its operation.

The largest single item for replacement is a million-pound magnet, a replacement that will cost more than $900,000, and expansion of the power equipment associated with the magnet—about $260,000. The only other major modification is the channel through which the gases are directed through the magnetic field and where the power is generated. Cost? An estimated $270,000. MHD—like other alternative energy sources—just doesn't come cheap.

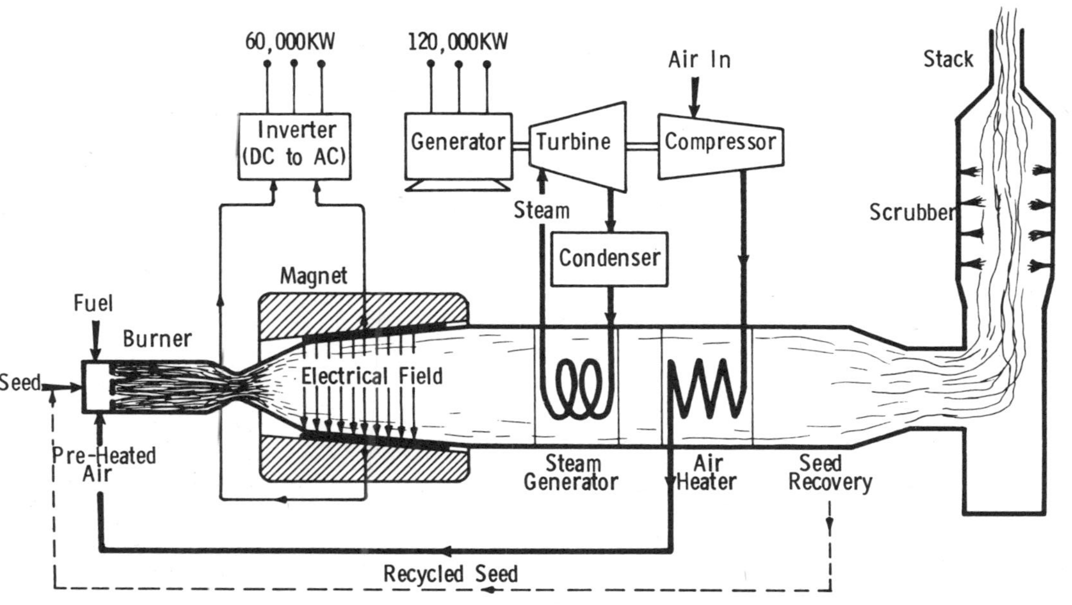

A MHD operation—such as this one at the Arnold Engineering Development Center near Tullahoma, Tennessee—is dependent on hot, fast-moving gases (such as the exhaust of a jet engine or a rocket motor) becoming ionized, or electrically charged. If this stream of ionized gas is passed through a strong magnetic field, an electric current is generated.

Chapter 6

WHY A REACTOR WON'T GO " POOF" IN THE NIGHT

Nuclear power reactors are now producing more than 7 percent of the nation's electricity by heat generated from the fissioning of uranium. It is anticipated that these reactors will supply more than 25 percent of our electrical generating requirements by 1985, and more than 50 percent when we top the 2000 mark. And, as of March 1, 1975, U.S. utilities—themselves involving a capital investment of $106 billion—stand committed to more than 220 nuclear plants. This means our national investment in nuclear power will stretch into hundreds of billions of dollars before this tired old century drops off into infinity.

Fine and good, you say, but what if one of the "things" blows its cool and goes "poof" in the night? Do we crawl in a ditch, head for the earth's center, or begin a very devout Act of Contrition?

None of these. Back-up systems for nuclear plants are like giant and very expensive safety pins. They are designed to take care of big "poofs" or little "poofs" all in a few seconds. An analogy might be drawn that's like what happens when the first minority family moves into a new neighborhood. Everybody twitters around for a while, scrambling and trying to figure out what to do about them. Then, one by one, they get to know their new neighbors. Then. Presto! "Why, there's nothing strange about these folks!"

Similarly, there's nothing mysterious or frightening about nuclear power plants when you understand the basic operating principles.

The basic elements of a nuclear plant are similar to those of a coal-fired one. The nuclear reactor and the coal-burning furnace are both sources of heat for producing steam. I found it helpful to go through what is nothing more than a very basic General Science explanation that goes like this:

Flame is applied to a flask. The water inside is then heated to boiling. The resulting steam, under pressure, begins to flow through a tube and strike the arms of a turbine, causing it to spin like crazy. This turbine is connected by a shaft and many turns of wire, each tightly wound together. As these wires spin within the magnetic field of an attached magnet, electric current flows through these wires to a meter which should register—on a home device built for demonstration purposes—a current of maybe eight-thousandths of an AMP.

Now, a power plant works on this same principle, but generating several million times more current. It has four important parts: a generator, a turbine that supplies steam under pressure, water in a boiler apparatus, and a heat source. A nuclear plant is no different: you will find these same components operating on the same principle as in a conventional plant. There is a generator, a turbine, and a cooler where the steam coming from the turbine is changed into water and then piped back to the boiler to be used over again. The heat to boil this water comes from a unit known as the heat exchanger, a name that aptly describes its function. The water is changed into steam on one side by the heat flowing on the other side. There is no connection between these two systems, and only the heat is exchanged from one side to the other. In other words, they share a duplex but never really get to say hello.

In a boiling water reactor, for example, the heat is created by the controlled nuclear fission of uranium fuel in the core. This uranium fuel consists of pellets that are small enough to fit in a large thimble. The fuel is a ceramic material resembling firebrick. The pellets are put inside metal tubes about a half-inch in diameter and twelve feet long, and these tubes are then assembled into bundles containing forty-nine rods each. These bundles—in groups of four—are then separated from other groups by control blades, which have to be withdrawn before any usable amount of fission can take place.

Some 800 of these bundles are installed in a large steel vessel that contains water. Heat is then supplied to the heat exchanger by the water circulating through the reactor. Subsequently, the nuclear reaction occurring in the fuel rods produces the heat. The process is called *fission* and takes place when the nucleus of an atom is split approximately in half. (Its opposite would be *fusion* which means to bring together.) Its output is regulated by control rods placed between the fuel rods: to increase the heat output, these control rods are raised or exposed to each other; to decrease the output, the rods are lowered.

As the control blades are worked, the fission process generates enough heat to boil the surrounding water and thereby make steam. This steam then leaves the reactor vessel through pipes and triggers an electricity-producing turbine.

After passing through this turbine area, the steam goes to a condenser, where it will be converted back to water by circulating around tubes containing outside cooling water. The water formed from the steam condensation is pumped back to the reactor vessel to continue its steam-making cycle, so in a real sense, its job is never finished.

Note here that the water used in a power system and the cooling water are separate and distinct at all times. The outside water which flows through a plant's condensers will *never* enter the reactor vessel, and the water in a reactor vessel will *never* come in contact with outside water.

Working reactors have proven to be fertile targets for critics who are fond of comparing these to potential atom bombs just waiting to be accidentally (or otherwise) detonated. One standardized phrase goes like this: "A reactor has enough fissionable material to make hundreds of atom bombs. Do you want *that* sitting in your back yard?"

Then there is the issue of sabotage and terrorism involving the theft of nuclear materials. The truth of this matter is that one bent on nuclear blackmail could cause more trouble to himself than anyone else. The heavy inert solids involved are less of a threat than so much dynamite.

As for do-it-yourself atom bombs, nuclear fuels have the wrong composition; the materials that move in industrial nuclear power channels are of little value to the would-be nuclear physicist. It is a separate question whether a highly sophisticated, heavily financed organization could acquire the array of special talent needed to seize, separate, and then purify plutonium to produce its very own personal bomb.

Sure, a brainy physics or chemistry major might be able to mix up a batch of bombs. But why go to the trouble of stealing industrial energy sources? As one scientist noted, this is a question that doesn't really touch the public utilities, since a truly intelligent person or group would know enough to go for weapons material, or the finished product might conceivably be stolen from places like Los Alamos or Oak Ridge, providing you can scale the elaborate security measures. Choosing any other route is, in effect, like stealing a paint-by-number set that has its numbers all jumbled.

I'm not trying to skirt this subject of terrorism and security or make it sound trivial. Three major assassinations (with an almost fourth) have taught Americans that there are indeed strange people and organizations floating around who feel they have a divine destiny to save mankind. In fulfilling this mission, such people often exhibit various degrees of violently anti-social behavior.

Nevertheless, to sabotage a nuclear power plant and cause release of radioactivity would be very difficult. First, it would require an intimate knowledge of the plant design and construction, its operational controls, and its protective systems. The fail-safe design and control interlocks of the plant would seemingly thwart all but the most expert attempts. It is, again, highly unlikely that a determined group of saboteurs could, in the limited time available, produce the chain of multiple failures necessary to release large amounts of radioactivity into the environment. A wise terrorist, or one bent on disruption, would find it far easier to attack food, water, or communication services, and with far less opposition.

As to the vulnerability of coastal plants to offshore enemy shelling, the many layers of containment shields involved would make penetration difficult. Not only is it difficult for fission products to get out of a reactor, but it is more difficult for missiles to penetrate the many barriers to get into the reactor core—and that's where the action takes place. First, the outer containment shell must be penetrated, then seven to ten feet of concrete must be ruptured, and finally, eight to ten inches of steel must be penetrated in order to cause damage severe enough to cause radioactive contamination to the environment. The actual containment shells are capable of absorbing substantial impact (as in large commercial aircraft) without danger of radioactivity release.

But to get back to the exact reason why a reactor cannot blow up like a bomb. There is only one major similarity between a nuclear power reactor and a bomb, and that is *uranium*. And there's so little of the right kind of uranium in a power reactor that it's just not possible for that big dome to go off with a big bang. It's as simple as that.

Here's the logic to that equation: To make a genuine bomb, two quantities of almost pure uranium must be brought together with a tremendous force. Explosives do this job for a real bomb, and the result is a chain reaction that really sets things moving. The heat in a nuclear power plant, however, is supplied by less than *four* atoms of pure uranium fuel for every *ninety-six* atoms of other materials.

These other ninety-six atoms have the exact effect of diluting the smaller quantity of uranium fuel. The other atoms also have the property of shutting off a reaction if the temperature does start rising. And speaking of temperature, remember that any unusual increase in either pressure or temperature triggers the control rods to drop, thus stopping a nuclear reaction. These are geared to operate manually should the automatic control fail.

Unless a potential bomb is set off in "clean" surroundings, free from neutrons, it will simply preignite and then shut itself off by thermal expansion. Vital Point: *A reactor always has neutrons present, which is definitely the wrong surroundings.*

There is one final factor to consider. A bomb-to-be must be fired by pushing its parts together in a few millionths of a second, and there is nothing in a reactor to give such speeds, even if it were possible to meet the other impossible conditions.

I still remember the words of a crusty old plant superintendent in upper-state New York who had been a U.S. Navy nuclear engineer.

"Honey," he said, "you learned the three *R's* in school. Right? Now, you got to remember the three *W*'s in respect to this bomb stuff. A reactor can't be a bomb because it has:

"*W*. Number 1: Wrong composition.

"*W*. Number 2: Wrong surroundings.

"*W*. Number 3: Wrong timing."

Satisfied, the ex-Navyman chewed on a well-worn cigar and waited for my response to his brief summary of why a reactor can't explode.

"Honey," I replied, "I can dig it."

Chapter 7

BACKUP SYSTEMS TO BACK UP THE BACKUP SYSTEMS

Nuclear plants don't just pop up out of the ground like spring dandelions . . .

Ask a scientist, juggling test tubes at Princeton, or another researcher, working in the lush hill country of Oak Ridge, and the answers are invariably the same: The finished product—a working reactor—is born of careful research and planning that spans many, many years. In addition, the Nuclear Regulatory Commission is a grandfather of the mountain-type personality, who decides just when a nuclear plant is ready for its world debut and roars when it isn't.

Granted, a newly evolving standardization process has speeded things up to a degree, but the emphasis on safety has not been touched. In fact, even more stringent measures have been applied, according to one commercial plant operator based on the West Coast. His widely agreed-upon anonymous comment: "Believe me, nuclear plants are the most kosher things going. The NRC acts as a Chief Rabbi who makes sure nothing impure messes up the works. It's orthodox all the way!"

I am convinced that airplanes are great for interviews. The main reason is that your victim is cornered and usually under the influence of alcohol mixed with altitude, which helps loosen a normally tight tongue. And then there's a relaxed atmosphere, reminiscent of a cozy cocktail lounge, unless a stewardess gets too pesky with trays and pillows.

On a flight from New York to Nashville, instinct told me to be brave and take an assigned seat in the "Smoking Section." The pickings here are usually different from "First Class" or "Non-Smoking," and on this particular occasion it was a wise, wise choice. Seatmate Number 18A turned out to be a tanned, six-foot-six, construction engineer whose speciality was—you guessed it—nuclear power plants. No kidding. This really happened. In fact, my initial reaction was, "One of us is a plant and it's not me." Until he started talking and sipping gin and tonic.

Apparently, it had been a long, hard year, and midway over Virginia, the engineer and I stopped mouthing polite chit-chat and got down to issues. There are several things griping America, he philosophized, then proceeded to name three: (1) bad, bad economy, (2) material costs skyrocketing, and (3) labor that's hard to find and expensive to keep.

So what else is new, I wanted to say. I didn't choose a two-hour flight in the cargo section, gagging smoke, just to hear that the United States is shooting "slag" rather than the pure stuff.

"Now, the root of the problem," the engineer continued, pointing a red swizzle stick toward me, "is the government. Simply screwing up the works. You can't get away with a shoddy job. How the hell do they expect you to make a buck? Inspectors are running around with magnifiers. And it's nearly impossible to pay off one of the bastards. Usually, he's got somebody standing over *him,* inspecting what *he's* doing. Damn! The country's in bad shape . . ."

Silence. The words cracked and jerked as the plane whizzed on. Finally, I looked out the window and then made a simple notation in my frazzled brown pad: God bless the inspectors who inspect the inspectors who inspect nuclear plants. They keep us poor but honest even when we don't want to be. Amen.

We do not, however, live in a germ-free society. In the past, accidents involving fission products have occurred, but all of them have been terminated and resolved without major release of radioactivity. And accidents are expected to happen in the future, but almost all of them will be terminated without any problem to the public. It has been estimated that rarely—perhaps once in every million reactor-years of operation—could a serious accident possibly occur. This could only happen if all the barriers are penetrated, which is not an easy task, even for an accident.

A Reactor Safety Study was commissioned in 1973 by the AEC to establish probability of various types of reactor accidents. The results were published in draft form in August 1974, in a document that has come to be known officially as the Rasmussen report. It was named for the chairman of the study group, Norman C. Rasmussen of the Massachusetts Institute of Technology. The final report was published in October 1975. The methods applied in the Rasmussen study have been used for several years in Britain to predict the probability of industrial accidents, but experience has shown that such predictions usually provide a higher frequency of accidents than actually occur. A basic aspect of the Rasmussen report is that the probability of a major release of radioactivity is about once in 100,000 reactor years. Such an accident would involve release of about half of the volatile fission products contained in the reactor. A release of that scale would have to be preceded by a meltdown of the fuel in the reactor, an event for which the report gives a probability of once in 17,000 reactor years. Finally, the report predicts that the water coolant from a reactor will be lost once in 2,000 reactor years but that in most cases a meltdown will be prevented by the emergency core coolant system.

Experience does provide at least some check on these estimates. Point One, there has never been a loss of coolant in 300 reactor years of commercial light water reactor operation. Furthermore, there has never been a fuel meltdown in nearly 2,000 reactor years of commercial and naval light water reactor operation. Therefore, if Rasmussen's estimate were wrong by a factor of 20 (in other words, if the probability of a meltdown were once in 850 reactor years), at least one meltdown should have occurred by now. Except that it hasn't.

Such safety prospects result from the way in which nuclear power plants are designed, licensed, regulated and operated. A bulk of the nuclear program is public safety, and, dollar-wise, it is in the economic interest of the utility companies to make sure plants and equipment behave. An elaborate system known as "Defense in Depth" includes highly reliable, fail-safe equipment, duplicate and triplicate systems and components, backup systems, diversity of operating principles for safety equipment, and the "design basis accident" analysis.

The facts are that any proposed nuclear power plant could very easily find itself "out at home" without ever swinging at the ball. The "ump" calling the signals is the Nuclear Regulatory Commission which has a rule book called "Licensing and Regulations." And this is one case, according to utilities and industry, where the book is followed page by page . . . chapter by chapter . . . volume by volume. The NRC gets its power in this regard under both the Atomic Energy Act of 1954 and the National Environmental Policy Act of 1969, as interpreted by the Calvert Cliffs decision of the U.S. Circuit Court of Appeals, Washington, D.C.

To license a nuclear plant, a utility must obtain sometimes as many as fifty-two different licenses or operating permits. In California, for example, the present number is thirty-two. Licensing is actually a two-step procedure. First, a construction permit must be obtained before any construction can begin. This application must contain a description of the plant, its site, and the utility's financial and technical qualifications, plus two vital documents: (1) the Preliminary Safety Analysis Report (PSAR), and (2) the Environmental Impact Report. These are not just several pages of loosely-stapled paper, but several thick volumes which are made available to the public for study and comment.

Once a utility has determined a need for a power plant, it then follows a selection procedure that has been developed from many years' siting experience with all types of plants—including ten years of nuclear work. Technological, economic, legal, environmental and public opinion factors are included. Specific site characteristics are evaluated through environmental engineering and scientific studies, including geology, ecology, water quality, meteorology, archaeology, and oceanography. Other areas considered include: topography, aesthetics, zoning, water supply, and transportation.

The Preliminary Safety Analysis Report (PSAR) is submitted prior to issuance of a construction permit. During the various reviews, a number of questions are raised by reviewers that must be answered in detail in writing by the applicant. The answers are usually submitted as supplements to the PSAR and normally add several more volumes to the total. Frequently design modifications, equipment additions, or both, are required to satisfy the safety requirements. After the plant is built, a Final Safety Analysis Report (FSAR) and an updated Environmental Impact Report are written that include all the supplements and changes, reflecting the "as built" instead of "as planned" aspects of the plant. These reports are submitted as part of the application for an operating license.

Then there is the time factor. Construction permits have taken from four to twenty-two months, averaging approximately twelve months for forty-one plants. Operating licenses have taken from nine to thirty months, averaging about twenty months for fourteen plants. The other fifty assorted licenses and permits required are usually obtained within the total time required for construction and operating permit processing. In addition, the Cliffs decision (Fall, 1971) initiated the environmental impact analysis which increases the time required to obtain both construction permits and operating licenses.

During the entire process, private citizens and interested public groups may present their views for consideration prior to the issuance of construction permits and operating licenses. Such individuals or groups may go before hearings held by the Atomic Safety and Licensing Board (ASLB). There is a minimum of two such hearings, one during review for the construction permit and another during review of the operating license. In addition, there are a number of public hearings conducted by a state public utility commission at which the views or individuals may be presented. Members of the public who may have a compelling interest in a particular plant hearing may ask the ASLB for, as the legal phrase goes, "leave to intervene" in the proceedings. As an intervenor, the person or group is considered part of the formal hearing, with the right to testify and to question the testimony of others. Decisions by the ASLB can be appealed to the Atomic Safety and Licensing Appeals Board and from there to the NRC.

In addition, there is the matter my airborne engineer groaned about: stringent quality control and quality assurance in the fabrication and installation of the equipment, plus detailed testing and inspecting procedures.

According to Floyd L. Culler, deputy director of the Oak Ridge National Laboratory, there are two "philosophical" approaches that are viewed in regard to reactor design. The first is the actual approach to the design and is referred to as "Design in Depth." Here the emphasis is on top quality building materials so that the possibility of even little accidents—links, valves, etc.,—are obviated by great care in the engineering design and execution.

The second link in this philosophical chain is "Defense in Depth," which means that the total operation, reactor et al, is designed in such a way that should an accident happen, it would occur in a benign way.

Since the beginning of nuclear reactor design, scientists and engineers have been asking themselves, "What if this happened? or that? or this and that?" Each time, the consequences of such hypothetical checkers are analyzed, and the plant is given an added stitch, if needed, to provide the acceptable level of protection for all concerned.

"Listen, there's a great amount of effort put into this," noted Culler, a member of the Scientific Advisory Committee to the International Atomic Energy Agency in Vienna. "Let's break it down a little more. You might say, for instance, 'suppose a valve doesn't close? What then?' That means we go back and re-analyze so that the failure mechanism is actually fail-safe. And that's relatively easy to do, with forethought."

In addition, there are numerous monitoring and detection systems to pinpoint leaks, locate potential "hot spots" and, in general, uncover any gremlins who show devilish potential. They get their fire cut off and quick. Furthermore, a third element assures that the system is equipped to handle even major failures deemed either improbable or at some very, very low order of probability. What we actually do," explained Culler, "is design an automatic layer of protection for accidents that we consider to be impossible."

There is even a final safety catch for those who are prone to visualize the ultimate scenario of molten radioactive lava rolling unleashed toward unsuspecting cities and towns. The Emergency Core Cooling System, tagged ECCS, is a combination third and fourth layer in the "Design in Depth" concept. Unofficially, it's the kind of guy you can count on in a dark alley when the hoods are ganging up and you left your trusty M-16 at home next to the grenade launcher.

The ECCS has a prime function—geared to operate either manually or electronically—to provide a means of recovery and protection for the so-called Class 9 Accident. If that sounds tough, relax. The experts say this hypothetical accident just isn't possible, but based on the supposition that it *could* happen, the Emergency Core Cooling System stands ready to guard and protect. You can look at it this way: here is a very expensive system designed for a job it will probably never have to perform. And yet, no American reactor can be licensed unless it is fully equipped with an ECCS.

Sounds good, doesn't it? Makes you feel confident, safe and secure. Everything's going to be all right. Right? Well, you can imagine my shock early one morning on March 30, 1975, when my eyes gradually began to focus on the *Tennessean's* front page. A headline under the masthead screamed: DA NANG FALLS TO THE REDS. This was bad, but at least it was somewhat expected. Two columns over, however, almostly directly beneath the fold, rested my shocker of the day. It appeared in the form of an ominous, squared-in box. *SYSTEM FAILED* was the headline, and I grimaced, knowing that "system" could only mean ECCS. Right.

My thoughts scrambled. But "you" can't fail. Or can you?

Figure 1 details the five-paragraph bulletin which appeared in the *Tennessean,* concerning a fire which broke out at the $815 million Browns Ferry Nuclear Plant in northern Alabama. Read it. Then, look at Figure 2, which details a reply to the article from Kenneth M. Clark, Field Public Relations Officer for the Nuclear Regulatory Commission. This NRC response went to press some ten days *after* the incident—by then the damage had been done.

The situation was a March 22, 1975 fire which was touched off by a testing flame—from a candle—that ignited solvent used to seal an air leak and burned up vital wiring in the cable spreading room, forcing the closedown of two giant reactors at Browns Ferry. This resulted in an immediate loss of 15 percent of TVA's total power output and required a daily shift to burning 10,000 tons of coal per unit at another installation.

So, thousands of Sunday morning readers—the *Tennessean* has wide readership—scattered throughout the mid-South area, learned at a glimpse that somewhere in their neck of the woods, the "system" had failed to operate. Only the grace of God (or luck) had prevented a *meltdown*—according to the article. Paragraph three is designed to send readers scurrying for Bibles or that second cup of sugarless coffee: "A MELTDOWN," it casually drops, "resulting from loss of coolant, would have turned the bottom of the reactor into a lava flow and released lethal radioactive gases throughout the plant and countryside."

How's that for a "Sunday morning coming down"? The odds are good that a sizeable number of lapsed, church-going Southerners got religion after reading that little hot particle (article).

Nor was the Nashville newspaper a stranger in the crowd. Walter Cronkite said over WCBS: "It has been revealed that there was a potentially serious fire last Saturday at one of the country's largest nuclear reactors, the TVA's plant at Decatur, Alabama. Electrical cables burned and made the plant's emergency control system inoperative. Experts say if the plant's nuclear core had overheated, automatic sprinklers would not have worked and a catastrophic accident would then have been possible. Nuclear critic David Comey of Chicago, said the fact that there was not a major accident was 'largely a matter of luck.' TVA officials said there was no problem with radiation and no damage to the reactors themselves. Repairs to the cables, however, could take months."

Here you have the type news report known as the "iffer". All sorts of possibilities are conjured up, on a level with claiming all ocean liners are unsafe because someone was killed going over Niagara Falls in a barrel. The final NRC transcript reveals that those systems directly concerned with maintaining water level in the reactor were operational. For the situation at Browns Ferry, one system was adequate to cool the situation, and this was just for shutdown cooling. There were two other systems operational and standing by. Note also that the *sole* purpose of the ECCS is to counter any recirculation pipe break, and there was no problem—at any time—with a break in this system.

Browns Ferry and its neighbors would have been in trouble *if* another fire, a tornado, an earthquake, a stray hurricane somewhat off-course, total loss of offsite power, plus a pipe break, *had all occurred simultaneously.*

"You know, we probably couldn't have made it then," sighed Plant Superintendent Green, a former Navy nuclear engineer. There then followed a weak smile, because Jim knew that even Mother Nature would have had trouble delivering that bill of goods.

Bleary-eyed and staggering from the strain of two sleepless nights, this native of Gallatin, Tennessee, had just undergone a baptism of fire that was far worse than steering a nuclear submarine through its paces under the Polar icecap. One sweaty hour before, he had faced some two dozen news reporters crowded in a small classroom at the plant. There were television cameras blaring their harsh white lights, photographers popping bulb after bulb, and insistent questions that pinched, skirted, and then drew blood.

Then, Nashville newswoman Jeri Lynn, of WSM, tossed out this question: "Mr. Green, were you at any time afraid during the course of this fire that things would get out of hand—and were we just lucky that they didn't?"

It was a good question and it got an honest reply.

"Was I at any time afraid? Yes, ma'am. Was I afraid things would get out-of-hand? No, not from the aspect of affecting the public, but just the damage to my plant."

". . . damage to my plant." Green, the sailor, had said his piece.

Later, rehashing events, the Tennessean shook his head and said: "Oh, and another thing, Jacque. We have tried to be truthful with everybody, whether it was the critics or anybody else. What more do they want?"

Your head on a silver platter, I thought, but told the tired man instead: "Sailor, hang in there. But *please,* no more candles to test for air leaks, huh?"

"Let me put it this way," he grinned, wiser for the experience. "You can bet that there will be a definite change in procedure from this point on."

Almost five months from the fire at Browns Ferry, the Nuclear Regulatory Commission, through its Office of Inspection and Enforcement, issued its first of three interrelated reviews. Additionally, there is a special technical group studying the circumstances and implications of the fire to determine where improvements in NRC policies, procedures, or technical requirements might be indicated.

This initial investigation has shown that a "basic cause of the fire was a failure to recognize the significance of the flammability of the materials involved . . . The immediate cause, then, was ignition of polyurethane foam sheeting used by construction workers during the sealing of penetrations. The men were using a candle to detect air leakage in a cable penetration between the Unit 1 and Unit 2 cable spreading rooms and the reactor building," according to Dr. Donald F. Knuth, director of the NRC Office of Inspection and Enforcement.

The agency also noted that both reactor cores "remained adequately cooled throughout this occurrence" and that the fire's "radiological impact . . . was no more significant than from the routine shutdown of the reactors."

TVA General Manager Lynn Seeber responded to the field report by noting that the utility's primary objective in any incident associated with its nuclear power plants is twofold: (1) preventing any abnormal release of radiation, and (2) protecting the public and plant personnel. "The Browns Ferry Fire was a serious incident," Seeber said. "But TVA met these primary objectives . . . I believe the action of TVA personnel in fighting and extinguishing the fire, and in safely shutting down both reactors, was commendable."

Thus, the Browns Ferry scenario—and press-wise that's exactly what it became—is only one example of the type fun and games too many journalists (media and print) are playing. Ignorance? Some. Deliberate distortion? Some. News management? Yes. Prostitution? Yes, on all counts. You can slug it: "Media Make." Sub-titled: "Minds for Hire."

Now let's leave fiery northern Alabama and, keeping the specifics in mind, return to Oak Ridge in East Tennessee. There, tapping his desk and eternally smiling, Floyd Culler seems to genuinely enjoy sharing his knowledge of the nuclear industry. He does not talk down or up, but rather, in "Cullerese" which, relating to safety mechanisms, goes something like this:

"You have a situation where there is a belt, tight pants, suspenders, *plus* a row of safety pins to make certain that nothing happens to your pants. Then, just to be absolutely sure, you put on an iron shirt so that the safety pins won't tear the weakest link. You may be over-cautious, but those trousers sure stay in place."

Satisfied, Oak Ridge's deputy director sits back and waits for his listener to react. Then, remembering something more, Culler's leather desk chair squeaks as he moves toward a large green blackboard and prepares to further define. This is just the philosophical approach, he notes. There is also what's referred to as "protection in depth" which covers those structures surrounding the reactor. Four principal areas have remained intact since this concept went operational nearly forty years ago, when nuclear power was thought of in terms of bombs rather than generating electricity.

"We recognized back in 1943 that the materials we were working with were dangerous," Culler recalled. "So, right then, basic work began on the properties of the materials and the possibilities of accidents that might occur in the reactor system."

For this particular chapter, you could have had quotes from people at Livermore, Princeton, Los Alamos, ERDA, the NRC, TVA, or even the People's Republic of China. I chose Culler's remarks and personality for this section because I liked what he said and the manner in which he said it. And he is a professional, respected in his field.

This may sting, but a major problem with today's critics of nuclear power is that the bulk are amateurs who simply don't know what they are talking about. It sounds impressive at parties to drop little tidbits on fancy-sounding terms like plutonium, wastes disposals, reactors blowing up, and speculation on terrorist activity. Suddenly, a hush descends on the room and the critic is the center of attention. Nice feeling. Unless somebody starts asking you probing questions and suddenly, things aren't so nice any more. I know because I've been there and back. Initially I bit anti-nuclear bait, hook, line and sinker, and even ran off with the line around my neck. As it began to tighten, I discovered there are just too many armchair scientists who mouth simple answers to complex problems. Again, you don't put two drops of energy in a one-gallon test tube and come out with instant electricity that will burn forever.

Well, anyway, on with the matter at hand . . .

A lot has happened since 1943 when man began to take this atom-splitting business seriously. Studies relating to the genetic effects of radiation, derived from experiments with mouse colonies, date back nearly thirty years. Steady, patient work like this has provided researchers with solid ideas on the genetic effects of penetrating radiation, and has initiated the evolution of separate and much-needed studies on the aging process, immunology, and carcinogenesis (origins of cancer). So, it's quite possible that without the early beginnings of nuclear power, there would certainly be a lot of catching up to do in these three main areas.

In the area of reactor safety, a legitimate question often centers on levels of radiation that are produced and how it affects living organisms. First, very, very low levels of radiation are produced in the coolants and process streams that cruise near the reactor. Officially, these emissions are called the reactor's Low-Level Constant Discharges and are produced when tiny pieces of trapped material are grounded within the fuel element. These discharges are pluton-activated and sent into the water where they are chemically exchanged and then literally kicked out.

Such emissions, however, are well in hand. The levels that are officially set as the lowest practical limits on reactor discharges allow doses that—at the maximum—are several thousandths of the natural dosage received from rocks or cosmic radiation. Yes, we are constantly subjected to doses of radiation from natural sources over which there is simply no control.

In fact, if Americans received all of their electric power from nuclear plants, instead of only about 7 percent, the additional daily radiation they would receive would total about one tenth of 1 percent of the amount coming from natural sources. And this figure represents less than *one-billionth of a fatal dose.* If you are really concerned about radiation, then stay out of airplanes, because ten hours in a jet can give as much radioactivity as a lifetime of living with nuclear power. Also, avoid X-rays, because just *one* X-ray examination can give up to 10,000 times as much radiation as you would get from living next door to a nuclear plant for one year. Watch those television sets, as well. And if you decide to simply stay home, better make sure it's not a brick house, because in six weeks brick gives as much extra radiation as a lifetime of living near a nuclear plant.

Basically, there are three types of radiation associated with nuclear energy that ordinarily affect humans: alpha, beta, and gamma radiation. Alpha particles (or rays) are the nuclei of helium atoms (2 protons and 2 neutrons), readily stopped by even a thin material like a sheet of paper. Beta particles (or rays) are high-speed electrons. Gamma radiation is a high-energy form of electromagnetic radiation (other types are ordinary sunlight and X-rays). Table 1 proivdes a listing of the estimated average annual doses from some of the alpha, beta, and gamma sources to which Americans are commonly exposed.

All three types of radiation (called "ionizing" because of its effect on matter) are produced in nature as well as by man's activities—regardless of whether there's a nuclear plant on the block or not. Such natural radiation in soil and water and from cosmic radiation is termed "background radiation."

In nuclear power plants, there is an additional type of radiation present which is due to neutrons. However, the only persons likely to be exposed even to low doses of neutrons are those working in the plant.

Also, coal-fired power plants emit measurable amounts of radiation due to the presence of naturally radioactive materials in the coal.

Table 2 explains how an individual can compute personal annual radiation exposure chart. Stripped of all the technicalities, the radiation measurement most meaningful to humans is the "rem," which is usually counted in thousandths, called "millirem," and shortened to "mrem." This unit of measurement takes into account the biological effectiveness of the three types of radiation. The National Council on Radiation Protection and Measurements (NCRP) has stated that no member of the general public should be exposed to more than 500 mrem per year of man-made radiation, excluding medical. The NRC allows additional annual doses to individuals living near current nuclear plants to be less than 5 mrem.

The second class of radioactive material release would come from the reactor in an accident situation. All of the "probable but not likely" accidents (with the exception of the "improbable one") have automatic systems which operate manually (meaning a technician must "push" a button and not hand-crank a wheel), if necessary, to contain radioactive material. This includes the massive containment vessel, designed to contain all stored energy—in the steam circuit, in the heat structure and in the nuclear reactor. So, what you have is an energy absorption system built into every reactor that will accommodate all the energy released in a maximum credible accident.

A secondary function of the containment shield is to provide another barrier in the event something happens to the radioactive material section of the reactor. There is, however, a first barrier. For you see, uranium and plutonium fuel is available in an insoluble oxide form. Such oxides are refractories—like magnesia or aluminum insulation found in houses—and are very carefully prepared as dense and grounded pellets. Contained within this heterogeneous matrix of oxide are essentially all fission products, with the exception of only a few.

Uranium and plutonium are mentioned as examples simply because, even in water reactors, there is significant plutonium production. In fact, wherever there's a reactor, you are going to find plutonium. It might be helpful to further define it in this specific manner, remembering that once you actually begin fissioning U-235, an equation such as this takes place:

$$
\begin{array}{lll}
 & \text{N} & \\
235 & \text{——FP} + 2.17 & \text{(neutrons per fission)} \\
\text{U} & (2) & \\
 & + \text{ energy} & \text{(plus energy)}
\end{array}
$$

This set of neutrons makes possible the initial chain of reaction in uranium-235. Consequently, plutonium-239 starts its production just as soon as the fission reaction process begins. You can safely figure, then, that the energy derived from the fuel that was initially uranium, comes from the fissioning of plutonium.

Providing that the neutrons are not slowed down by water or other materials that have atoms of the same size as neutrons (referred to as moderators), it is possible to use enough of these plutonium-spawned neutrons and its fission products to make much more plutonium from the uranium-238. Scientists, in fact, are trying to produce up to 1.3 times more than is burned. A different type of reactor is required, however, to achieve this effect, because the resulting neutrons move fast and can't be slowed down. Also, in order to maintain such a tempo, hydrogenous material like water cannot be used.

The consequence of using a hydrogeneous material is that the neutrons simply move too slowly to produce a high yield of plutonium-239 in the secondary process. Therefore, sodium (preferred) or some other liquid metal that will not slow down these neutrons must be used as a coolant.

As a matter of review, remember that the first barrier is not the massive containment shield, but rather the tiny fuel pellet that is put into the reactor. Contained within its matrix are most of the fission products and all of the plutonium. Actually, you might compare it to a rather large, brownish vitamin pill which is safe enough to hold.

Now, let me explain what is meant by "most of the fission products." There are some which diffuse out of this oxide pellet. These are officially known as Fission Product Gases (i.e., xenon, krypton) and serve the function of providing a pressure of fission gas over the pellet. This, in turn, then creates the necessity of designing a second barrier to contain the fission products and to protect the fuel from water or sodium coolant. Again, remember, this is in addition to the large containment shield. Fast breeder reactors (detailed in Chapter 11) utilize a very carefully prepared, high-grade stainless steel cladding and, in the case of water-cooled reactors, a zyconium one is used.

We can now proceed to Barrier Three, which consists of long pins of fuel that crudely resemble those metal cylinders country folks sometimes use to draw well water. Normally, a reactor fuel rod contains about 217 pins (slightly more or less) with a U-235 concentration of from 2 to 3 percent. Such "bundles" of fuel pins are stacked together—in proper geometry—to allow a critical mass, a critical reaction, to proceed. Insertion is used to control the reaction's tempo.

Reviewed, Barrier Three is essentially a reactor pressure vessel that fulfills three functions:

- Contains coolant
- Protects the core
- Protects the primary system from radioactive material release

Note: "Primary system" refers to the reactor pressure vessel and all pipes hooked to it, plus circulation of water through the core to it.

What you should be visualizing now are three layers of protection that are *inside* the reactor. As a final guardian, the containment shield still hovers over the entire operation.

One of the places I visited at Oak Ridge concerned some ten years' worth of brittle fracture experiments geared toward this specific area of pressure vessels. Without exception, their studies, verified by visual and written reports, prove that there is no way to burst these devices by any type of built-in or automatic failure. And the only manner in which coolant can be lost from a pressurized water reactor or a boiling water reactor is for the pressure vessel to rupture or for a main water pipe to break. The conscensus of technical experts is that it is virtually impossible for the high-quality, eight- to twelve-inch thick steel pressure vessel to rupture. Therefore the break could only occur by rupture of the piping. In analyses, no mechanism for causing this break is postulated; it is simply assumed that in the worst case, a two and one-half foot diameter pipe with two-inch thick steel walls breaks completely apart and the broken ends separate in such a manner that all of the cooling water and steam can be lost from the system through these ruptures.

Now, here is the interesting aspect. Thus far, estimates are based on experience with non-nuclear, high-pressure piping systems. Failures by a major rupture of a pipe have occurred about once in 10,000 years of systems operation. However, nuclear systems operate at lower pressures—1,000 to 2,500 pounds per square inch (psi) instead of 3,000 to 4,000 psi—and are built to much higher quality standards both in materials and workmanship. Then there's quality control and inspection procedures that are equally tough. Therefore, the actual chances of failure are much, much less than once in 10,000 years. The exact figure is not known, simply because it has never happened. A major pipe rupture in a nuclear system has been estimated to occur between once in one million and once in ten million operational years. This is not to say that leaks and breaks of lesser degree, with lesser potential consequences, will not occur more frequently through time and use. The system, however, is equipped to deal in a number of ways with such matters.

Now, around Barrier Three and in addition to the final shield, there is a relatively pressure-tight concrete structure that is a good *sixteen feet thick*—and sometimes more. This then is the Fourth Barrier and it is not absolutely pressure-tight: if a few gases leak out, it's possible these might ooze through where the pipes come in. It is, however, practically pressure-tight and holds water without any difficulty. This is essentially the shielded cavity area. Here there is space for the primary pumps, the primary heat exchanges and piping, plus auxiliaries that function as part of the Emergency Core Cooling System (ECCS). Are you now beginning to get a better idea of the ECCS and its unique seating arrangement in the chain of command? In the case of a Westinghouse reactor, this might be a large ice chamber, or in the case of a General Electric model (boiling water reactor), it means a great doughnut of water in the bottom.

The fifth physical barrier is the actual containment vessel. It is designed for two purposes: controlling the maximum heat release and storing heat, plus containing any of the shut-down reactor heat, which is about 10 percent of its rated power. Culler explained that you lose a factor ten of this in ten minutes and another factor of ten within one hundred days, which implies that heat is being generated by decaying fission products.

Until just recently—while reactors were small—it was accurate to say that no matter what happened, there was insufficient energy in the systems to allow for breaching a containment vessel. This remained valid up to about 500 to 750 megawatts. As reactors have grown, it could not be argued, on the basis of analytical models, that it was impossible to breach the containment barriers. More importantly, says Culler, as the power density went up, it became conceivable that, starting at Barrier One, the beginnings of a potential meltdown could occur. He depicts the scenario like this:

The oxide would melt and interact with the ziron alloy. It would then get hot enough to react with the water. Such an effect would place a small amount of hydrogen in the vessel. Then, the sequence of events would thrust a reactor into the "big time" known as the China Syndrone. . . . This is the term used by nuclear scientists to describe a hypothetical chain of events resulting in total failure of a commercial nuclear electric power plant. Highly radioactive material generates great heat, even at rest, because of the decay of radioactive particles. The heat is greatly intensified during the fission process that occurs in the core of a working nuclear reactor. Thus, it must be cooled at all times and water is commonly used in most U.S. "burner" plants.

ECCSs differ depending upon the type of reactor. Current pressurized water reactors (PWRs) have a number of systems which flood the reactor core with water and prevent its heatup. Or, boiling water reactors (BWRs) have both a system which sprays water on the core from its top and another system which floods the core by hitting it with water from the bottom. And the ECCS is not likely to fail because it is not a single system. So, failure of a single subsystem or piece of equipment does not mean cooling capacity is lost. This was one aspect that was so erroneous about reporting of the Browns Ferry fire. Take the current PWRs, for example: the primary ECCS has duplicate and independent pipe systems, each with two different water pumps, each taking electrical power from any one of three different supplies and two different sources of water. Also, there is another system which will automatically inject water stored in tanks into the core using stored air pressure. So, you would have to have several successive failures in parallel equipment in order to prevent coolant from getting to the core and doing its job.

Even further speculating. In order for a failure to take place, a guillotine-like, double-ended pipe break would have to occur in the largest cooling line. This is no easy task. We are talking about a hunk of metal (remember the brittle fracture tests) that is not only cracked at one end, but *totally split—sliced like a carrot—simultaneously at both ends.* All the water that flows through this pipe would then no longer be available to the core and would be soon dumped by the system's stored energy.

Based on this evidence, it is safe to say that actual chances of a loss of coolant accident and total ECCS failure are minuscule. But perhaps you are still thinking ". . . what if." Herein lies controversy. It is frequently asserted by critics of nuclear power that a catastrophe would result on the level of those derived from a 1957 AEC Study called WASH-740. This little fellow has become quite popular lately and details a maximum hypothetical accident from a gory and out-of-date vantage point. In this study, it was assumed that half the fission products in the core were thrown (remember: reactors *can't* explode like atom bombs) into the air as dust particles, to be blown under the most adverse wind and weather conditions and deposited on the landscape and waiting population. As a result, it was estimated, the maximum would be 3,400 killed, 43,000 injured, plus $7 billion in property damage. Any way you slice it, such a catastrophe would mean quite a clean-up job. That was back in 1957. Present nuclear plants do not resemble in any way the study design contained in the WASH-740 report, and yet, it is frequently quoted as an outstanding example of what's in store for us should so-and-so or this-and-that happen. Actually, modern plants employ a series of engineered safeguards that were not included and probably not even thought of during the WASH writing. It is now believed that an accident with consequences even remotely resembling the severity of those indicated in the 1957 study is virtually impossible. So, sniff twice the next time somebody puts a musty smelling WASH-740 under your nose.

There is another frequently quoted report which details failure of what is alleged to be an ECCS during a test run at the National Reactor Testing Center in Idaho. Again, this did not involve a representative test of ECCS capability. In fact, no full-scale tests have been done, nor are any planned, mainly due to their being impractical and very expensive. Carefully designed smaller experiments, plus appropriate mathematical models, provide sufficient data on which way the cards are stacked. The test referred to in Idaho involved a nine-inch diameter pressure vessel with one set of inlet and outlet pipes. In this experiment, a break in the inlet vessel was simulated and the attempt was made to inject water into the pressure vessel to cool the electric heaters simulating the core. The water was forced to enter against a residual steam pressure as steam and water were being expelled through the break. Obviously, it did not work. In contrast, reactor vessels are from 14 to 22 feet in diameter and have two or four sets of inlet and outlet pipes, each about 2½ feet in diameter. So, as you see, the total picture takes on a different coloring.

Even with the Emergency Core Cooling System, it was necessary to decide just how fast and whether the coolant could go in to prevent the temperature of the fuel from getting higher than its melting point or higher than it should be for any reason. Culler said that a temperature of 2300° Fahrenheit has been tagged as what the core can take and still remain well below the melting point of uranium and plutonium. As a result, all systems are now designed to hold fuel temperatures at

2300° F. Don't confuse this with the interior core temperature, where there is an allowance of 3000° F., which is still well below the melting point of plutonium and uranium (about 5000° F.).

Here then is a review of the much-discussed China Syndrone: It proposes that the fuel overheats and goes directly to the melting point of plutonium and uranium (about 5000° F.). The metal shielding is gone and the fuel sulks into the reactor base. And, because it is no longer in an extended array, there is no water to cool it easily. It gets hotter and hotter. There's no dispersement. In short, it's time to dial "maintenance" because there's fireworks coming. It's a radioactive blob, a little blob but a very mean one. Finally, yearning to move out, this mass begins to melt its way through the steel, eventually concrete, and conceivably all the way to China.

Don't laugh. This scenario—officially the Class Nine Accident—is one of the most quoted on record. It is the hell-fire and brimstone approach which critics use to bring the masses of unconvinced down front. Here you have the final punch line to the last stanza of that popular invitational hymn: "Save Our Country." And there are mass conversions until some of these "saved" take a closer look at the fine print.

The Brookhaven Report is another example of earlier evaluations that were so horrible and yet are still quoted as accurate by critics. Brookhaven speculated that from 10 to 50 percent of all fission products in plutonium would be released from fuel all the way through the circuit. New experiments have reduced these dosages considerably, and a source turnoff has been calculated for *every* hypothetical reactor accident.

Studies have also been made to determine Ecological Reduction/Concentration in the event of a low-level, routine release of radiation. This means that scientists now know basically what the biological pathways would be in the area of human consumption and environmental effects. For example, it is now known just where most of the radioactive material will settle down in the body and how long it will stay, plus what type dosage each organ can be expected to receive. How? Such data was collected from studies of organs from cadavers obtained throughout the world. Researchers were able to specifically pinpoint what concentration of isotopes could be anticipated. Incidentally, this research began in 1946 and was vigorously pushed by the AEC, beginning in 1952 and continuing today.

You can safely conclude that based on the data and assuming *all* layers of protection—plus the ECCS—failed to work, the released dosage of fission products would be less than the presently allowable five million rems per year. Here you have the belt, suspenders, shirt, safety pin, and tight pants approach: a back-up system that backs up the back-ups.

As Culler noted: "We do something that no other system in technological history has done. We assume that the system is going to fail—although, at the same time, we think it's impossible—and have made reasonably adequate provisions to protect against that eventuality."

Reasonable people, he continued, are "relatively competent and don't think that we are trying to put on to mankind a problem that is just not controllable. Those series of events, taken in logical form by rational people, provide a degree of protection not morally provided in other exposures."

Success then of the China Syndrone hinges on total penetration of all physical barriers, plus failure of all systems mentioned in the philosophical approach. They are redundant—there are at least two of everything—and all must fail for the improbable to succeed. We can assume, then, that it would not be a very logical occurrence.

You got a full dose of the most improbable accident first so that these other hypothetical "heavies" can be viewed in the proper perspective. Also, remember that the safety mechanisms that are locked in on the China Syndrone are equally tied to other operations in the system. As far as a reactor is concerned, no part is an island unto itself.

Tornadoes, earthquakes and such "acts of God" and, occasionally, man, have been viewed with legitimate concern, especially when a utility seeks licensing through its initial Environmental Impact Statement. A reactor would obviously not be placed on a known fault or near areas of turbulent weather patterns. Unofficially, this is known as common sense . . .

QUESTION: Will a reactor fall apart if it is struck by a tornado or earthquake?

The answer is no, because of numerous automatic control devices and systems that scramble either automatically or manually to shut down a reactor in seconds. The reality of this "scram" system was given a severe test in April, 1974, when tornadoes and high winds ripped through North Alabama and headed for . . . Browns Ferry. Wind tossed power lines like match sticks and created general havoc for miles around, but the nuclear plant—perhaps to the dismay of critics—shut down automatically just as it had been designed to do. There was no panic and as far as the engineers were concerned, the system moved like clockwork.

Another example: Southern California's strongest earthquake in thirty-five years struck on a quiet morning in February, 1971. The San Onofre Nuclear Power Plant near San Clemente rode it out beautifully and for a very good reason. The plant had been designed and site selected to guard against damage from quakes, lightning, offshore storms and other such acts of nature and man.

Earlier, in April, 1968, an earthquake of 6.5 Richter magnitude centered near Borrego Springs, California. It also shook San Onofre with a maximum force of 5 percent of its weight (i.e., an acceleration of 0.05 g). The plant experienced no difficulty and continued to generate electricity without interruption, according to Plant Superintendent Hans Ottoson. He told me that this plant, located directly on the Southern California coast, is even protected by a wall twice the height of the largest tsunami (tidal waves spawned by earthquakes), even though the plant base elevation is above the tsunami. There is also protection, he said, against entry of water through the cooling water pipes going ino the sea.

A fault is a break or crack in the earth, along which the ground on one side moves permanently either vertically, laterally or both, with respect to ground on the other side. Criteria from the Nuclear Regulatory Commission stipulate that a fault can be considered active if it has moved in the last 35,000 years. Construction crews are required to dig at thirty- to forty-foot depths and at various distances from a fault to determinc if there are vertical or horizontal offsets in the soil or rock. These are even measured in inches. And, just to keep the plant crew honest and as an extra measure of safety, independent geological consultants then go in and review the entire survey. Then, both parties must agree, according to Gayle Hunt, a geologist who studied the first commercial sites for nuclear power plants in California. Federal regulations, he said, require that each plant be designed to withstand the maximum earthquake motion that could be expected at its site and be able to shut down safely. Initial licensing processes zero in on such extensive evaluation of the site and plant design.

It is interesting to compare, or perhaps contrast, survey preparation normally required in the construction of large apartment and office buildings. A contractor might invest between $20,000 and $30,000 to survey a thirty-story office building. (Put a quote on "might," because this is not always done.) Not so with a nuclear plant. The average survey runs about one million dollars or more, and the bulk is contained in a massive report, mentioned earlier, called the Environmental Impact Statement.

Still wondering how the experts can possibly know that an earthquake fault couldn't split a nuclear plant right down the middle? First, remember that plants are *never* sited in fault zones. Such areas are defined by mapping previous fault lines near each potential site. Since faulting usually occurs along previous lines, the chances of one intersecting a reactor are almost negligible. Incidentally, this aspect of site selection is unique to nuclear plants. Most vital structures, including freeway overpasses, aquaducts, hospitals and schools are constructed near known faults, and some buildings even sit right on top of them!

Obviously, it is not currently possible to predict the exact size or timing of an earthquake. It is possible, however, to estimate—based on historical records, current knowledge and measuring techniques—the largest quake that might occur at a particular site. Information of this type is then used to determine the Operating Basis Earthquake and the Safe Shutdown Earthquake. The former is the largest which might occur at least once during the life of a plant and during which the plant is designed to be able to keep going safely. The Safe Shutdown Earthquake is the most severe that could be expected at a specific site. An earthquake of this magnitude may never have been experienced and may never occur, but regardless, it determines the design of the plant's safety systems. At this level, some components, such as the turbine—that are considered non-safety related—could conceivably be damaged. However, all safety-related components—reactor and associated equipment—can not be damaged during a major re-settling of the earth.

How? Such facts are ascertained by a computer printout of a mathematical model that is subjected to rigid runs and re-runs before any construction is permitted on the actual plant. A combination of structural analysis and testing also determine that structures and equipment will survive. Then, vibration tests are conducted to confirm analyses or the exact point where the analytical complexity introduces that little bug called "uncertainties." Tests are routinely performed on the first models of individual components, including piping, fuel elements, pressure vessels, pumps, valves and full-scale reactor structures. In fact, whole reactor buildings are tested using mechanical shakers attached to the structure, and high explosives are detonated nearby to simulate strong quakes.

Here's another thought to ponder. Fossil fuel—coal, oil, and gas—power plants are not designed to the same exacting quake resistant standards as their nuclear cousins, but are constructed to resist forces only 50 percent greater than ordinary building and highway structures, which are covered by a catch-all term known as "uniform" or general building codes.

Only nuclear plants are designed to withstand a major earthquake motion, and because of this, only nuclear plants can be expected to operate during even a moderate quake. Moreover, should a major seizure take place, that dome-shaped reactor building could be expected to remain operational and supply badly needed power, while in all probability, conventional plants will be knocked out of commission. In fact, it has been suggested that seismic evaluation is long overdue for fossil fuel plants and that these would certainly be an area of grave concern should a natural disaster occur.

In conclusion to this chapter, there seems to be only one fault against which the nuclear plant is totally defenseless, and that is, the sheer fantasy of the human mind—verbalized as "what if?" *

*The Price-Anderson Act, first passed in 1957 but due to expire in 1977, set a ceiling of $560 million on the damage claims to which a nuclear licensee could be held liable. Since at that time private insurers would offer a licensee no more than $60 million coverage, the act stipulated that the federal government indemnify licensees for the remaining $500 million. Now, private insurance of up to $125 million is available, but the $560 million ceiling remains in effect, with the government indemnifying licensees for the difference of $435 million. The intent of the act in 1957 and of a 10-year renewal bill that already has moved through the House, was two-fold. On one hand, it sought to provide more public protection than private insurance made available, but, on the other, to limit the liability of licensees so that the potential threat of tremendous liability claims would not retard the growth of the nuclear power industry. Industry and government spokesmen say that the renewal bill allows the ceiling to "float" upwards as more reactors are built. And if the nuclear power industry is to be able to find the financing it needs to expand, they add, it must continue to have protection against claims.

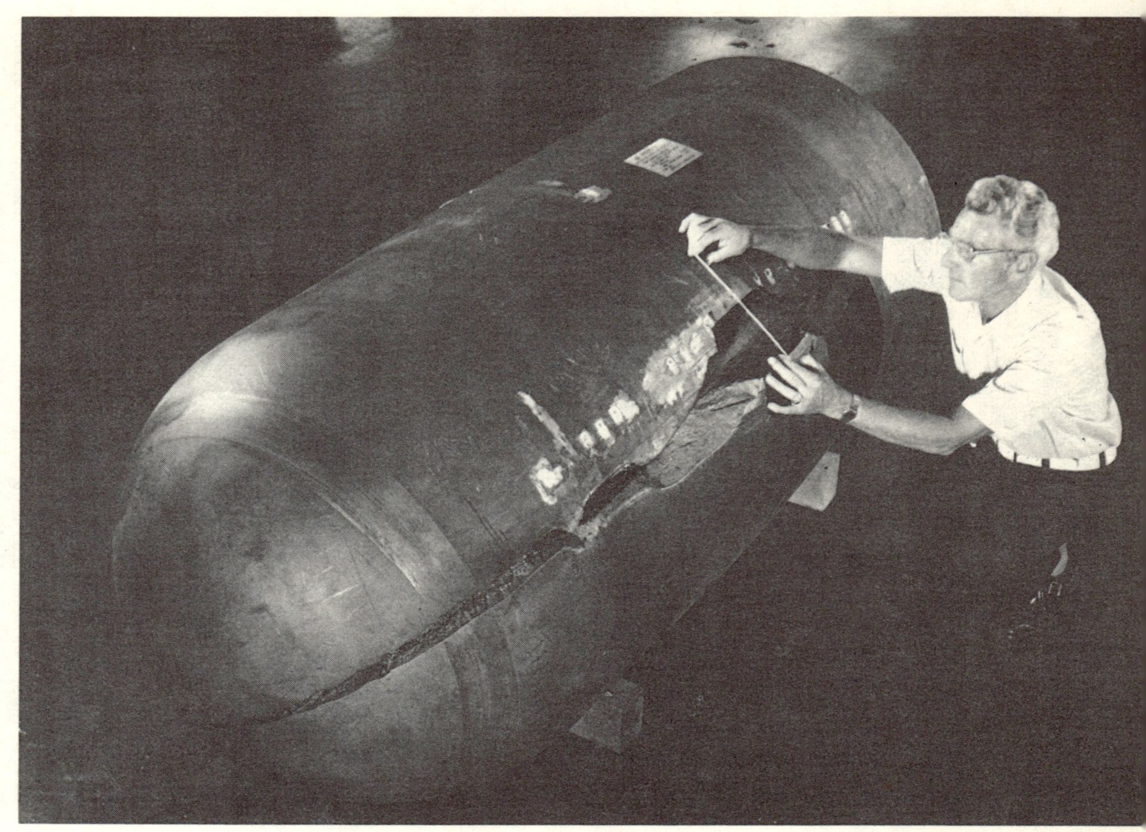

A series of tests to reach a failure point on thick-walled, carbon steel vessels which are being conducted at the Oak Ridge National Laboratory in support of the Energy Research and Development Administration sponsored Heavy Section Steel Technology Program. The purpose of this program is to evaluate failure characteristics of test vessels representative of the design, material, fabrication and inspection requirements for current light water reactor vessels. After a flaw is deliberately machined at external or internal locations on a vessel, it is subjected to intense, internal liquid pressure until it bursts. This one failed at approximately 31,000 pounds per square inch, which represents about three times the design pressure of the vessel. Such results demonstrate that reactor-grade steels have a remarkable capability to tolerate even very large defects, thereby verifying the margin of safety in operating vessels. (4178-74)

Dr. Floyd L. Culler, deputy director of the Oak Ridge National Laboratory (center), with Nobel prize-winning physicist Eugene Wigner (left), an ORNL consultant and Dr. Alvin M. Weinberg, former director of the ORNL and now director of the Institute for Energy Analysis at Oak Ridge.

Final construction work on the proposed Sequoyah Nuclear Plant located near Chattanooga, Tennessee, and under supervision of the Tennessee Valley Authority System.

System Failed

The emergency core cooling system (ECCS), designed to automatically shut down a nuclear reactor in case of fire, failed to operate as planned when fire broke out at Browns Ferry Nuclear Plant March 22, the Nuclear Regulatory Commission has disclosed.

However, manually operated backup systems managed to cool down the core of the reactor containing uranium-235 fuel and prevent a meltdown, the NRC report said.

A MELTDOWN, resulting from loss of coolant, would have turned the bottom of the reactor into a lava flow and released lethal radioactive gases throughout the plant and countryside.

"The ECCS behavior (at Browns Ferry) is under specific investigation," said the NRC report.

An in-depth report on the fire, a press conference, a tour of the plant, a TVA board meeting held Thursday and the NRC preliminary investigation is written by *Tennessean* staffer Dolph Honicker.

Figure 1

A front page "News Account" of the Browns Ferry fire caught more public attention than a more accurate account which appeared several weeks later as a Letter to the Editor (Figure 2).

Letters to The Editor

An Explanation Of the ECCS

To the Editor:

In the Sunday, March 30, 1975, edition of *The Tennessean*, you mistakenly identified an Emergency Core Cooling System (ECCS) for a nuclear reactor on your front page.

Your newspaper states that the ECCS is "designed to automatically shutdown a nuclear reactor in case of fire..." in reference to the recent problem at TVA's Browns Ferry nuclear plant.

The ECCS has nothing to do with shutting down a reactor and is not related to fire. It is designed to cool the core, which contains the nuclear fuel, in the event of a sudden loss of water accident. A reactor is shut down by inserting the control rods which absorb neutrons and stop the "nuclear reactor."

You also refer to a "meltdown resulting from loss of coolant" in the third paragraph of your front page Browns Ferry piece.

There is no connection with the hypothetical circumstances your story describes and what happened at Browns Ferry. As our information indicates, the reactor core there was under water at all times; and the ECCS was not required at any time to maintain the fuel in a safe condition.

KENNETH M. CLARK
Field Public Affairs Officer

U.S. Nuclear Regulatory Commission
Directorate of Regulatory Operations
Region II
230 Peachtree St. NW, Suite 818
Atlanta, Ga. 30303

Figure 2

Table 1: This illustrates estimated average annual doses from some of the alpha, beta, and gamma sources to which persons in America are exposed. As a point of reference, the NRC requires that annual doses to individuals living near nuclear power plants be less than 5 mrem per year . . .

Type of Radiation	Some Sources	Average Annual Dose mrem/yr)
Alpha	Natural radioactivity (uranium) in soils, rocks, minerals	30
Beta	Natural radioactivity (e.g., potassium-40)	20
	Television (an average of 1hr/day)	1-2
	Luminous dial wrist watch	2
	Natural radioactivity in the air (e.g., tritium)	2
Gamma	Medical and dental x-rays	50-100
	Cosmic radiation at sea level	40

Table 2 **COMPUTE YOUR OWN RADIATION DOSAGE**

We live in a radioactive world. Radiation is all about us and is part of our natural environment. By filling out this form, you will get an idea of the amount you are exposed to every year.

	Common Source of Radiation	Your Annual Inventory
RE YOU	Location: Cosmic radiation at sea level Add 1 for every 100 feet of elevation Typical elevations: Pittsburgh 1200; Minneapolis 815; Atlanta 1050; Las Vegas 2000; Denver 5280; St. Louis 455; Salt Lake City 4400; Dallas 435; Bangor 20; Spokane 1890; Chicago 595. (Coastal cities are assumed to be zero, or sea level).	40 ___
	House construction (¾ time factor): Wood 35; Concrete 45; Brick 45; Stone 50	___
	Ground: (¼ time factor): U.S. Average	15
T YOU DRINK EATHE	Water, Food, and Air: U.S. Average	25
YOU	Jet Airplanes: Number of 6000-mile flights ______ x 4	___
	Television viewing: Black and white Number of hours per day ______ x 1 Color Number of hours per day ______ x 2	___ ___
	X-ray diagnosis and treatment: Chest x-ray ______ x 100-200 Gastrointestinal tract x-ray ______ x 2000 Dental x-ray ______ x 20	___ ___ ___
	Compare your annual dose to the U.S. Annual Average of 225	Sub Total ___mrem
CLOSE IVE EAR T	At site boundary: Annual average number of hours per day ______ x 0.2 One mile away: Annual average number of hours per day ______ x 0.02 Five miles away: Annual average number of hours per day ______ x 0.002 Over 5 miles away: None	___ ___ ___
		Total ___mrem

One mrem per year is equal to: **Moving to an elevation 100 feet higher**
Increasing your diet by 4%
Watching one additional hour of black and white TV per day
Taking a 4-5 day vacation in the Sierra Nevada Mountains.

Chapter 8

THE MOST TOXIC SUBSTANCE KNOWN TO MAN?

Plutonium is the love child of an ultimate alchemy:
It can reproduce itself.
—Rolling Stone

Two dozen green, smiling marigolds shake their fuzzy heads in the warm summer breeze. Nearby, a plot of snapdragons add a color chorus of their own, flanked by several rows of country-looking zinneas. Satisfied, the children and I wash the spades and brush away all traces of dirt and weeds. Our summer garden is truly a thing of beauty and we are pleased.

Tragedy strikes two days later. A team of hideous insects strip the tender young stalks, leaving only rows of gawky stubs. Our summer garden has become a thing of terror and we are saddened.

Angry and wreaking revenge, we rush to a nearby garden center. The children amble around, sniffing and poking as the situation is explained to a specialist. Nodding, the clerk smiles at my two preschoolers and then produces an innocent can labeled "Rose and Flower Garden Spray." It is manufactured by ferti-lome and sells for $2.31 plus tax.

Satisfied and without reading the label, we return home and prepare to attack the choppers. The insecticide is left on the patio table and remains there for several hours. The children play and laugh near the cylinder and the baby even juggles it until the can is pulled from her clutching fingers. A battle ensues then as my third grader tries to jerk the canister from his younger brother. There is laughter and shouting until finally the can rolls off to a safe corner. The label still has not been read, and I'm unaware of the monster that has come into our home.

Finally, as darkness settles, I retrieve the spray and prepare my assault. It is then that the very small print catches my attention. Here is a partial reading:

> WHAT IS THIS PACKAGE FOR? This package makes insect and disease control easier. It contains insecticides, miticide, and fungicide that will fill most of your rose and flower garden needs.
>
> HOW DO YOU USE IT? Shake well before using. Remove protective cap. Hold container upright. Point nozzle in direction indicated by arrow . . . Point spray away from people and pets . . . Do not use indoors or in confined areas . . . Do not attempt to spray against the wind.
>
> HERE ARE THE RESULTS YOU MAY EXPECT! By providing fast, economical insect and disease control, ferti-lome Systemic Rose and Flower Spray makes it easy and pleasant to grow more beautiful roses and flowers.

And now for the clincher that made me want to get a pair of tweezers and put the can in a lead-lined vault:

> *CAUTION: May be fatal if swallowed, inhaled or absorbed through the skin. Do not breathe fumes or spray mist. Do not get in eyes, on skin, or on clothing. Wash thoroughly with soap and warm water after handling. Wash clothing with soap and hot water before reuse. Keep children and pets away from treated areas until spray deposit has dried on treated areas. Do not handle treated plants until thoroughly dry. Do not contaminate feed or food stuffs.*
>
> *This product is toxic to fish and wildlife. Birds feeding on treated area may be killed. Keep out of any body of water. Do not contaminate water by cleaning of equipment or disposal of waste. Do not use on plants used for food or forage. This product is toxic to bees and should not be applied when bees are actively visiting the area. Do not re-use container for any purpose. If poisoning occurs obtain prompt medical aid.*

One or two cans of this stuff and you could easily wipe out a neighborhood—insects, people, birds and bees—poof! Just like that and all for under $5. This should never fall in the hands of our enemies, I have to muse. Just think of what it could do to a water supply. Or what happens one morning when the office prankster gets bored and decides to level the ultimate joke? . . . The possibilities are as endless as the human mind.

It seems incredible that such a spray can be bought on the open market. Remember, the manufacturer says that mere skin contact may be fatal . . . not to mention what will happen to the unsuspecting birds and bees.

Unfortunately, such products are not really the exception—although this may qualify near the top—because the market is crowded with cleansers, dusters, detergents and other little goodies that, mixed together, will produce lethal gases and liquids that kill their unsuspecting victims, sometimes in seconds.

Another example. You may have noticed that manufacturers of a very popular drain opener have switched to plastic containers. The reason? A rash of consumer lawsuits from widows and widowers because the metal canister kept blowing up on the cupboard shelf.

This is not meant to be a long way around to the so-called hazards of plutonium, but in fact, I would probably have been much safer sprinkling plutonium pellets in my garden rather than using that insecticide. What you have is an analogy that vividly illustrates just how many so-called harmless substances there are floating around for the eager consumer to reach out and take home. None have the stringent restrictions that surround plutonium.

Webster's *New World Dictionary* defines "plutonium" as (1) a form for the planet Pluto and (2) a radioactive, metallic chemical element. This chapter will deal with the latter definition and hopefully expand on it just a little, although no criticism is intended toward Mr. Webster, who was limited by space and time . . .

Ask most people about this element and most who have done a little reading will categorize it with this response: reactors create plutonium and it can be used to make a bomb. This covers a lot of unanswered ground. If, however, you are talking about present and future non-military applications, then power production is the name of our game. Its prime value lies in the fact that recovered plutonium can be recycled into light water reactors' "reload" fuel. All but the older LWRs have such a capability. In this capacity, plutonium produced during reactor operation is separated and then added to the new fuel, thus contributing about one-third of the total heat output. This means a very significant long-term savings of U-235, even though large amounts of new U-235 are still required in the uranium-plutonium process. At this point, you have the important fast breeder reactor (see Chapter 11) which can increase the energy output of a pound of uranium by as much as seventy times. That's right, the figure is seventy times. This allows lower grade ore to be mined more economically, which, in turn, extends the nation's nuclear energy resources.

Plutonium has other uses that are not directly related to the fuel cycle and which must be specially produced. Another isotope, Pu-238, has been used in various devices: one is a thermoelectric generator such as the SNAP-27 device, used in the space program to provide power for experimental stations on the moon; other Pu-238 heat sources are providing power for orbiting weather and navigational satellites. Another potential use involves plutonium for batteries in human circulatory assist devices, such as the heart pacer.

This is interesting background, you think, but the real question is the validity of a popular impression that there are a lot of plutonium-damaged humanoids walking or crawling around, most with cancerous bodies afflicted by large boils and open wounds. Plutonium has been presented (almost with relish) as a fiendish demon, "the most toxic or poisonous material on earth and the most hazardous material man has ever faced." Grim scenarios—I even drew one in my first article—have been painted in which every man, woman and child eventually falls victim to this creeping death. Ring a familiar bell? Again, in a matter such as this, I went to the experts: scientists at Los Alamos Scientific Laboratory, Oak Ridge, and private companies that have handled the stuff for years. They offered not only their own expertise but also documented reports, dating back to the very early forties.

For example, Lynn Wallis, a health physicist employed by General Electric, provides some very credible insight on this question of environmental and biological effects. His reference point includes an extensive background of experience as a Health and Safety Officer at Hanford and at GE's Vallecitos Nuclear Center—installations that handle plutonium on a daily basis. Wallis has stated quite frankly that plutonium is neither the worst nor the only hazardous substance that man has had to cope with on the industrial level. Nor is its radioactive hazard rating even unique from an environmental standpoint. A reference work here is the International Atomic Energy Association Technical Report Series 15, which shows that natural-occurring thorium and radium isotopes fall into the same hazard classification as plutonium.

Also, Wallis explained, the common idea of plutonium's chemical toxicity is not directly applicable to the biological effects of radiation. You might compare it to a taste—less than a drop or fraction of a teaspoon—of nicotine, parathion, TEPP or many other pesticides, which are purely fatal. These are the ultimate trip with no pretty dreams to bid you farewell. Ingestion of the same quantity of plutonium, however, would *not* be fatal.

Other statements charge that "in terms of total weight of material that represents a lethal dose, plutonium-239 is . . . 1000 times more toxic than modern nerve gas." Actually, the median lethal dose of nerve gas "Sarin," which is stocked by the U.S. Army as GB, is 1000 micrograms, and death is guaranteed within only a few hours. No matter how one defines the lethal dose of plutonium, it is not more than a factor of four more toxic than that. Moreover, there are biological agents which are in orders of magnitude more toxic. Botulism, for example, is lethal in sub-microgram amounts, and anthrax spores are some ten times more toxic than plutonium (J.H. Rothchild, *Tomorrow's Weapons,* 1964, McGraw Hill).

In some cases, as little as two minute drops of certain pesticides in contact with the human skin are lethal. In such extremely low concentrations, there is an almost instant and fatal response to such materials. In other words, the bugs will go, but there's a good chance you'll be crossing over with them . . .

Plutonium, however, in direct contact with the skin is not harmful. In fact, if you are willing to prove a point, you could even go so far as to hold an ingot of it next to your heart or brain and still fear no adverse effects. In one facility, I held a few pellets in the palm of my own hand. Their appearance was like polished rabbit food or large, brownish vitamin pills.

Also, the effects of plutonium have been compared to botulin (a food poisoning toxin), but it has been found that botulin, gram for gram, is 270 times more hazardous than plutonium. That's *270 times more hazardous.*

Let's clarify on just a few points. Plutonium is rendered generally in one of three forms: metal, nitrate, or oxide. The oxide is a fluffy, yellow-green powder, and this is the form which is generally used as reactor fuel. It can be fine enough to be inhaled and therein rests the controversy. Never, never inhale plutonium or you could suffer serious, even fatal consequences. Here a key factor is the amount of dosage. It turns out, depending on how the calculation is done, that there are some five to eight tons of plutonium dispersed in the atmosphere just from nuclear weapons testing. We put it there. The Chinese put it there. The Russians put it there. We also know that most of this material has long since returned to the earth's surface. In 1963, there was a limited test ban treaty which essentially halted atmospheric weapons testing. Mankind was simply prohibited from setting off any more of his big firecrackers above ground. Instead, we now blow the earth's insides to bits. Regardless, the point being made is that everybody in the world has a quantity of plutonium in them. It can be measured by means of very sensitive instrumentation. There have been studies done for years which involve not only occupational workers after death, but also non-occupational exposure cases. The U.S. Transuranium Registry continues to attempt to correlate postmortem findings with body plutonium measurements.

Studies of bone samples, lymph node tissue, etc., show that this material goes into the lungs and then slowly travels to surrounding tissue, after which some is eventually excreted. Figures show that there are roughly three to five pCi of plutonium in everybody on the face of this earth. If there was the super toxicity alluded to by critics, then there would be some indication that something indeed has gone afoul. But it just hasn't happened to the extent that there is anything documented and directly attributable to plutonium. And that's what we are interested in—official documentation, not hearsay based on rumors and emotion.

A question has also been raised as to what might happen if plutonium were released into a water supply. First, the introduction of any extraneous agent into a water supply is undesirable, especially if the element is as radioactive as plutonium. The problem, however, would not be so great as popularly imagined.

Insoluble—and the key word here is insoluble—plutonium exhibits a collodial behavior. That is, the material tends to precipitate and fall to the bottom as sediment; thus the substance is quickly put out of circulation. In addition, man would not be endangered by either soluble or insoluble plutonium in water: for one thing, in either form it passes through the gastrointestinal tract almost entirely unabsorbed. Precisely, only .002 to 0.5 percent, depending on the form of the material, is absorbed into the body.

In short, one of the best ways to do a job on somebody is to make certain they inhale microscopic particles of plutonium. Experiments with laboratory animals have proven that it is the level of radioactivity of the plutonium that is important and not the quantity inhaled, as in the case of a chemical poison. Regardless, the radioactive hazard is high once plutonium is actually absorbed into the body—.6 microgram of plutonium-239 has been established by medical authorities as the maximum permissible dose over a lifetime, and an amount approximately 500 times greater is believed to lead to cancer.

Inhalation is virtually the only way that it can be effectively absorbed into the body, and about 15 percent of such particles are likely to be retained in the lungs where they may cause cancer. However, there is little danger if plutonium is ingested in food or drink since it passes unchanged through the digestive tract and only about one part in 30,000 enters the bloodstream. Therefore effective plutonium poisoning of the water supply or agricultural land is virtually impossible.

It is often stated that one reason high level radioactive wastes will have to be so carefully maintained for hundreds of thousands of years is due to the extreme toxicity of plutonium-239. Again, this must be viewed in the proper light. In a preliminary draft, Dr. Bernard L. Cohen for the Institute of Energy Analysis writes that if all our present electric power was derived solely from nuclear plants, the Pu-239 in the wastes would not total 160 Kg per year, or 16 tons per century. And, once plutonium becomes part of the soil, half the battle is over, because its hazards become comparable to those of other poisons in proportion to their relative toxicities and relative abundance. As Dr. Cohen speculates, even if all this plutonium was distributed into a plowed soil used to grow grains, it would be responsible for fifteen cancer deaths per century. Now, if this material, through authorized wastes disposal, is buried some 1000 feet below ground, there is no way it can all become concentrated into the top layer of soil.

Water contamination? No real problem since plutonium does not move readily through soil with ground water. Its vertical movement is typically about 0.5 cm per year and its lateral movement is much less. Here's some further reinforcement: In arid Western states where wastes disposal sites are planned, the distance to the nearest creek would typically be at least 5 km, so it would take plutonium at least 10^6 years to reach it. And, during this time its activity would be greatly reduced (by a factor of 10^{12}), making its effects completely negligible. Note here also that all precautions are taken to select burial sites not accessible by water, which voids speculation as to this "what if" at the start.

There is another rumor circulating which says that land once exposed to plutonium can never be reclaimed. This is incorrect, and the best example of large-scale decontamination occurred at Palomares, Spain, where plutonium had contaminated a small village, a hillside and some irrigated agricultural land producing alfalfa and tomatoes. The whole thing happened when a nuclear weapon was accidentally dropped and its conventional TNT explosive detonated, splattering the landscape with generous amounts of plutonium and giving the villagers a bad scare.

Decontamination was accomplished by first, cleaning and removing the topsoil and vegetation. Then it was plowed. Extensive follow-up ecological studies, documented LA-DC–544 (1968), indicate that no health hazard now exists from the ingestion of agricultural produce grown on the land where plutonium was dispersed. Updated in 1975, the official record still shows the same analysis.

Palomares proved a perfect controlled study that was produced by accident. It was a nightmare that has thus far come through with a happy ending.

Who or what watches over such things and records them?

In the United Kingdom, there is the National Radiological Protection Board, which corresponds to our National Council on Radiation Protection. The British are keenly aware of the controversy over this hot particle question, and since it has been their experience to inherit problems if messy situations aren't cleared up immediately, if not sooner, the English don't waste much time. They also have a way of saying things as they are without worrying too much about embarrassing people.

In the recent writings of T. B. Taylor (*Nuclear Theft: Risks and Safeguards,* Ballinger [Cambridge] 1974), an issue is clearly made of the possibilities of plutonium dispersal by terrorist or criminal groups. There has also been debate over the chance of accidental dispersal in various operations of the nuclear fuel industry, to which Taylor makes reference. Actually, plutonium would be a very poor choice as a terrorist weapon. Point One, cancers generally occur from fifteen to forty-five years after initial exposure (lethal dosage can be verified at 260 micrograms through inhalation), and in most victims the increased risk is less than 10 percent. Compare this to the normal risk of cancer death in the United States, which is 16.8 percent (this varies according to geography).

When reports first reached the American public on the cancer risk in cigarettes, millions of people didn't simply turn their backs and run in the opposite direction. Research showed a 10 percent increased risk in cancer death from cigarette smoking. No mass panic. Sales are still great. The young, especially, are puffing away. "You're gonna get it. You're gonna get it," they reply. There's even little sign of terror among the high risk group of heavy smokers who run a 50 percent chance of cancer.

So, the point is, a smart terrorist would employ a better psychological weapon than inducing a risk of cancer among the populace that would not be fulfilled for at least fifteen to forty-five years. Terror-wise, that's just not where it's at. Empires have fallen in such short times. Nor would much public sympathy be generated for a terrorist running around dabbing plutonium here and there like the Easter Bunny. If you're into guerrilla warfare, there's one basic rule to learn: your campaign will only be so successful as to the degree of support that can be generated among the masses which you seek to "liberate." And it's not likely that the populace would be turned-on to mass indiscriminate murder.

Still not convinced that plutonium isn't the quickest way to be king of the mountain? Okay. If you are a terrorist or a newcomer to the business and you sincerely want to indiscriminately murder a few thousand people, then here are a few viable options:

No. 1 Release a poison gas into the ventilation systems of large buildings. (i.e. Time-Life Building).

No. 2 Discharge a load of gasoline or napalm (either by airplane or tank truck) on the spectators at a football or baseball stadium—lighted cigarettes would do the rest.

No. 3 Blast open a large dam. There are situations like this which could kill more than 200,000 people in one massive splash.

No. 4 Poison city water or food supply.

No. 5 Attack a tanker carrying liquified natural gas as it passes near densely populated areas. Estimates on this range up to 50,000 fatalities.

Better still make up your own list! All it takes is a little imagination and you are into the terrorist business. Just remember: nuclear plants are not the best way to go—stick with the tried and proven methods of death and destruction.

As to the distinct impression that many people have been hurt or damaged by plutonium, the facts clearly show otherwise. There have never been any adverse biological effects of any kind ever reported in the handling of plutonium. Perhaps most representative are the remarks contained in the conclusion of a paper written by Drs. G. B. Schofield and G. W. Dolphin titled: "U.K. Experience on the Medical Aspects of Radiological Protection of Workers Handling Plutonium." Schofield is a physician with British Fuels Ltd. and Dolphin, a member of the NRPD. A section of this report reads:

> It's true to say that after 30 years experience in the U.S.A. and twenty-two years in this country (United Kingdom) no disease attributable to plutonium toxicity has been diagnosed in any worker concerned with the production and methodation of plutonium.

You will notice that both workers and the general public are referred to in this statement.

An immediate response might be to question how this British document can be verified. These answers plus a copy of the actual report were provided by a scientist who is regarded as a foremost authority on plutonium. He is Dr. Chester R. Richmond, now associate director for Biomedical and Environmental Sciences at Holifield National Laboratory. (The only other American who might have qualified in this area of expertise was a colleague of Dr. Richmond's at Los Alamos, Dr. W. H. Langham, who has since died in an airplane crash.)

Prior to joining Oak Ridge in 1974, Dr. Richmond was associated with LASL in New Mexico, almost continuously since 1955. There he had a major role in initiating projects which have had a general impact on radiation protection aspects of plutonium production and use in the nuclear energy program. In addition, he has either authored or co-authored numerous technical reports on this subject.

Soft-spoken, precise and to the point, Dr. Richmond is a scientist who does not cloud his words in a kind of mysticism. His approach is cautious optimism coupled with a firm and lasting faith in the scientific method. A hope that he frequently voiced during our interview was that the general public sift through "sound data, documented reports" on this aspect of nuclear power.

"People can be very selective," he explained, "in picking things totally out of context." Then again, he smiled, there is the frequently-voiced thought that the experts are just "too close to the industry and have got too much at stake. Very often this is not only misleading but also an actual error in fact."

Another recommended document is one which Dr. Richmond co-authored with Drs. L. H. Hempelmann, Department of Radiology, School of Medicine and Dentistry, University of Rochester, Rochester, New York; W. H. Langham (deceased), and G. L. Voelz, of the Health Division, LASL. It is titled: "Manhattan Project Plutonium Workers: A Twenty-Seven Year Follow-Up Study of Selected Cases."

This Manhattan Study, as it's referred to, documents the findings on twenty-five men who worked during World War II at what is now Los Alamos, building the first nuclear weapons. During that hectic period—in the excitement, in the crunch to build "the bomb" and hopefully end World War II—the project was top secret and the men worked under extremely tight time restraints. Yet, the same guidelines or restrictions that are now enforced on the nuclear industry were then applied to plutonium handling and fabrication facilities.

"The longer you study," Dr. Richmond explained, "the greater are the chances you have enough time elapsed for something to show up. One of the things a lot of people are now harping about is: 'you can't believe the human data because you haven't waited long enough.' Very few, if any, however, will argue with thirty years."

An abstract of the Manhattan report states:

> Twenty-five male subjects who worked with plutonium during World War II under extraordinarily crude working conditions have been followed medically for a period of twenty-seven years [now 30]. Within the past year, twenty-one of these men have been examined at the Los Alamos Scientific Laboratory and three more will be studied in the following year. In addition to physical examinations and laboratory studies (complete blood count, blood chemistry profile and urinalysis), roentgenograms were taken of the chest, pelvic, knees and teeth. The chromosomes of lymphocytes cultured from the peripheral blood and cells exfoliated from the pulmonary tract were also studied. Urine specimens assayed for plutonium gave a calculated current body burden (excluding the lungs) ranging from 0.005 to 0.42 uCi, and low-energy radiation emitted by internally deposited transuranic elements in the chest disclosed lung burdens of probably less than approximately 0.01 uCi. To date, none of the medical findings in this group can be attributed definitely to internally deposited plutonium . . . Disease and physical changes characteristic of a male population entering its sixth decade were observed.

And perhaps the most important finding in this abstract:

> Because of the small body burdens on the order of the maximum permissible level in these men so heavily exposed to plutonium compounds, we conclude that the body has protective mechanisms which are effective in discriminating against these materials following some types of occupational exposures. This is presumably explained by the insolubility of many of its compounds. Plutonium is more toxic than radium if deposited in certain body tissues, especially bone; however, from the practical point of view, plutonium seems to be less hazardous to handle.

Additional material on this subject can be obtained in a rather lengthy Los Alamos informal report, LA-5148-MS (1973), by the same authors. Titled "A Twenty-Seven Year Study of Selected Los Alamos Plutonium Workers," it provides additional data on the operations and procedures (including flow charts), and chest counting plus urine assay procedures. Keep in mind these are not fabrications but rather documented medical case studies covering a long thirty-year period.

"In a sense, you are opening Pandora's Box," Dr. Richmond sighed. "Because as soon as you show changes—fully realizing that some individuals who smoke a great deal will show changes—we have to be certain if they all showed these or just one or two."

Note: Exposure to tobacco smoke and other toxic materials is known to alter the normal cytology of bronchial cells. This makes observed effects of plutonium exposure difficult to interpret and subjects have been asked to voluntarily quit smoking. However, there has been no way to fully ascertain if this is being done.

Early in 1944, the potential danger of exposure to plutonium was recognized by its discoverer, Dr. G. T. Seaborg. He was aware of the similarity of the radioactive properties of plutonium and radium and of the extreme toxicity of the latter elements which caused bone cancer in man after deposition of microgram quantities in the body.

In an effort to learn more about the biological and metabolic properties of plutonium (and, hopefully, thereby to avert another disaster such as happened in the radium dial painting industry in the late twenties), Dr. Seaborg gave about 10 mg. of plutonium, out of the first half-gram produced, to Dr. Joseph G. Hamilton of the Crocker Radiation Laboratory in Berkeley, California. It went for biological experimentation. Thanks to Dr. Seaborg's foresightedness, many basic facts related to the biology of plutonium were known when those first milligram samples of the element arrived at Los Alamos in the spring of 1944.

Nevertheless, as Dr. Richmond noted, to be aware of the potential biological hazards of plutonium and to protect against them were two entirely different matters. Safety regulations could be established on the basis of experience in the radium dial paint plants, but protection problems in the two cases differed greatly.

It became apparent in the early 1920s that the young women painting radium dials on watches were developing bone cancer. Then milligrams of radium in watch plants were subjected to simple mechanical operations, while at Los Alamos, kilogram quantities of plutonium were involved in complex chemical and metallurgical manipulations. All work with plutonium was carried out in a wooden Chemistry and Metallurgy Building, called simply, "D Building." Air entering the facility was cleansed of dust by electrostatic precipitation methods, but air leaving the building was mostly unfiltered since few, if any, of the exhausts which contained hoods, even those from plutonium, had filters. Suffice it to say that working conditions were quite primitive compared to those in effect now.

Regulations included: (A) a complete change of street clothing on entrance to the contaminated areas with two changes per day of freshly laundered coveralls, canvas bootees, and surgical caps (all persons showered on leaving the building); (B) use of surgical rubber gloves and respirators during all chemical procedures involving plutonium; and (C) whenever possible, use of closed systems, which initially were homemade dry boxes with rubberized canvas sleeves attached to surgical gloves.

Sometimes it was impractical to use a closed system and chemical hoods had to be used. In the early days, these were not even made of stainless steel. All workers were fully appraised of the hazards of plutonium, as they were then known, and required to sign a statement agreeing to abide by the safety rules. With rare exceptions, all cooperated to the best of their ability, although during the tension and feverish activity of developing the first atomic weapons it was difficult to avoid some short-cuts in observation and enforcement of safety rules.

When milligram quantities of plutonium first became available to the New Mexico bomb makers, efforts to live with what was considered to be safe contamination levels were hampered by the fact that portable alpha counters and continuous air samplers hadn't yet been developed.

However, again due to the urgency of the times, work with plutonium had to proceed, and improvised methods of monitoring and decontamination were unbelievably primitive by today's standards.

One example was the use of stationary counters which were the only means of detecting alpha particles from plutonium. It was necessary to make "swipes" with lightly oiled filter paper of the laboratory surfaces considered to be most likely for contamination and to bring these paper strips to the Alpha Proportional Counter for counting. This was done systematically on a daily basis (or after an accident) in each laboratory containing plutonium (six swipes per laboratory).

Any "swiped" area (approximately 6 x 6 inches) having an activity of more than 500 counts per minute required decontamination. Similarly, in the absence of air samplers for radioactive dusts (first available in the fall of 1944), both nostrils of each plutonium worker were "swiped" routinely with rolled, moistened strips of filter paper at the end of each working day, or when accidental pulmonary exposure was suspected. These samples were also counted in a Stationary Alpha Particle Counter. If the nose swipe exceeded fifty counts per minute, the subject was questioned carefully about possible accidental inhalation or negligence of safety regulations. For example, it was quite possible for a worker to contaminate himself by scratching his nose.

Progressively changing words to arrive at a direct statement of fact or conclusion is often used in connection with plutonium. There is one case of alleged contamination resulting from a puncture wound that is extremely interesting and frequently quoted. It has been interpreted by some as resulting in a "precancerous lesion." These quotes come from a report titled: "Radiation Standards for Hot-Particles: A Report on the Inadequacy of Existing Radiation Protection Standards Related to Internal Exposure of Man to Insoluble Particles of Plutonium and Other Alpha-Emitting Hot Particles." It was prepared by A. R. Tamplin and T. B. Cochran for the Natural Resources Defense Council (NRDC) in Washington, D.C.

In this reference, released in 1974, the following statement is made: "This precancerous lesion indicates that a single plutonium-239 particle irradiates a significant (critical) volume of tissue and is capable of inducing cancer."

This entire matter stems from a worker who allegedly got some plutonium on a puncture wound in his hand while on a loading dock back in the early sixties. Later, a tumor developed there and a suspicion has been raised that it was induced by the hot-particle. Depositions from physicians attending the case, however, state specifically that there is no connection between the alleged contamination and the tumor.

Information on this case was originally published in 1962 by C. C. Lushbaugh and J. Langham, in a report titled: "A Dermal Lesion from Implanted Plutonium." It appeared again in 1967 under the new heading. "Histopathologic Study of Intradermal Plutonium Metal Deposits: Their Conjectured Fate."

The authors (Lushbaugh and Langham) stated:

> Although the lesion was minute, the changes in it were severe. Their similarity to known precancerous epidermal cytological changes, of course, raise the question of the ultimate fate of such a lesion should it be allowed to exist without surgical intervention. Although no malignancies of the skin of man have ever been shown autoradiographically to be associated with such alpha-emitting foreign bodies, the changes would seem to indicate that the development of such a lesion is possible.

There is a key phrase here, and I sincerely hope that you will take the time to plow through some of this medical terminology. It's well worth the effort, and it's the only way that, we, the American public, are going to get at the truth of this entire controversy. The key phrase is ". . . similarity to known precancerous epidermal cytological changes."

Individuals studying this situation acknowledged this wording in the first reference of the Lushbaugh and Langham (1962) report. From this point on, however, there begins the steady progression of subtle word changes. Refer back to the earlier quoted 1974 reference by Tamplin and Cochran and the wording has been altered slightly to read ". . . precancerous changes in tissue . . ."

In a third reference, dated 1974, it becomes ". . . particle (referring to plutonium) induced cancer . . ."

Again, what has occurred is the old trick of progressively changing words to arrive at a direct conclusion. This may have been the basis of a recent statement by R. Gillette in *Science* magazine ("Plutonium and the Hot Particle Problem: Environmental Group Proposes a Draconian Answer"). It states: "Only one human cancer case is clearly linked to plutonium exposure." *Actually no human cancer case has ever been "clearly linked" to plutonium exposure.* Cytologic changes have been described in cells in the vicinity of embedded plutonium particles in man, but the malignant cellular transformation required for the diagnosis of actual cancer has never been found next to the "hot-particles" in human tissue.

Again, the primary reference point here is Lushbaugh and Langham, 1962, and Lushbaugh's 1967 statement. Similar results have also been reported for several animal experiments designed to study the biological effects of hot-particles by Dr. Richmond and his colleagues. Specific report titles are listed in the index.

Granted, it can be stated that under certain deliberate exposure conditions, plutonium has been found to be an efficient cancer-producing agent in *experimental* animals. The key wording here is "certain exposure conditions" and "experimental." There are, however, no human cancer cases "clearly linked" beyond a doubt to plutonium and, yet, you repeatedly hear the phraseology: ". . . plutonium is known to cause cancer. Plutonium has caused cancer in men. A pound of plutonium means death." Etc. Etc. Etc.

Such statements of "fact," noted Dr. Richmond, are often made, not by kooks but rather by "respected individuals who have looked seriously at this problem." Nor is concern limited just to the United States. "It's of international interest and should be," he said, adding that the French have made some interesting observations and research applicable to the hot-particle problem. "The Russians have also made some comments which I think are very interesting in terms of how they interpret hot-particle effects," he said. These show an interesting parallel, he added, to findings contained in the Manhattan Study.

The United States has processed large quantities of plutonium during the past thirty years with no indication that any member of the public has been adversely affected. Even in an industrial context where exposures have taken place, some even exceeding acceptable standards, no cancers have been identified or attributed to plutonium. I have not intended to imply in this chapter that the element is not hazardous or that it does not require considerable control. But we have and can continue to control its release to the environment and man. Remember that the release of small quantities of plutonium does not ipso-facto mean a threat to man's health. If every one in the world were to receive the maximum permissible dosage of radiation from plutonium, roughly 200,000 metric tons of it would have to be uniformly dispersed in the atmosphere. Not an easy task.

Plutonium, then, should be viewed in the same framework as other highly toxic and carcinogenic material utilized in this society and compared on its merit.

"It hasn't, as some people would have you believe, destroyed the human race," Dr Richmond concluded. "There are much larger concentrations—much, much higher—involved in individual exposure cases (occupational) and we still don't see effects. So, you wouldn't anticipate effects on levels that are much, much lower."

Dr. Chester R. Richmond

From the Office of
C. R. Richmond

6/24/75

Jacque Srouji:

Attached is a list of references related to the hot particle question that I thought you might be interested in. I cannot remember what reports I gave you when you visited with us earlier this year. If you need any of the attached information, please let me know.

Chet Richmond

Figure 3. On subjects like plutonium, I sought the advice of experts like Dr. Chester R. Richmond, formerly of Los Alamos and now Associate Director for Biomedical and Environmental Sciences, Oak Ridge National Laboratory, Tennessee.

Chapter 9

GARBAGE-DUMPING ON A GRAND SCALE

Garbage is as old as man, but getting rid of radioactive wastes from a nuclear reactor is a little more complicated than simply tossing a sack in your neighborhood Dempsey-Dumpster.

Right now, as you read and wonder about this chapter, there is probably an archaeologist kneeling reverently somewhere and gently brushing aside dirt from a crude artifact discarded by some nameless being thousands of years ago. Such fragments of earthenware and chipped stones, resting in the ashes of ancient fires, tell their own unique story. These are wastes, leftovers, from a now-extinct human culture.

Today, in the rolling wheat fields of Kansas, tons of straw left behind by threshing machines must be disposed of each year by baling or by burning. These straw wastes may never be seen by future archaeologists, but other wastes of the twentieth century will intrigue them: things like piles of slag from steel mills, city dumps under tons of soil, mountains of rusted automobiles, and perhaps a stray Coke bottle or two that has somehow withstood time's onslaught.

Most acts of men lead to wastes of one sort or another: hunters' ejected cartridges, smoke from power plant chimneys, waste paper in school rooms, or the curls of childhood lost forever on some barbershop floor. The nuclear industry is not an exception, and those who manage these installations voice concern about wastes handling, processing, and disposal. After all, the earth is their home too.

The actual volume of nuclear wastes is a significant aspect of this entire question. Mention "wastes" and the immediate image is one of a huge garbage dump, with trucks moving in and out and those piles of death growing higher and higher. Eventually, you might fantasize, such radioactive piles turn into monstrous blobs of amoeba-like substances that roll out to absorb anything in their path. Now, that's pure fiction, but it makes good copy, guaranteed to catch a reader's attention.

The real story is much different and not quite so dramatic. The physical volume of the solidified high-level wastes of a single, 1000 megawatt electric nuclear plant for a year's operation, could be placed in nine canisters, each one foot in diameter and ten feet long, with each holding about six cubic feet of wastes. Simplified even more, that means it's just about enough to fill a nice-sized clothes closet . . . a little smaller than one of those walk-in jobs.

Let's scale that down even more, because the message here is quite clear: the amount of high-level wastes to be handled in the future is actually miniscule. Each American uses an average of 7,000 kilowatt-hours of electricity per year, which takes about one gram of nuclear fuel (the weight of one aspirin tablet). "Burnup" of this fuel produces about one gram (three-hundreths of an ounce) of fission fragments. After being processed and solidified, the gram of waste has grown in volume by about three times. The total high-level wastes generated by one person's energy consumption in his 70-year (average) lifetime, is less than one-half pint (in solid form). The total volume of such solid wastes generated in the United States is less than 1,000 cubic feet per year. If this were in one solid pack, it would require a room ten feet by ten feet in an area that was ten feet high.

Remember, however, that solid radioactive wastes aren't stored in a solid pack. They are packed in steel canisters about ten feet long and one to two feet in diameter, so that today's total annual consumption of nuclear energy produces about 130 canisters (ten feet by one foot) of such wastes per year.

One final point. Suppose we take the maximum predicted growth rate of nuclear power, which says that a thousand, one-million-kilowatt power plants will be operational in the United States by the year 2000. The total solid wastes generated during that year would fill about 8,000 canisters, and the total nuclear wastes accumulated up to the year 2000 would fill about 80,000 canisters. The largest estimate of storage area required is about twenty-five square feet of floor area for each one-foot diameter canister. Thus, *by the year 2000, the total land area needed for storage of all wastes ever produced by nuclear power generated in the U.S. would involve just about fifty acres.*

Surprised? I was.

These aren't just speculations. They are the official estimates from people like Congressman Mike McCormack, who, as you may remember, serves as chairman of the powerful House Energy Committee.

As Representative McCormack said, long-term storage does involve an acceptable final disposal strategy. Some of the top contenders are: deep burial in geological formations, such as salt or granite; and burial at sea. There is also a rumor circulating that wastes may be shot into space, but this particular strategy has received official frowns and is, at this point, not considered very practical or safe. Frankly—and maybe I'm reacting as a layman—I really can't buy shooting that stuff into space. It's like garbage-dumping on a neighbor's lot without asking their permission: somewhere in the universe there just might be intelligent life that might not appreciate our using the sun as a big Dempsey-Dumpster . . .

Salt beds appear to be the most feasible and practical alternative. These formations are usually dry, impervious to water, and not associated with unstable groundwater sources. The ability of salt to change shape under pressure causes rapid closure of fractures, which is a good point to consider since any tear would be instantly sealed. Salt is sufficiently strong that large cavities formed by mining will not collapse. Also, salt deposits underlie some 400,000 square miles of the United States. The estimated volume of high-level wastes solids to be produced by the year 2000 would occupy less than 1 percent of the volume of salt now being mined each year.

I find this interesting and reassuring, because to be honest, when salt beds were first mentioned, all I could see was a shaker of tiny, white granules used for seasoning. Nothing clicked. Now, it does, and here are the specific advantages of depositing nuclear wastes in salt beds:

- Good radiation shielding properties
- High heat conductivity (to remove decay heat)
- Low earthquake potential: beds have been stable for more than 200 million years . . . a long time any way you notch it.

- Free of circulating ground water (no danger of wastes being carried away)
- Impermeable rock isolation from aquefiers
- Easy to excavate

An Oak Ridge National Laboratory program has included studies of the effects of radiation on salt—on its physical properties, the possible production of chlorine, and the effects of movement under pressure and plastic flow when heated. Such information is necessary in the design of actual disposal facilities.

A nineteen-month demonstration disposal of high-level radioactive wastes solids was carried out in a salt mine at Lyons, Kansas. Spent reactor fuel was used in lieu of actual solidified wastes. During the field demonstration, the feasibility and safety of handling highly radioactive materials in an underground environment was clearly demonstrated. Studies also provided valuable data on the stability of salt under the effects of heat and radiation, as well as the creep and plastic flow characteristic of salt.

ERDA is now conducting engineering studies to determine the location and subsequent establishment of such a facility. Sites in Kansas and New Mexico are being studied for suitability. The Kansas Geological Survey determined that the Lyons site was not safe, since it had been penetrated by oil exploration drill holes and there was a possibility—although remote—that water might be able to seep through these. So, to be extra safe, Lyons is disqualified.

Radioactive wastes vary greatly in the concentration of their radioactive materials. It is convenient to classify them according to their potential hazard, as indicated by their level of activity. Three levels have been defined, somewhat arbitrarily, as follows:

- *Low-Level Wastes:* These have a radioactive content sufficiently low to permit discharge to the environment with reasonable dilution or after relatively simple processing. They have no more than about 1000 times the concentrations considered safe for direct release. In liquid form, low-level wastes usually contain less than a microcurie of radioactivity per gallon. This represents one millionth of a curie.
- *Intermediate-Level Wastes:* These have too high a concentration to permit release after simple dilution, yet they are produced in relatively large volumes. Their radioactivity is approximately 100 to 1000 times higher than that of low-level wastes. In liquid form

they may contain up to a curie of radioactivity per gallon. A curie, you may remember, is a measure of the number of atoms undergoing radioactive disintegration, that is, the rate of decay in one gram of natural radium.

- *High-Level Wastes:* These are the sticky ones. They contain several hundred to several thousand curies per gallon in liquid form and result from chemical processing of irradiated nuclear fuels. High-level wastes pose the most severe potential health hazard and the most complex technical problems in management. But we are working on it . . .

Now, to define a little further the wastes produced from reactor operations. These are of two types: relatively short-term and relatively long-term. The toxic life time of the short-life wastes called fission products is approximately 500 to 1,000 years. The long-life wastes, including plutonium, while initially not the dominant toxic component of radioactive wastes, would come under this category because their toxic life is more than 250,000 years.

What bothers many people is that the toxicity scale of these wastes extends far beyond the stability of known governments and, in fact, most human institutions. Therefore, the question is just how can we assure future generations adequate protection from the release of large quantities of radioactive wastes? It is a legitimate question for those who will go after us. Just what kind of legacy are we prepared to leave them? The value judgment here is on a much broader scale than simply selling your old homeplace and leaving a closet full of dirt and debris for the new owner to sweep away.

Although the quantity of nuclear wastes will increase in the future (see Table 3) as nuclear power expands, the problems of disposal are not related to volume or weight. In fact, one of the decided advantages of nuclear energy is the small amount of waste in relation to the energy obtained. An excellent illustration of this is to compare nuclear wastes with those from coal-fueled power plants, as illustrated in Table 4.

Radioactive materials are produced in solid, liquid, and gaseous forms. Liquid wastes include corrosion products, some fission products and tritium; while gaseous ones contain some gaseous fission products and tritium. It should be remembered that gaseous and liquid wastes from nuclear power plants contain very low levels of radioactivity which are carefully monitored with amounts determined by the Nuclear Regulatory Commission (NRC).

If the term "tritium" causes a little "gas," you might want to remember that Chapter 12 (Fusion) goes into greater detail about this radioactive form of hydrogen. Reactors produce it in several ways, and it can be removed from the system as either a liquid or a gas. Because it can become part of water molecules and because of its long half-life (12.3 years), tritium represents a potential biological hazard, but its main advantage is that it disperses (i.e., evaporates) quite rapidly.

In the fission process, uranium splits into two new and lighter atoms. These are called fission products, most of which are stable (non-radioactive) while others are highly radioactive. Only the latter are hazardous. The fission products build up in the fuel during its use and are retained there until the fuel is removed for reprocessing.

Sometimes fission products escape from the fuel cladding, usually through pinhole-size defects in the fuel cladding, into the surrounding coolant, in which case they are collected and removed as either liquid or gaseous wastes, depending on the reactor type. For example, xenon, krypton and iodine are fission products which are normally retained entirely within the fuel elements but occasionally escape, to be collected by the reactor coolant.

Where do we go from here? Highly radioactive wastes—separated from the cooling water—are of very small volume and are sealed in special containers for shipment to licensed disposal areas. Low-level wastes are stored temporarily, permitting some of the radioactivity to decay, and are then diluted to harmless levels with large quantities of water and discharged into lakes, rivers, and oceans. And I know exactly what you are thinking at this point, because I asked the same question at places like Browns Ferry in northern Alabama and San Onofre on the West Coast: What about thermal discharge? Just what kind of effect does it have on marine life? The simplest answer, based on reports I was given plus first-hand observation, is: nothing detrimental. Often, it is to the contrary. Again, you might want to refer to the section on San Onofre (Chaper 10) detailing a very exacting "Thermal Effect Study" that was a joint venture by the Environmental Quality Analysts, Inc., and the Marine Biological Consultants, Inc. Their findings can be substantiated through physical data and are quite reassuring. If in doubt, I would again strongly suggest referring to this chapter since it does clear up a lot of muddy water.

Radioactivity in liquids discharged to the environment is limited by NRC standards to very minute quantities. The maximum allowable levels were reversed downward in 1971 from 100 to 20 picocuries/liter (the curie is a measurement of radioactivity; pico means one trillionth). The revised standard reflects the actual levels achieved in nuclear power plant operational experience since 1967. As shown in Table 5 the range of radioactivity level experienced is comparable to, or lower than, the levels found in ordinary (tap) drinking water and other common liquids.

Such environmental study of areas near nuclear energy installations is viewed as an important aspect of wastes management research. Objectives are to:

- Determine the fate of radioactive materials released to waterways;
- Determine the mechanisms of their dispersion;
- Evaluate the hazards associated with disposal practices;
- Evaluate the usefulness of waterways for disposal;
- Recommend appropriate long-term environmental monitoring procedures.

Principal studies in the United States have been made at the Clinch-Tennessee River system near Oak Ridge; the Columbia River below Richland, Washington; and the Savannah River near Aiken, South Carolina. Similar studies have also been made at the Ottawa River in Canada, the Thames River in England, and the Rhone River in France.

The Clinch-Tennessee River Study, for example, is a thorough probe, on a multi-agency level, to evaluate physical, chemical and biological effects caused by the disposal of low-level radioactivity. Water-sampling stations have been established at seven sites, located as far as 125 miles from the home station at Oak Ridge. Probes have included dispersion tests with dyes and radioactive tracers, determination of the distribution of radioactivity in bottom sediments, methods of incorporation of radioactivity into the bottom sediments, and detailed analyses of the biological distribution of released radioisotopes and the resulting dosages to human and animal life in the area.

Dr. William R. Bibb, chief of the Research and Development Branch for the Division of Research and Technical Support at Oak Ridge, told me that studies at Clinch River have provided valuable thermal tolerance data for the Liquid Metal Fast Breeder Reactor that is being constructed at Oak Ridge.

As nuclear fuel is "burned" in a reactor, the uranium fuel material is slowly converted by nuclear fission into products often referred to as "nuclear ashes." Except, they really aren't that at all. They collect as a contaminant in the individual fuel rods and may be either solids, dispersed uniformly through the ceramic (uranium oxide) type of fuel material in the fuel rods, or as gases trapped within the fuel rod. Such fission products are highly radioactive. After the fuel material is partially exhausted and the fission products have built up to a certain level in the fuel, the fuel rods are withdrawn from the reactor and shipped to a fuel processing plant where the remaining fuel material (the nuclear fuel does not change its physical form) is chemically separated from the fission products. The recovered fuel is then recycled back into the nuclear power program.

At this point, it should be fairly clear in your mind that high-level nuclear wastes are those highly radioactive waste fission products which are stripped out of the fuel elements in the chemical extraction cycle used in the fuel processing plants. The wastes are initially in a liquid (aqueous) form. The liquid wastes may be placed in tanks to "cool," i.e., to lose some of their radioactivity through the process of radioactive decay.

During such cooling off periods, the "flasks" produce a bluish-green glow called the Cerenkov Reaction (after the Russian who discovered them). I stood high above a cooling tank (half the size and about as deep as an Olympic pool) and stared down at the water. There sat the containers, stacked side-by-side, and each glowing with a greater or lesser intensity, depending on its confinement period. Strange, I told Dr. Bibb, that something so beautiful can be so deadly. He agreed.

So, through this Cerenkov Reaction, the fission products decay, give off their radioactive energy, and become less radioactive. This loss of radioactivity continues with time. After a storage period as liquids in the plant (up to five years), they are then solidified into beads, granules or chunks. Again, these are highly radioactive because they contain the fission products from the spent fuel. The solid wastes may be then further stored at the plant—up to a total of ten years from initial production of liquid wastes—prior to shipment to a federal waste repository. Note: High-level wastes will always be transported in solid form, because a federal repository will not accept it any other way.

Fission products stay in place until the fuel assemblies are removed from the reactor. They are called, appropriately enough, spent fuel assemblies, because they have worked hard. Fission products are pesky little critters who have the knack of absorbing neutrons needed in the nuclear chain reaction, thus slowing down the entire operation. This means that they must be removed or they eventually stop the reactor's entire works. About two-thirds of the fissionable uranium (uranium-235) in the fuel is consumed before the assemblies must be removed for reprocessing.

More than 90 percent of the fission products are stable, which means non-radioactive. Of these products, some decay in a few seconds while others take a longer time, ranging sometimes up to thousands of years. Some fission products have an economic potential in the near future, possibly in medicine. Others show promise.

The most common type of fuel in use today consists of bundles or assemblies of sealed metal tubes, about one-half inch in diameter, containing pellets of uranium dioxide, each about one inch long. In reprocessing, these tubes are chopped into small segments, the fuel pellets are dissolved in strong acid, and the fuel and fission products are then chemically separated. These extraction processes are quite similar to the techniques used by industry with, for example, pharmaceuticals. The basic difference is that the spent nuclear fuel rods are highly radioactive, and from the time that they leave the reactor, they are handled with extreme care, specifically by remotely operated machinery called master slaves, and are always kept under water in shielded containers or behind thick concrete walls of shielded "hot cells." In a real sense, they get the VIP treatment all along the line.

The liquid and sludge remaining after the uranium and plutonium have been "mined" from their spent fuel elements, can now be correctly referred to as "wastes." These high-level radioactive materials include not only the fission fragments themselves—the ashes of nuclear fissioning—but also some of the chemicals used to extract the leftover fuel. The liquid wastes may be stored in double-walled stainless steel tanks at the fuel reprocessing plant for up to five years. They must then be solidified and, within another five years, shipped to a government plant for disposal.

You might wonder about the feasibility of fuel reprocessing. Why not just throw the entire mess away and begin anew? Point one, the spent fuel is reprocessed to recover valuable fissionable material contained therein, namely uranium-235 and plutonium. Point two, recovery of uranium reduces the cost of power production by about 5 percent or $3 million per year for a 1000-MW plant, which not only conserves a valuable natural resource, but also reduces (by 20 percent) the amount of mining necessary to replace it. The cost of waste disposal has been estimated to be from 1½ percent to 2 percent of the cost of nuclear power.

Thus far, very little high-level waste has accumulated from the generation of commercial nuclear power. Also, a relatively minute amount of such waste is created by each plant, since a full-size (one million kilowatt) facility will normally produce only about sixty cubic feet of high-level solidified wastes during a year's operation. Until recently, a single reprocessing plant has been able to handle these amounts: a West Valley, New York, facility which is owned and operated by Nuclear Fuel Services (NFS). This plant was closed in 1972 for modifications and is expected to resume work by 1977. Two other commercial plants are also designated for this responsibility: General Electric's Midwest Fuel Recovery Plant near Morris, Illinois, and the Allied-General Nuclear Services Plant near Barnwell, South Carolina, which is a joint venture by Allied Chemical and General Atomic. There is also a total of four government plants in operation and in addition to these, thirteen reprocessing plants worldwide—mainly in Europe.

Much of the high-level radioactive wastes generated to date by nuclear reactors are still stored in their "interim" liquid form in large tanks. All these wastes will ultimately be converted to solids, reducing their volume considerably, and then turned over to the government for storage or disposal. Note: There is also a much larger volume of liquid wastes which have been generated by federal operations. These are now stored in tanks on large government reservations at Hanford, Washington, and the Savannah River in South Carolina, and the National Reactor Testing Station in Idaho.

Storage safety is another valid question. Again, the record is optimistic. After two decades of storage of both commercial and government-generated liquid wastes, not a single member of the public has been over-exposed to radiation from stored high-level radioactive wastes.

True, there has been leakage from the old carbon-steel tanks in the government installations during the years, but there has been no detectable increase in radioactivity of local ground water and no over-exposure to radiation by anyone in the area, as was demonstrated by extensive measurements in welds around the leaking tanks. Nevertheless, this method of storage is unquestionably obsolete and has no place in future plans for storage of wastes from commercial nuclear plants. At Hanford, for example, fifteen of 151 tanks have developed leaks at weld areas over a period of about twenty years. This has not been a problem, and massive tank failure, resulting in large flows and a total loss of contents, has just never happened and is not expected.

At commercial reprocessing plants, wastes in liquid form are collected, concentrated by evaporation, and then piped into underground tanks on site for the up-to-five-year cooling-off period before being converted to solid form. These tanks (see Figure 4) are now constructed of stainless steel and are designed to last more than fifty years. In the event of leakage, there are standby tanks available for immediate transfer. Such leakage would be detected almost instantly by built-in monitoring instruments located in the liner around each tank. As a further measure of safety, the gravel bed under the tank also has detectors buried in it and is equipped with standby pipes through which any wastes which might leak out during the transfer operation can be pumped to the surface for storage in the new standby tank. The tank is designed to withstand earthquakes, as well, although sites for commercial reprocessing plants are selected for geological stability.

After the wastes are solidified, they will be shipped to the government for "retrievable surface storage" or "permanent disposal." It is important to distinguish between these two terms.

Storage is designed to be temporary, even on a time-scale of decades or centuries. A storage facility must be monitored and maintained as long as it is operated, in a manner described by TVA Director Aubrey Wagner, like the dikes and dams. Retrievable surface storage, often called "engineered storage," is considered a temporary expedient to be employed until the best permanent disposal method and site are selected which will virtually eliminate the burden of monitoring and maintenance.

The first storage facility is now being designed to handle 5,000 to 7,000 canisters, covering all storage needs until about 1995. During this period, storage capability will be extended, plus ample testing done to evaluate just where we go from there.

This chapter has covered wastes in reactor situations, as well as actual storage and maintenance. The next logical question should be: Great, but how do you get from Point A to Point C? How and what is Point B, and how safe is it?

First, consider what shipping casks are like (Figures 5 and 6). These are probably the most carefully designed shipping containers known to man. They are made to contain fuel elements in a very strong steel shell about one inch thick. The shell is surrounded by 8–12 inches of lead and finally by another 4½ inches of steel. Radiant cooling fins are welded to the outer steel cylinder to remove the heat generated by the radioactivity in the fuel elements. The ends of the outer cylinder have massive flanges designed to protect the finned sides.

Perhaps, you are still not convinced. Just how do the experts really know that these casks would survive an accident? Designs are analyzed and the casks tested by subjecting them to each of the following accident conditions, in sequence:

- A 30-foot free drop onto a flat, unyielding surface with the cask oriented so that maximum damage will occur.
- A 40-inch free drop onto the top end of a six-inch diameter, eight-inch long steel bar with the cask oriented so that maximum damage will occur.
- Exposure to a 1,475° Fahrenheit fire for a thirty-minute period, with no artificial post-fire cooling.
- Immersion of the entire cask in water to a depth of three feet for at least eight hours.

After a cask has been through these four conditions, it is tested for leakage and must come through leak-tight. The test results are very carefully reviewed to assure that the established performance criteria are met before an okay is given.

If a cask gets up and walks away from this type of torture test, it receives that gold star which is officially known as the Government Transportation Permit.

The significance of such tests is tied to the "design basis accident" concept that goes back to the old "pants and suspenders" approach described by Dr. Culler on reactor safety. It goes like this: A shipping cask is supposedly aboard a train passing over a freeway and is thrown off when the train derails. The cask lands thirty feet below on the freeway, where a tank truck carrying jet aircraft fuel just happens by, crashes into it and promptly bursts into flame. The cask is then enveloped in the flames which burn for exactly thirty minutes. Then it promptly rolls off the road into a lake where it is submerged for exactly eight hours before it is finally recovered.

Okay, granted the darn things are mighty strong and no accident is going to push it around . . . what about sabotage and those ever-present terrorists lurking in the bushes?

Relax, cask designers know about these things as well and have prepared the devices to successfully withstand any type of assault you might envision. These include high-power rifle fire, even with armor-piercing bullets, and explosive attacks, such as one might expect with a satchel charge (several sticks of dynamite), or small bombs that might be surreptitiously placed on the vehicle or dropped from an overpass. Even small thermite bombs would have no serious effects.

With lids and closure plugs weighing as much as five to ten tons, it is not likely that anyone will be "peeking inside" to see what's going on. Also, with the casks themselves weighing between 20 and 100 tons, theft of the entire thing is very unlikely. It would take more than a U-Haul-It to cart one of those things away.

All kidding aside, there is no doubt in my mind that wastes management is serious, even killing business, if we aren't extremely careful. There comes a time, however, when the child must walk . . . when the child must deal with the real world in a rational manner. The nuclear industry is at that stage. Waste management has been around for some 30 years already and it's one of those bitter pills that we must swallow and learn to accept until a permanent solution is found. And someday it will be. Mankind has yet to be dwarfed by that which he does not fully understand, and garbage-dumping will not be the first to defeat our best efforts.

Table 3: Reprocessing Requirements and High-Level and Alpha Waste

Year Ending	1970	1980	1990	2000
Installed nuclear capacity, MW†	6,000	150,000	450,000	940,000
Fuel reprocessed metric tons/yr.	55	3,000	9,000	19,000
Solid High-Level Waste Annual *	0.17	9.7	33	58
Solid High-Level Waste Accumulated *	0.17	44	290	770
Alpha Waste				
Annual *	1,000	1,000	1,000	1,000
Accumulated *	1,000	11,000	21,000	31,000

† 1 MW = 1 megawatt, or one million watts
* Volume in thousands of cubic feet

Table 4: Fuel Consumption and Waste (1000-Megawatt Power Plant)[1]

	Hourly	Daily	Annual
Fuel Consumption			
Coal	683,000 lbs.	8300 tons [2]	3,000,000 tons
Uranium	0.3 lbs.	7.4 lbs.	about 1 ton
Waste Production			
Coal (ashes)	68,300 lbs.	830 tons [3]	300,000 tons
Uranium (fission products)	0.3 lbs.	7.4 lbs.	about 1 ton

[1] 1000 megawatts is enough electricity for a city of about 1 million
[2] equivalent to a 100-car trainload every day
[3] equivalent to a 33-car trainload every day (not including airborne wastes)

Table 5: Liquid Radioactivity Levels

(Picocuries/liter)*

Typical Nuclear Power Plant Radioactive waste discharge			1-10
Domestic Tap Water	20	Whiskey	1200
River Water	10-100	Milk	1400
4% Beer	130	Salad Oil	4900
Ocean Water	350		

* A curie is a unit used to measure radioactivity
A picocurie is a trillionth of a curie
A liter is approximately one quart

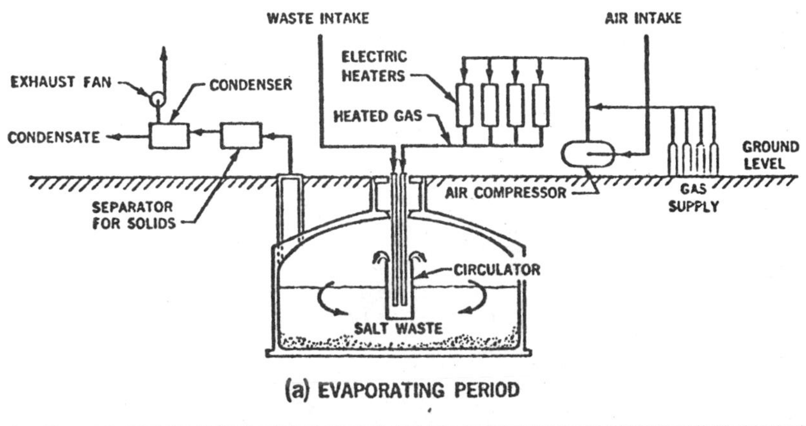

(a) EVAPORATING PERIOD

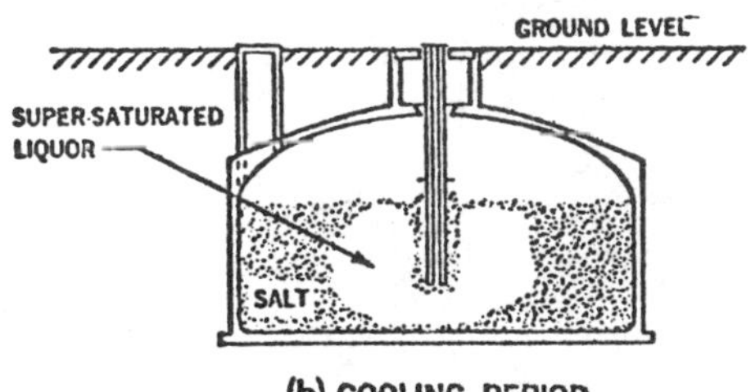

(b) COOLING PERIOD

Figure 4. Proposed in-tank solidification of high-level wastes at the Hanford Plant. More waste can be added as tank level drops during evaporation.

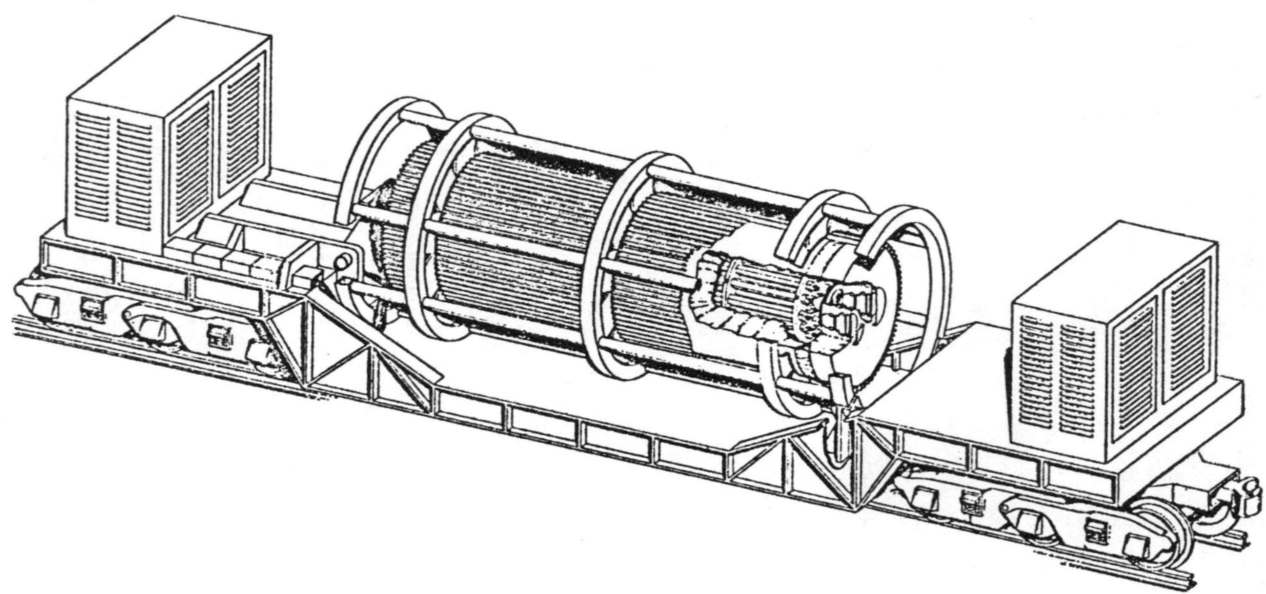

Figure 5. This cask is designed to ship 18 Light Metal Fast Breeder Reactors (LMFBR) spent fuel assemblies.

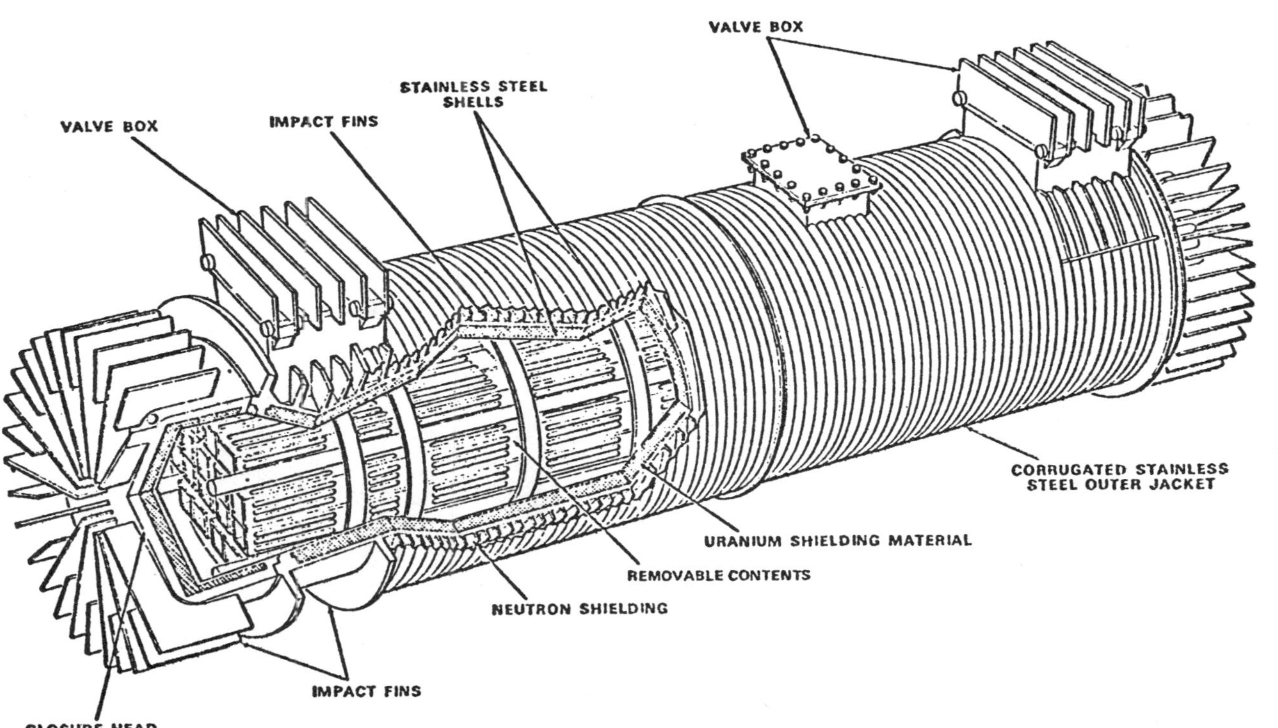

Figure 6. The IF-300 spent fuel cask was designed by General Electric and has been licensed for use by ERDA's licensing staff.

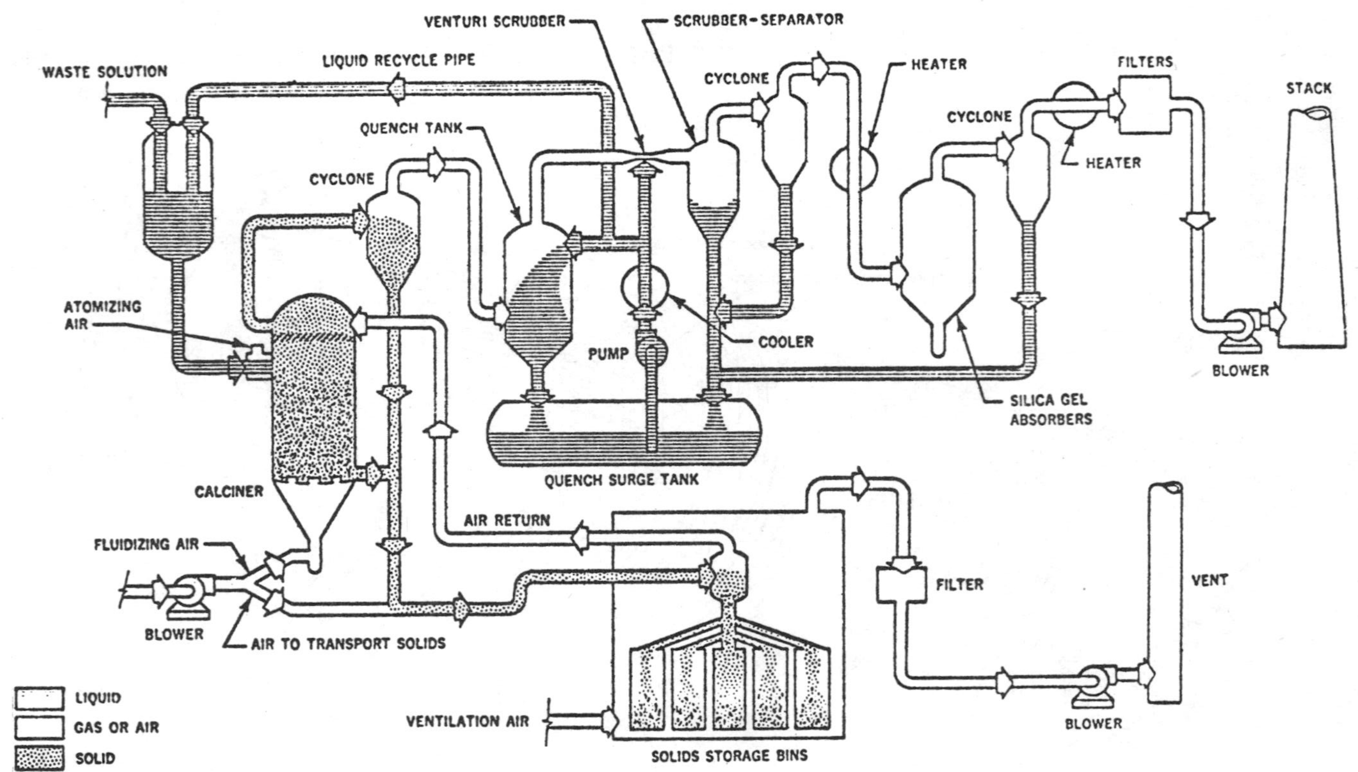

Figure 7. Flow diagram of the waste calcining facility at the Idaho Chemical Processing Plant.

Table 6: Typical Radioactive Wastes and Disposal Methods

Type of waste radioactivity	Source of waste	Form of waste	Typical isotopes	Type of radiation	Disposal methods
Natural activity	Mining of uranium ores	Solids	Uranium-238	α, γ	Pile in open
		Liquids	Thorium-230	α, γ	Seep into ground
		Gases and dusts	Radon-222	α	Ventilate mine
	Fuel Fabrication Plants	Solids	Uranium-238 Uranium-235	α, γ	Decontaminate
		Liquids (acid)	Uranium-238 Uranium-235	α, γ	Neutralize, concentrate, and bury residue
		Dusts	Uranium-238 Uranium-235	α, γ	Ventilate, filter, and disperse to air
Fission-product activity	Fuel irradiation and processing	Solids (from purposeful solidification)	Strontium-90 Cesium-137	β β, γ	Encase in container and store permanently (-600 years)
		Liquid (with strontium and cesium removed)	Technetium-99 Ruthenium-103 Cerium-144	β β, γ β, γ	Store in tanks for several years; then solidify in place
		Gases	Iodine-131	β, γ	React with chemicals to bind in solid, e.g., silver iodide
			Krypton-85	β, γ	Disperse to air
Activation-product activity	Reactor materials unavoidably irradiated during operation	Solids	Aluminum-28 Manganese-56	β, γ β, γ	Package and ship for land burial
		Liquids (dissolved material)	Cobalt-58	$\beta+, \gamma$	Evaporation or ion-exchange; bury residue
		Gases	Nitrogen-16	β, γ	Hold for decay (very short life); then disperse to air
	Purposeful irradiation to produce useful isotopes	Solids	Cobalt-60	β, γ	Ship for burial when no longer useful (long life)
			Phosphorus-32	β	Store for decay to safe levels (short life)

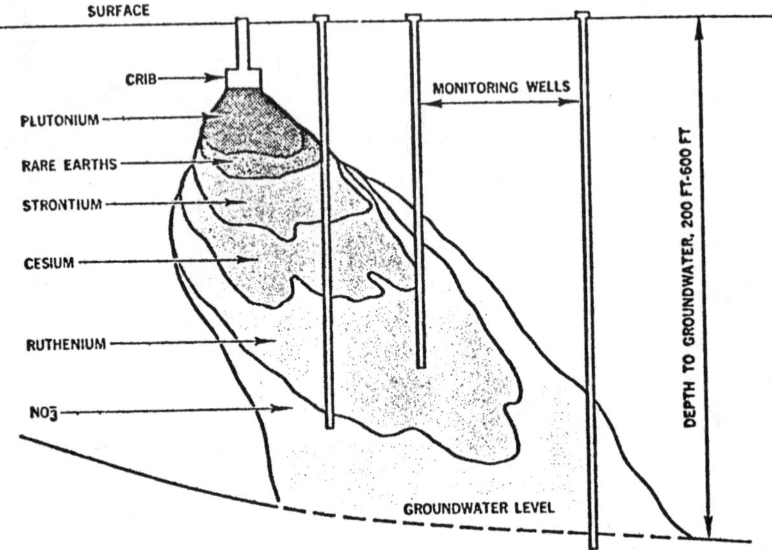

Figure 8. Disposal of liquid wastes into the ground at the Hanford Plant using an underground crib. As shown, the various materials found in these wastes are kept in the soil.

Chapter 10

BEACH BALLS BY THE SEA

When the swallows come home to Capistrano, they sometimes pause to nest or sun themselves on the sea cliffs around the San Onofre Nuclear Generating Station. The bluffs jutt out toward the Pacific and form a sort of semi-breaker near the containment bubble that shields the reactor.

Located near San Clemente, the plant is one of ten major electrical generating stations in Southern California that are located near coastal waters. Of these, it is the only one that uses nuclear fuel for electrical power generating. The others require fossil fuels such as gas and oil.

San Onofre has several things going for it. One is its unusual location within Camp Pendleton, which is one of the largest Marine bases in the United States. The plant officially began operations in 1967 and is owned jointly by Southern California Edison and the San Diego Gas & Electric Company.

Pioneer Unit 1 is a single electrical generating facility that uses a uranium fuel nuclear reactor to provide thermal energy for steam generation to drive electrical turbine generators. Actual thermal energy output of the reactor is 1350 megawatts. Electrical generation output is 430 megawatts net.

Southern Cal's PR men half-jokingly advised San Onofre engineers that the "neatest" thing they could do for the sphere-shaped containment shield was to paint it patriotic red, white and blue—like a gigantic beach ball. And driving in from San Diego, that's the first impression a motorist gets as the car winds up a slight hill and comes in full view of the plant. It's either that or perhaps a giant egg, left by some confused sea serpent who somehow got off-course on her way to the islands . . .

"They didn't quite buy our beach ball idea," sighed Milton H. Transchel, a former CBS producer who now directs Public Programs for California Edison. "Still, we think it's a cute idea . . . maybe later."

Two more "beach balls" are being constructed, each designed for twice the generating capacity of the existing station. Each of the proposed Units 2 and 3 will have a thermal energy output of 3,410 megawatts and an electrical output of 1,140 megawatts. As with Unit 1, most of the heat energy not converted to electricity is to be discharged into the marine environment through a cooling water system. Power operation for the second unit is tentatively scheduled for 1980, with the third section expected to achieve commercial capability some eighteen months later.

Enter here, Memo Number 1 on the nature of public relations people and things . . . In my fifteen years as a journalist, Southern Cal's Transchel has to be the most vivacious and intelligent PR director I have yet encountered. Publicity oriented people are too often obnoxious types who do more damage than good in their contacts with the press. In fact, some specimens are no less than living examples of the famous "Peter Principle," and beautiful exceptions are just that—beautiful. Two other top contenders would have to be TVA's Charlie McBride, a sedate individual who always seemed to come through even if I dialed at the last minute and requested a priority picture or interview. Another guy who knows his trade is Wayne Range, an all-seeing media eye for Oak Ridge. This man could cut through red tape and provide access to key people like a McCulloch Chain Saw operating on "high," and usually without stirring far from his cluttered desk.

Now, enter here, Memo No. 2 on public relations and the psychology of sales techniques . . . During college I attempted to sell "Great Books of the Western World," and at our first sales meeting, all new recruits were "Dale Carnegied" with one basic fact: No one will buy your product unless you are convinced that it is the best on the market. The most polished presentation in the world is no good unless you are sold on what you are selling. That's a fact that carries over, regardless of whether you are hawking books or reactors. You just have to believe in yourself and your product.

Enter here, then, conclusions to Memos No. 1 and 2: As far as I'm concerned, Southern Cal has a man who knows what he is doing. Or maybe that's the gusto way things are done on the West Coast. It was, however, a very pleasant surprise.

Another invigorating experience was encountering the security chief for San Onofre, who provided an interesting diversion from reality as I had come to know it. From the beginning, this man flustered me so much—after flying 2000 miles—that I teased, "What are you, on some kind of ego trip or something?"

My antagonist grinned and proceeded to control the situation even more. "I'm very sorry," he said, "but you can't take either your camera or tape recorder inside the plant. You'll also have to remove the religious medal around your neck. . . ."

Now, confiscating my material goods was one thing, but snipping the silver chain of my religious faith was another matter . . .

"Jesus has to stay," I persisted meekly, thinking what an ungodly place this was to be taking a stand for the Christian faith. The medal had carried me through uprisings in places like Vietnam, the Dominican Republic, and even domestic crises in the United States. And now, just as I was about to enter a nuclear plant and what critics allude to as "the shadow of death," I was being told to part with all and enter on faith alone . . .

"Sorry, the rules," the man answered in a firm but polite manner. He then explained that since the metal detector was neither Catholic, Protestant nor Jewish, it could not make exceptions and would begin clicking in a very unholy manner. . . .

Exasperated, I turned to Milt Transchel and said, "Is he for real? I've come all the way from Nashville, Tennessee, to make San Onofre famous with my camera and tape recorder." On my other side stood a tall Swede named Hans Ottoson, plant superintendent, who was watching the entire episode in apparent disbelief.

Undaunted, the security chief continued his trek and said, "Your watch. I'll have to check it. Also, empty your pockets and leave the contents in this plastic bag."

Holding the slender gold band of my watch, the man probed thoughtfully and then observed: "This is a very unusual instrument. There's no ticking or movement."

Ready, I quipped, "That's because it's a laser death ray and I've just switched it on SWAT."

Hans continued to stare, even ventured to shake his head. Meanwhile, Milt began apologizing and said that he couldn't understand just what was going on and he would certainly make amends as soon as possible. Bitter glances were then exchanged between the various levels of command.

Despite the harassment, I was beginning to admire my opponent in blue. Here was a fellow who took his job as security director of a major nuclear plant quite seriously. More power to him. In this business, you can't take chances and he wasn't. Of course, I wouldn't have let him known this. . . .

Our swords—bearing "he" and "she" markings—were lifted once more, and the officer shifted to a less penetrating stance. It was obvious that he didn't believe the watch was a death ray, yet, there was no need to take a chance. . . .

At this point, I shot for the jugular in our war of words and inquired if it was also necessary for me to relinquish my M-16 which was taped on my back . . .

"Only the ammo clip," he answered without hesitation.

Touché. Later, I learned that Governor Ronald Reagan had received (minus a joke here and there) the same treatment. In fact, Transchel finally had to phone the head of Cal Ed and get him out of a board meeting just to get clearance for my equipment. All this, despite the fact that my visit was arranged several weeks previously.

Regardless, that security man is worth his weight in gold. He was another very pleasant, though somewhat exasperating, surprise. His name? Oh, he asked that it not be used due to security reasons. . . .

Americans, it seems, have a tendency to be simply too trusting. We base value judgments on appearance and instinct more often than following exact procedure that allows for no exceptions. On the other hand, you have to maintain some sort of balance and be able to filter at the appropriate times. The question is when, and it's another one of those gray zones where there is no precise answer except, maybe: when in doubt *don't.* In fact, you might even want to erase that and put: when remotely suspicious, *don't.*

Here's a case in point before we return to sunny California and that beach ball by the sea. Following a rash of bombing threats, demonstrations and assorted trivia, the Pentagon has become a very tight unit with guards stationed at every exit and entrance. Without the proper credentials, an outsider is politely refused entry to any section other than the lower level visitor's shopping mall. Unless . . .

I happened to be in Washington during the height of the 1975 Cambodian fishing boat crisis. One of my stops was at the Soviet Embassy to pick up some photographs on USSR nuclear plants that their scientific attaché had ordered for me. Then I dashed over to the People's Republic of China Liaison Office to get some background material on energy usage in Communist China (anyone tailing me was probably thoroughly confused as to my ideological leanings). Finally, with several hours before another interview—this one at the Nuclear Regulatory Commission—I decided to visit the Government Printing Office. Finding the North Capitol store crowded, I simply drove on to the Pentagon in Virginia.

(Of course, this was prior to April-May 1976 when I was assigned to Navy duty in the office of Admiral David Cooney, Chief of Naval Information, known as CHINFO and based at the Pentagon, Washington, D.C.)

Parking was another problem, so I walked and then walked some more, finally arriving at the Potomac entrance. No dice. No way, lady. Clearance was necessary and I lacked the right badges. Several other entrances and the same response from friendly but firm guards. I kept being directed toward some hazy visitor's entrance that loomed further and further as I walked more and more. Sweating and tired, it finally became a sheer challenge just to gain entry into that bastion without going through public access—if indeed a point actually existed.

Walking and watching, I noticed a woman in her mid-forties, sitting quietly on a bench. It was clear that she was on some type of lunch break, either beginning or ending, and instinct told me that here was a key. Smiling and innocent, I moved toward her and said, in my best Tennessee drawl, "Can you *please* tell me how to get in this place? It's *endless* . . ."

It turned out that she was from Virginia, even had friends or relatives in my part of the South, and after a few minutes of general chatting, was satisfied that I represented no threat to the Free World. My manner and appearance told her that here was one tired girl who was a-okay, which, in that instance, was a true evaluation. However . . .

"Look, I have to go back in," the lady said kindly. "You come with me and we'll go through on my pass. Then I'll show you where the bookstore is, leave you there while I go to the bank, and then take you back out. It'll save you more walking."

The woman was being kind, thoughtful and considerate to a visitor. In any other situation such a response would be commendable. Here it was, security-wise, nothing short of tragic.

Her key phrase was "leave you there," which she did and returned some fifteen minutes later. It had been obvious to me as we moved past the guards and gained entry to the Pentagon that my new found friend carried a top secret security classification. High ranking officers saluted or smiled as she passed them in the hallways. In fact, I learned —almost dazed—that my escort was no less than the administrative assistant to one of the key people in that eight-sided building. She even showed me a tiny glimpse of her classified domain, which was stirring like a hornet's nest as news of the Cambodian incident flashed and flared.

So, there I went, clutching a large, sealed manilla envelope that no guard searched or even questioned once my escort's badge was flashed. The woman had known me for only thirty minutes, and yet, her trust provided me with easy entry into one of the most sensitive areas of our military complex.

Her name isn't being used because I don't want to intentionally cause problems for this kind-hearted person. Still, I couldn't help but be bothered as I thought of the two places I had just left—the Soviet Embassy and the Communist Chinese Liaison Office—and now here I was strolling down Omar Bradley Hall with a very respected and helpful government worker. Empires have fallen for just such carelessness. Bombs have also been planted and assassinations completed. Instinct is not always the best judge of character, unless you just happen to be a lady reporter trying to get into the Pentagon.

But back to San Onofre, and no more detours.

Southern California lies within a climatic region that can be broadly defined as Mediterranean, characterized by short, mild winters and warm, dry summers. The climate is influenced chiefly by the East Pacific high-pressure area, whose center lies to the northwest between Hawaii and Alaska. In summer, this pressure system intensifies, resulting in mild, dry weather and a blockage of storms. In winter, the high pressure area weakens and moves to the south and west. Cold fronts occasionally cross the Southern California coast, resulting in cooler weather and occasional rain.

The generating station is situated on the very edge of a narrow coastal plain that extends from the coastline to a range of low hills some two miles inland that have a maximum elevation of 1,725 feet above sea level. The plain ends at the beach in a line of wave-straightened cliffs that extend sixty to eighty feet above a narrow sandy beach. Numerous ravines are cut into the cliffs as a result of erosion by storm runoff from the coastal plain.

The natural beach cycle in the San Onofre vicinity consists of growth in the summer and erosion in the winter. During summer, low-height, long-period waves transport sand onto the beach from offshore, creating a very pleasant and sandy beach cover. Then high, short-period waves, common during winter, wash away sand, exposing underlying rock and cobble in some areas.

Beyond the surf zone, a gently sloping bottom reaches a depth of thirty feet below mean approximately 3,000 feet offshore. Patchy distribution of bottom sediments inshore of the thirty-foot depth includes stretches of sand interspersed with areas of cobble and boulders. Most of the cobble near shore areas shows evidence of periodic burial, while those farther offshore are apparently free of sandy cover throughout the year. The active sedimentary material is mainly composed of quartz, feldspar and a sand known in geological circles as biotite.

Due to its steep sea bluffs and limited public access, the seacoast around San Onofre has remained a natural, relatively undeveloped area. Water contact sports like swimming and surfing have been sparse. The U.S. Marine Corps has maintained relatively tight security in the area, and swimming and surfing activities have been largely restricted to military personnel and members of the San Onofre Surfing Club. Since the establishment of a state beach, however, this area is rapidly becoming popular, particularly during the summer months.

Biological studies of the aquatic environment at San Onofre began in 1963 and were used to develop a program of periodic monitoring that started the following year. Studies, according to Alek Ronald Strachan, senior marine biologist for Southern California Edison, included probes of intertidal, benthic and plankton communities. The latter term refers to the passively floating animal and plant life found in a large body of water.

At the request of the California Regional Water Quality Control Board, San Diego Region, the thermal effect study of San Onofre Unit 1 has consisted of review of the monitoring program results. Where possible, reviewers of these data were to be selected from persons who had been or were currently engaged in the monitoring program, but, in addition to these people, separate and independent reveiwers were brought from throughout the academic community. A listing of such individuals and their affiliations is found in Table 7. Specific topics included kelp, phyto-plankton, zooplankton, fishes, and effects on the food chain, in addition to intertidal and benthic ecology.

Since San Onofre went operational, there have been some twenty marine surveys, according to Strachan, and none of these have shown any significant change in marine environment. The only noticeable difference, he added, has been an increase in the number of fishing boats that tend to hang around the outfall of Unit 1. This is attributed to water movement and slightly warmer temperatures (briefly) which serve to attract fish.

For a while, environmentalist groups attacked San Onofre on the basis of possible damage to valuable kelp beds. Then a healthy new kelp bed sprang up directly opposite the outfall and solidly squelched any arguments in that direction. Among other things, kelp provides food for fish, which in turn provide food for fishermen.

I know the kelp is there because Strachan, as a group of us were eating lunch at the San Clemente Inn, made the mistake of telling me about the beds. Traditionally, he explained, kelp in Southern California has generally declined since the first surveys were made as far back as 1911. This decline was accelerated during abnormally warm water periods between 1957 and 1959 (both pre-nuclear). It was estimated during this span that perhaps 90 percent of the beds were gone. In recent years, however, a change took place. Beds in the San Mateo Point kelp subregion have increased (with fluctuations) to 0.4 square miles since their apparent disappearance. Beds in the Barn kelp and San Onofre kelp regions have undergone natural restoration since their reduction during and following the warm water years. And since 1963, the subregions nearest San Onofre, which are most likely to be affected by thermal release from the generating station, have displayed higher percentages of kelp increase than the two control subregions further away.

Sipping a cool glass of lime-tasting liquid, I listened as Strachan detailed his work as a marine biologist, first with the California Department of Fish and Game, and later as a deck hand on a commercial tuna boat called the "Ventura." At forty-one, here was a Californian who lived and genuinely loved his work with sea life.

Strachan confided to me that one result of the twenty marine surveys has been observation of a sea invertebrate that has decided to attach itself—make a nest—right on the outfall structure. "We have consultants who have looked at this matter and scrutinized it for some time," he explained. "It's definitely a filter feeding organism, and it seems to be growing faster than at the control stations."

There are three causes of "mortality" which could possibly produce losses to life like zooplankton (belonging to the plankton organism) in the San Onofre area due to continuous and simultaneous operation of Units 1, 2, and 3. These are: (1) Intake and passage through the cooling water system, (2) Entrainment into the discharge plume from the immediately surrounding water, and (3) Transportation through the thermal field by the natural currents.

Temperature-wise, we are talking about 105° F. maximum down through 97° F. minimum, and the total heat treatment occurs during a three-hour period. The first system utilized two hours; the new system uses three to see if an even lower temperature can do the job.

Overall, the most reasonable estimate of mortality from the combined operations of the three generating units is based on a 25 percent intake mortality, a 10 percent entrainment mortality and a zero percent plume mortality. Again, analysis of species composition of the local plankon and their geographic ranges toward thermal tolerance, indicates no appreciable damage. Another report, mentioned briefly, is an independent analysis of zooplankton data which was obtained during a monitoring period and shows significant increases in population which "appear to be the direct result of the operation of the San Onofre Nuclear Plant."

There is one hypothesis which helps explain such enrichment and it involves changing the distribution of organisms in the following manner: (1) cooling water is intaked about 3,200 feet offshore at a depth of approximately 15 feet below the surface, (2) the discharge is located 600 feet inshore of the intake, and (3) six of the nine organisms examined showed clear trends toward greater abundance offshore.

Translated, this implies that with the intake and transport of plankton from an area of relatively high concentration to one of relatively low concentration, there results an apparent increase in the discharge area. In short, the plankton are moving out to the suburbs.

Strachan is a very gracious name-dropper, and when he stumbled over "muricea california" (a sea fan that is thriving like everything else under the shadow of San Onofre), I interrupted and said I wanted to see "muricea" the anemone, cystoseira (a sea plant known as a feather boa) and all the rest of that gang out there in the blue—especially the kelp beds.

At first, my host didn't fully grasp my plans, so I was more specific: "I want to dive off the area you have been describing."

Poor Strachan needed a little convincing that I had actually used scuba gear, but soon we were on our way to the warm Pacific which turned out to be chilly despite our dry suits (scuba term for rubber suits). A brief adjustment to the regulator and it was bon voyage for a few precarious minutes in a world of gentle beauty. That dive was a rare treat for this one mother of three and I enjoyed every minute. And it was all there, just as he had described it . . . the monitoring markers, everything.

The dive ended as we pushed against the waves and moved toward the shoreline in a dark Navy raft. Gradually, two forms took shape on shore. One was the PR director and the other I recognized as the security officer who had apparently picked up our watery presence as some form of amphibious assault from the sea. He peered intently through heavy black binoculars and waited as we sloshed nearer.

"We come in peace," I said reverently, waddling over to the two men and holding aloft one slimy glob of dark kelp. There really had been beds of the stuff under the restless mass of bluish-green water. I was feeling salty, wet, sandy and totally at peace in the California sun. The few times like these compensate for the long hours a writer must spend huddled over a typewriter and praying for inspiration. It makes up for months of research that must be sifted and then re-sifted some more to find the most palatable pieces. Alas, however, the moment is too brief for the work that must come.

Carefully lowering the binoculars so that no grain of sand would scratch the lens, the security officer squinted for just a second and then produced two very large plastic bags. . . .

"This is for your gear. I'm sorry, you can't go into the plant like that. Rules," he smiled, adding, "I cleared you for the plant—not the Pacific Ocean."

As I said, the man does his job.

During the dive, we observed such fish as barred sand bass, opal-eye, California sheephead, kelp bass and sargo. The most common species was the barred sand bass, which would swim around us, gawk in a fishy way, then dart off. Four fish—Pacific bonito, Pacific mackerel, jack mackerel and northern anchovy—dominate the commercial catch in this heavily trawled area of Southern California. And, while it seems that the San Onofre plant has had no direct effect on commercial fishery, its operation has influenced sport fishery. Catch statistics indicate a marked increase in both catch and angler efforts since 1962.

Some fish have been entrained in the cooling water intake during normal plant operation, but the number removed from the ocean is limited by a velocity cap fitted over the intake structure. The amount of fish removed averages about fifteen tons per year for normal plant operation and about three tons per year for periodic heat treatments (every five to six weeks).

A conclusion reached from analysis of the fish collection records is that neither normal operation nor heat treatment, as presently conducted, poses a significant threat to the fish or community stability. In fact, there are two direct effects expected from the operation of Units 2 and 3. These are: (1) entrainment of fish into the cooling water system, and (2) attraction of fish to the intake and discharge structures. To cope with this, the proposed cooling water system for Units 2 and 3 will have a fish diversion louver and by-pass system to transport entrained fish back to sea with no exposure to temperature excesses or mechanical damage. It is estimated from experimental work that the system will be at least 80 percent effective. Therefore, the maximum anticipated loss of fish is about 36,000 pounds per year, an insignificant loss when compared to California's sport and commercial catches. The cost of the new system will run in the neighborhood of about $1 million.

Only a few weeks prior to my visit to the West Coast, a very curious seal found herself moving rapidly through the cooling water intake system. Eventually, the confused seal dropped, flappers and all, into a large concrete basin designed to keep nosey sea creatures and other types from ending up as fuel for the reactor. It took a busy work crew nearly five hours to raise the slippery seal—via a crane—and transport it back to the sea. As with the new and costly fish screening system, this is an example of procedures on either a large or small scale which can accommodate even a stray seal or two . . . even though there's no official rule book which dictates "you gotta do this."

San Onofre is unique in other ways. It represents the first run of the third generation power plants. First, there were the prototype reactors such as the one which began operating in December, 1957, at Shippingport, Pennsylvania. Located near Pittsburgh, it was the first large nuclear power electric generating plant in the United States and utilized a pressurized water reactor. Now, the net electrical kilowatt capacity for the Shippingport Atomic Power Station is 90,000.

Plants may have different types of reactors, and their power output can also differ by substantial amounts. Several factors influence this. Some plants may have been built as demonstration units to determine relative advantages of one reactor-type compared with another and were built to the size requirement of a particular power company or system.

The Humboldt Bay Power Plant, located near Eureka, California, is about 200 miles north of San Francisco. It utilizes a boiling water reactor designed by the General Electric Company, with a 68,500 net electrical kilowatt capacity. Owned and operated by Pacific Gas and Electric Company, it began commercial operation in 1963.

The same type reactor can be found at Big Rock Point Nuclear Plant near Charlevoix, Michigan, about 200 miles northwest of Detroit. This plant started commercial operation in 1963, has a 70,400 net electrical kilowatt capacity, and was also designed by General Electric. Consumers Power Company owns and operates the system.

The Peach Bottom Atomic Power Station Unit Number 1 is a helium-cooled reactor with a 40,000 net electrical kilowatt capacity that was designed by Gulf General Atomic. It started commercial operation in 1967 and is located on the Susquehanna River near Peach Bottom, Pennsylvania, sixty-five miles southwest of Philadelphia. Peach Bottom is a demonstration reactor featuring high temperatures and a gaseous coolant. It is operated by Philadelphia Electric Company.

There are various types of reactor systems now in use or being developed. Chief among these are the "light water" reactors and the gas-cooled ones. There are two classifications belonging to the latter: the boiling water reactor (BWR) and the pressurized water reactor

(PWR). In the first, water, heated to steam in the reactor, turns the turbine directly. In the PWR, water passing through the reactor is kept under sufficient pressure to keep it from boiling in the reactor vessel. From the reactor it goes to a steam generator, where another water system is heated to form steam to run the turbine.

Gas-cooled reactors operate in much the same way as the PWRs, except that the coolant, or working fluid, is a gas instead of water. It is generally agreed that credit for this particular expertise goes to General Atomic, who have pioneered with the High Temperature Gas-Cooled Reactor (HTGR). The HTGR is often referred to as a converter reactor, because it generates a certain amount of its fuel from thorium as it operates, thereby conserving valuable uranium fuel resources. Using helium as the coolant, this system produces steam in the 1,000° F. range. There is an even larger—330,000 kilowatt—HTGR plant, the Fort St. Vrain Nuclear Generating Station, located near Denver.

The HTGR operates on a basis whereby more of the system's energy is converted to electricity and less heat is rejected into the environment. Less cooling water is required and therefore smaller cooling towers or ponds, resulting in less environmental impact and greater siting flexibility. In addition, the HTGR has proven its efficiency both in Europe and the United States. Attendant problems are discussed in greater detail elsewhere in this book.

Now, among the so-called second generation systems were those in the 200,000 megawatt range such as the Dresden Nuclear Power Station, Unit Number 1, located about fifty miles southwest of Chicago. It is designed by GE and began commercial operation in 1960, utilizing a boiling water reactor. Dresden Unit 1 was the second largest nuclear power plant to be built in the United States. It is owned and operated by Commonwealth Edison Company and has a 200,000 net electrical kilowatt capacity.

The fourth largest nuclear station built in this country was at the Indian Point System of Consolidated Edison Company of New York. Located on the Hudson River, Indian Point is about thirty-five miles north of New York City and uses a pressurized water reactor. It has a capacity of 265,000 kilowatts and the projected capacity once all three of its units are operational is more than 2 million kilowatts.

The Connecticut Yankee Atomic Power Plant, located at Haddam Neck on the east bank of the Connecticut River, is about twenty miles southeast of historic Hartford. It was designed by Westinghouse Electric Corporation and has a pressurized water reactor with a 462,000 kilowatt capacity. The plant, owned by Connecticut Yankee Atomic Power Company, went operational in 1967.

A more detailed "Summary of Operating U.S. Power Reactors" is contained in Table 8.

Thinking nuclear and military doesn't always produce weaponry, according to the Department of the Navy, headquartered in the Pentagon. In fact, their agency maintains that the U.S. Navy was the first organization to adapt nuclear power for essentially peacetime uses. Putting it to practical use in ships was first begun by Captain (later Vice-Admiral) Hyman G. Rickover when he requested the AEC to undertake the design, development and construction of a nuclear reactor for use in a submarine. That was September, 1947.

Consequently, in five years and seven months, the Westinghouse Mark I Reactor went into the "critical stage"—and that's a good term. Constructed inside a submarine hull surrounded by a shell of water, the Mark I reached its full power—critical stage—capacity and started a simulated submerged crossing of the Atlantic. The 96-hour test was completely successful. From this experiment, the navy's Bureau of Ships awarded a contract for the production of a nuclear submarine incorporating the Mark I reactor design. President Harry Truman laid the ship's keel in June, 1952, and in September, 1945, Mrs. Dwight D. Eisenhower broke the ceremonial bottle of champagne against the U.S.S. *Nautilus* as its lithe body slid into the Thames River at New London, Connecticut.

Why nuclear submarines? With almost unlimited cruising range and endurance limited only by the crew, such submarines are capable of extended submerged operations in the international waters of the world —which cover about 70 percent of the earth's surface. Free of the need to surface, they remain hidden by an ocean curtain, their locations hard to pinpoint.

Nuclear power was one of the United States Navy's highest national priority items contained in fiscal year 1976 and 1977 authorization requests, according to Admiral James L. Holloway III, U.S. Navy Chief of Operations who succeeded Admiral Elmo R. Zumwalt on July 1, 1974.

Holloway, incidentally, is the first nuclear power-trained officer promoted to the navy's top uniformed position. After instruction in the nuclear reactor program under Admiral Rickover, then Captain Holloway was assigned commanding officer of the USS Enterprise (July 1965) which was then the navy's first nuclear powered carrier.

Back on land, nuclear power plants first became economically competitive with the conventional form of steam power about 1965. Prior to this, it was considered very expensive to obtain usable electric power from atomic energy. The combined efforts of government and industry, however, have switched this. The nuclear industry officially began with the Atomic Energy Act of 1954, which permitted industry to participate in the development of peacetime applications of atomic energy.

A utility company spends money on two principal items in order to produce electricity for its customers: the power plant itself and the fuel which this plant consumes. Right now it costs more to build a nuclear plant than it does a conventional one, but this is really only half the story, since fuel costs for the nuclear facility run much lower.

Once all three units of the San Onofre system are fully operational, officials say outright that 3.5 million gallons of oil *per day* will be saved. That's oil which could otherwise be refined to gasoline for the family car.

One idea being pushed by industry is a concept referred to as "one stop shop." It refers to state and federal level clearance required in the building of nuclear plants. There are now some eight different agencies involved in clearance—often acting on the same point and creating in most cases, duplication, worn nerves, and tons of paperwork.

"It gets ridiculous," says Southern Cal's Transchel. "You can have eight different agencies trying to clear you on the same point and going through the same duplicating investigations, which is what's now happening. This is why it takes ten years to go to a nuclear plant. I'm talking about thirty-five different regulatory permits which are required before you can go ahead."

For example, it took four years for clearance to be given for construction of Units 2 and 3 of San Onofre.

The old AEC was looking closely at this situation until it collapsed and died, to be speedily replaced by the Nuclear Regulatory Commission and the Energy Research and Development Administration. The system thought most practical would involve standardized components that have been previously cleared as safe or adequate and would not necessitate re-examination each time construction is applied for on a nuclear facility. Those in the industry agree that this would greatly loosen a heavy log jam and streamline clearance procedure.

This is one aspect of the nuclear industry that is like churning butter on a hot July evening. It takes time and your arms begin to ache and ache, but eventually the whole mess begins to settle and you can get on with the business at hand. Most of us find it's easier to simply go to the supermarket and buy a box already prepared.

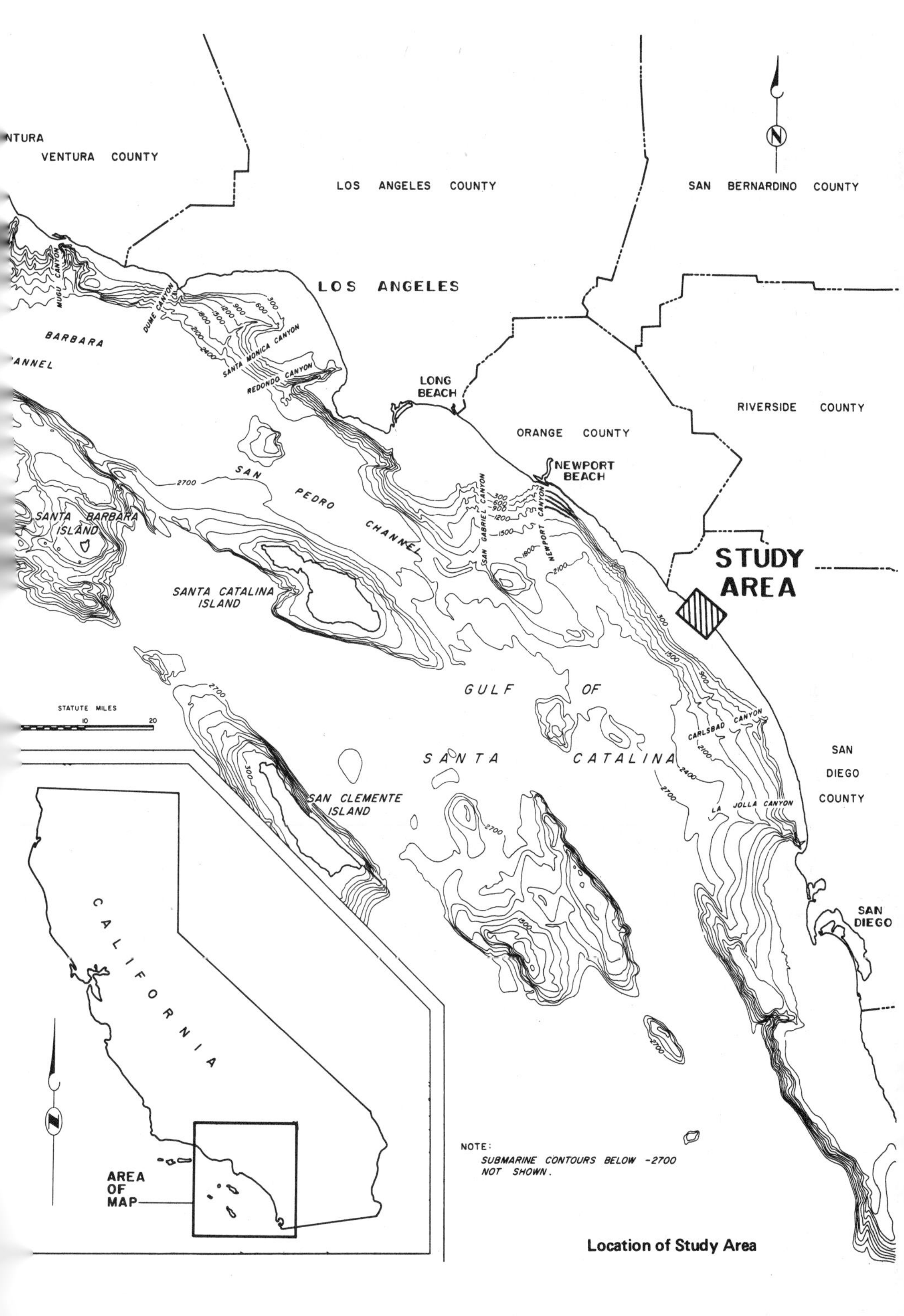

Location of Study Area

The USS Long Beach *(CGN-9) just before cruising into Pearl Harbor Channel, Oahu, Hawaii. The square box-like object on the top deck is the reactor for this nuclear-powered Navy ship. One priority project scientists at the Naval Research Laboratory are working on deals with small pulsed power systems capable of supplying energy needs for ships like the* Long Beach.

The nuclear-powered fleet ballistic missile submarine USS Daniel Boone *leaves port at Roosevelt Roads, Puerto Rico, enroute to the Mediterranean.*

Iron workers install giant steel reinforcement bars at the base of the first two new nuclear power units now under construction at the San Onofre Station. When completed in the early 1980s, each of the new generators will produce 1,140,000 kilowatts of electric power for Southern California residents. The expansion project is being undertaken by Southern California Edison Company and the San Diego Gas & Electric Company.

The San Onofre Nuclear Generating Station near San Clemente recently produced its 20 billionth kilowatt-hour of electricity. The energy produced during the eight years of operation by this nuclear facility is equivalent to more than 30 million barrels of expensive imported fuel oil. San Onofre is owned 80 percent by Southern California Edison Company, and 20 percent by San Diego Gas and Electric Company.

An artist drawing of the proposed Sequoyah Nuclear Plant located near Chattanooga, Tennessee, and under supervision of the Tennessee Valley Authority system.

Workmen in the control room of the San Onofre Nuclear Plant in Southern California keep the "big machine" under a watchful eye.

Thermal effects studies such as those run by Southern California Edison and San Diego Gas and Electric companies, represent a joint venture between environmental quality analysts and marine biological consultants.

The long-range purpose was to determine just what effect a nuclear plant has on organisms. The results showed growth encouragement rather than discouragement.

…e San Onofre …uclear Generating …ant located near San …emente in Southern …lifornia.

…he San Onofre …uclear Generating …ant in Southern …lifornia is a high-…wered 'beach ball' … the Pacific.

Author Jacque Srouji and Hans Ottoson, superintendent for the San Onofre Nuclear Plant near San Clemente, California, narrating the job of keeping a major nuclear plant healthy and happy . . .

Panasonic and note pad in hand, writer Jacque Srouji and Superintendent Hans Ottoson, of San Onofre, listen as Alek Ronald Strachan, senior marine biologist for Southern California Edison, details biological studies of the aquatic environment at San Onofre which began in 1963.

Pressing for more details on thermal effects, writer Jacque Srouji listens as Alek Ronald Strachan, senior marine biologist for Southern California Edison, breaks down studies on a year-to-year basis.

Plant Superintendent Hans Ottoson takes Author Jacque Srouji and Marine Biologist Alek Strachan, to the new construction site at San Onofre, designed to generate twice the capacity of the existing station. Each of the proposed units (2 and 3) will have a thermal energy output of 3,410 megawatts and an electrical output of 1,140 megawatts.

Southern California Edison's Senior Marine Biologist, Alek Ronald Strachan explains a thermal effects study involving some twenty marine surveys along the California coast.

Author Jacque Srouj listens as an analyst details how marine life such as kelp and plankton has flourish on test plates locate around the outfall o Unit 1 at the San Onofre Nuclear Ger erating Plant.

Alex Ronald Strachan, senior Marine Biologist for Southern California Edison, shows Hans Ottoson, San Onofre Plant Superintendent, and writer Jacque Srouji how grills protect fish and other marine life from plunging through intake valves.

Hans Ottoson, San Onofre superintendent, offers writer Jacque Srouji a glimpse of the main control room at the San Onofre Nuclear Generating Station.

Laboratory Marine Analyst Jed Springer notes how sea life such as kelp and plankton have flourished on test plates located in and around the outfall of Unit 1 at the San Onofre Nuclear Generating Plant.

Author Jacque Srouji takes notes as Hans Ottoson, Plant Superintendent for the San Onofre Nuclear Generating Plant near San Clemente, California, provides operating procedures and other data.

San Onofre's Director of Security—who refused to be identified for security reasons—and his assistant offer stiff resistance to those seeking unauthorized admittance to the nuclear plant.

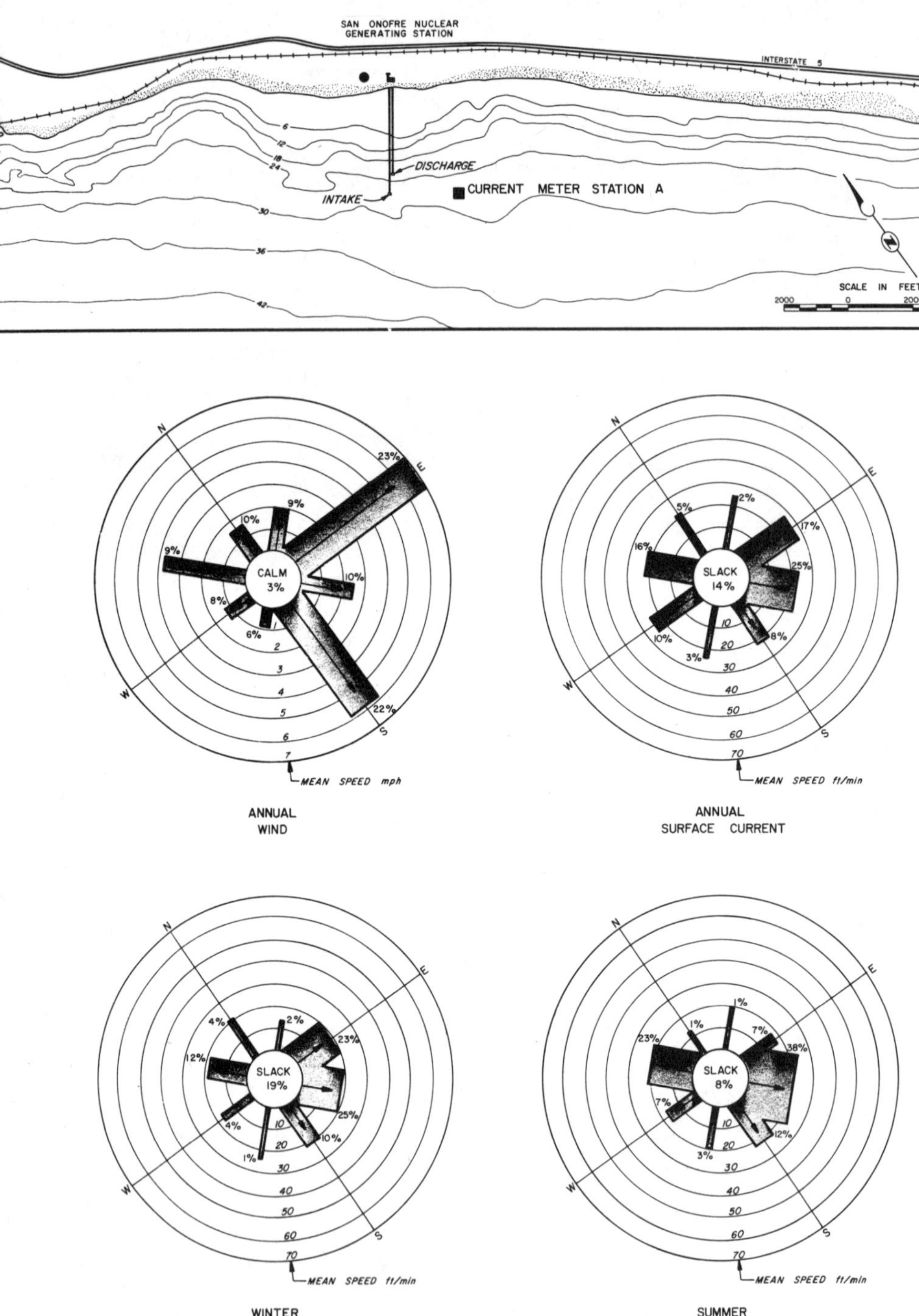

Frequency of Wind and Surface Current at San Onofre

Table 7: A sampling of the professional expertise available during a thermal effect study of the San Onofre Nuclear Generating Plant which was requested by the California Regional Water Quality Control Board, San Diego Region.

Subject	Reviewer	Affiliation
Primary Reviews		
Kelp	Dr. Wheeler J. North	California Institute of Technology
Phytoplankton	Anne Dodson	University of California San Diego Scripps Tuna Oceanography Research
Zooplankton	Arthur M. Barnett	Scripps Institution of Oceanography
Intertidal Ecology	Terry Parr	Marine Ecological Consultants
Benthic Ecology	Dr. Thomas B. Scanland	Marine Ecological Consultants
Fishes	Harry Hickman	Marine Biological Consultants, Inc.
Food Chain and Productivity	Frank Alverson Gordon Broadhead	Living Marine Resources, Inc. Living Marine Resources, Inc.
Independent Reviews		
Plankton	Dr. James Enright Dr. John McGowan	Scripps Institution of Oceanography Scripps Institution of Oceanography
Beach and Intertidal Ecology	Dr. Robert Given	University of Southern California
Benthic Ecology	Dr. Robert Given	University of Southern California

Summary of Operating U. S. Power Reactors as of Nov. 1, 1974

Table 3

Name and location owner operator	Docket No.	Reactor type (reactor designer)	Design power		Operating license No. (date)	MWh gross* (Sept.–Oct. 1974)	10^3 MWh cumulative† through Oct. 1974	Recent actions
			MW(t)	MW(e)				
ARKANSAS NUCLEAR ONE 1 Pope County, Ark. Arkansas Power & Light Co.	50-313	PWR (B&W)	2568	850	DPR-51 (5-21-74)	177,551	184	In power testing stage
ARNOLD Cedar Rapids, Iowa Iowa Electric Light and Power Co.	50-331	BWR (GE)	1658	569	DPR-49 (2-22-74)	451,561	883	
BIG ROCK POINT Charlevoix County, Mich. Consumers Power Co.	50-155	BWR (GE)	240	75	DPR-6 (8-30-62 POL) (5-1-64 OL)	91,820	4,151	
BROWNS FERRY 1 and 2 Decatur, Ala. Tennessee Valley Authority	50-259 50-260	BWR (GE) BWR (GE)	3293 3293	1065 1065	DPR-33 (6-26-73) DPR-52 (6-28-74)	921,160	4,346	Some outage for maintenance of Unit 1; Unit 2 was recently licensed
CALVERT CLIFFS 1 Lusby, Md. Baltimore Gas & Electric Co.	50-317	PWR (CE)	2560	845	DPR-53 (7-31-74)			Recently licensed
CONNECTICUT YANKEE Haddam Neck, Conn. Connecticut Yankee Atomic Power Co.	50-213	PWR (West)	1825	575	DPR-14 (6-30-67 POL)	816,053	26,472	
COOK 1 Benton Harbor, Mich. Indiana & Michigan Electric Co.	50-315	PWR (West)	3250	1060	DPR-58 (10-25-74)			Licensed for fuel loading and testing
COOPER Nemaha County, Nebr. Nebraska Public Power District	50-298	BWR (GE)	2831	778	DPR-46 (1-18-74)	753,350	1,334	In power testing stage
DRESDEN 1, 2, 3 Grundy County, Ill. Commonwealth Edison Co.	50-10 50-237 50-249	BWR (GE) BWR (GE) BWR (GE)	700 2527 2527	200 809 809	DPR-2 (9-28-59 POL) (10-14-60 OL) DPR-19 (12-22-69 POL) DPR-25 (1-12-71)	31,421 405,404 858,672	13,317 16,262 13,032	Some downtime of Unit 1 for refueling and maintenance; some downtime of Unit 2 for repairs

FITZPATRICK Oswego, N. Y. Power Authority of State of N. Y.	50-533	BWR (GE)	2436	821	DPR-59 (10-17-74)			Licensing authorized pending plant completion
FORT CALHOUN Washington County, Nebr. Omaha Public Power District	50-285	PWR (CE)	1420	457	DPR-40 (5-24-73)	532,176	2,787	
FORT ST. VRAIN Platteville, Colo. Public Service Co. of Colo.	50-267	HTGR (GGA)	842	330	DPR-34 (12-21-73)			Recently licensed
GINNA Ontario, N. Y. Rochester Gas & Electric Corp.	50-244	PWR (West)	1520	490	DPR-18 (9-19-69 POL)	676,497	13,063	
HANFORD 1 Richland, Wash. USAEC Douglas United Nuclear, Inc.	‡	Graphite (GE)	4000	850	‡	810,210	27,372	
HATCH 1 Baxley, Ga. Georgia Power Co.	50-321	BWR (GE)	2436	786	DPR-57 (8-6-74)			Authorization for full-power operation was issued in Oct. 1974
HUMBOLDT BAY Eureka, Calif. Pacific Gas & Electric Co.	50-133	BWR (GE)	240	65	DPR-7 (8-28-62 POL) (1-21-69 OL)	67,449	4,163	
INDIAN POINT 1 and 2 Buchanan, N. Y. Consolidated Edison Co. of New York	50-3 50-247	PWR (B&W) PWR (West)	615 2758	265 873	DPR-5 (3-26-62 POL) DPR-26 (10-19-71)	304,650 789,300	13,557 2,877	
KEWAUNEE Carlton, Wis. Wisconsin Public Service Corp. Wisconsin Power & Light Co. Madison Gas & Electric Co.	50-305	PWR (West)	1650	541	DPR-43 (12-21-73)	327,900	1,517	Down for maintenance and inspection
LA CROSSE Genoa, Wis. USAEC Dairyland Power Coop.	50-409	BWR (A-C)	165	53	DPRA-5 (7-3-67 POL) DPRA-6 (10-31-69 POL) DPR-45 (8-28-73 POL)	23,944	1,159	Down for maintenance and inspection

*From *Nucleonics Week*, **15**(43): (Oct. 24, 1974) and **15**(47): (Nov. 21, 1974).
†As of the end of Oct. 1974. From *Nucleonics Week*, **15**(47): (Nov. 21, 1974).
‡Not required.

(Table continues on the next page.)

Summary of Operating U. S. Power Reactors (Continued)

Name and location owner operator	Docket No.	Reactor type (reactor designer)	Design power		Operating license No. (date)	MWh gross* (Sept.–Oct. 1974)	10^3 MWh cumulative† through Oct. 1974	Recent actions
			MW(t)	MW(e)				
MAINE YANKEE Lincoln County, Maine Maine Yankee Atomic Power Co.	50-309	PWR (CE)	2440	790	DPR-36 (9-15-72)	275,180	6,939	On line following downtime for refueling
MILLSTONE POINT 1 Waterford, Conn. Northeast Nuclear Energy Co.	50-245	BWR (GE)	2011	652	DPR-21 (10-7-70 POL)	0	12,064	Down for refueling and maintenance
MONTICELLO Monticello, Minn. Northern States Power Co.	50-263	BWR (GE)	1670	545	DPR-22 (9-8-70 POL)	698,770	11,117	Hearing to be held on conversion of provisional operating license to a full-term license
NINE MILE POINT 1 Oswego, N. Y. Niagara Mohawk Power Corp.	50-220	BWR (GE)	1850	625	DPR-17 (8-22-69 POL)	794,170	14,717	ACRS issued favorable report on full-term license Sept. 10, 1974
OCONEE 1, 2, and 3 Oconee County, S. C. Duke Power Co.	50-269 50-270 50-287	PWR (B&W) PWR (B&W) PWR (B&W)	2568 2568 2568	886 886 886	DPR-38 (2-6-73) DPR-47 (10-6-73) DPR-55 (7-19-74)	751,224 457,092 76,949	6,586 1,279 77	Unit 1 is down for refueling; Unit 2 is down for repairs; Unit 3 was recently licensed
OYSTER CREEK Oyster Creek, N. J. Jersey Central Power & Light Co.	50-219	BWR (GE)	1930	640	DPR-16 (4-9-69 POL)	769,822	19,070	
PALISADES Covert Township, Mich. Consumers Power Co.	50-255	PWR (CE)	2200	700	DPR-20 (3-24-71 POL)	91,510	4,502	Operating under limited (90-day) authorization; down for repairs
PEACH BOTTOM 1, 2, and 3 York County, Pa. Philadelphia Electric Co.	50-171 50-277 50-278	HTGR (GGA) BWR (GE) BWR (GE)	115 3293 3293	40 1065 1065	DPR-12 (1-24-66) DPR-44 (8-8-73) DPR-56 (7-2-74)	37,391 1,161,760 267,290	1,385 4031 325	Unit 1 was shut down Oct. 31 for decommissioning; Unit 3 is in power testing stage
PILGRIM 1 Plymouth, Mass. Boston Edison Co.	50-293	BWR (GE)	1998	664	DPR-35 (6-8-72)	795,260	6,378	
POINT BEACH 1 and 2 Manitowoc County, Wis. Wisconsin–Michigan Power Co. Wisconsin Electric Power Co.	50-266 50-301	PWR (West) PWR (West)	1518 1518	497 497	DPR-24 (10-5-70) DPR-27 (11-16-71)	663,200 534,050	12,267 6,652	Unit 2 is down for refueling

PRAIRIE ISLAND 1 and 2 Red Wing, Minn. Northern States Power Co.	50-282 50-306	PWR (West) PWR (West)	1650 1650	530 530	DPR-42 (8-9-73) DPR-60 (10-29-74)	104,250	858	Unit 1 is in power testing stage; Unit 2 was recently licensed
QUAD CITIES 1 and 2 Rock Island, Ill. Commonwealth Edison Co. Iowa–Illinois Gas & Electric Co.	50-254 50-265	BWR (GE) BWR (GE)	2511 2511	800 800	DPR-29 (10-1-71) DPR-30 (3-31-72)	818,660 540,590	10,371 11,088	Unit 1 is down for maintenance; Unit 2 is on line after repairs
RANCHO SECO Sacramento County, Calif. Sacramento Municipal Utility District	50-312	PWR (B&W)	2772	913	DPR-54 (8-16-74)	11,456	11	In power testing stage
ROBINSON 2 Hartsville, S. C. Carolina Power & Light Co.	50-261	PWR (West)	2200	700	DPR-23 (7-31-70)	837,033	15,657	Hearing to be held on increase of power from 2200 to 2300 MW(t)
SAN ONOFRE 1 Camp Pendleton, Calif. Southern California Edison Co.	50-206	PWR (West)	1347	430	DPR-13 (3-27-67 POL)	594,000	19,184	
SURRY 1 and 2 Surry County, Va. Virginia Electric & Power Co.	50-280 50-281	PWR (West) PWR (West)	2441 2441	788 788	DPR-32 (5-25-72) DPR-37 (1-29-73)	754,500 110,545	7,593 6,366	Unit 1 is down for maintenance; Unit 2 is down for repairs
THREE MILE ISLAND 1 Three Mile Island, Pa. Metropolitan Edison Co. and others	50-289	PWR (B&W)	2535	819	DPR-50 (4-19-74)	1,004,525	1,534	
TURKEY POINT 3 and 4 Dade County, Fla. Florida Power & Light Co.	50-250 50-251	PWR (West) PWR (West)	2200 2200	693 693	DPR-31 (7-19-72) DPR-41 (4-10-73)	338,656 784,942	7,261 5,177	Unit 3 is down for refueling
VERMONT YANKEE Vernon, Vt. Vermont Yankee Nuclear Power Corp.	50-271	BWR (GE)	1593	514	DPR-28 (3-21-72)	397,884	4,904	Down for refueling and maintenance
YANKEE Rowe, Mass. Yankee Atomic Electric Co.	50-29	PWR (West)	600	175	DPR-3 (7-9-60 POL) (6-23-61)	266,158	15,533	
ZION 1 and 2 Zion, Ill. Commonwealth Edison Co.	50-295 50-304	PWR (West) PWR (West)	2760 2760	893 893	DPR-39 (4-6-73) DPR-48 (11-14-73)	885,435 375,622	3,416 681	Some downtime of Unit 2 for repairs

*From *Nucleonics Week*, **15**(43): (Oct. 24, 1974) and **15**(47): (Nov. 21, 1974).

†As of the end of Oct. 1974. From *Nucleonics Week*, **15**(47): (Nov. 21, 1974).

Chapter 11

ABC'S OF LMFB'S

A school-sized poster, looking every bit like a sheet torn from a child's primer, has been widely distributed by the old Atomic Energy Commission. Scrawled in bold crayon print, its message goes like this:

"Johnny had 4 truckloads of plutonium.

"He used 4 of them to light New York for 1 year.

"How much plutonium did Johnny have left?

"Answer: 5 truckloads."

No trickery here. The answer is just that simple: The appeal of the Liquid Metal Fast Breeder Reactor (LMFBR) is a unique ability to produce more fuel than it consumes. It goes without saying that this makes the breeder, as it is called, a very desirable body in a fuel-conscious society.

In fact, the nuclear battle has shaped to the point that you either love the breeder or you hate it. Both sides do agree on one specific: the LMFBR figures as a major element in the U.S. energy strategy. Since early 1975, it has been subjected to the scrutiny of four major reviews, including one by a panel of the (Congressional) Joint Committee on Atomic Energy, headed by Representative Mike McCormack.

Interest in the breeder reactor concept blazed very early in the nuclear program because of an obvious need for efficient utilization of limited uranium resources. Since the mid-1940s, numerous research projects have studied this technology, working toward the actual development and commercialization of a LMFBR.

Its use will stretch uranium from decades to centuries, because the breeder pulls a neat little trick of converting uranium-238 into nuclear fuel. It takes an otherwise useless piece of "slag" product—which is usually stockpiled—and makes it all new again and ready for action. Thus, plutonium is contained in a cycle without end and can be used to fuel other breeders or light water reactors.

A hard-working LMFBR is geared to extract more than 60 percent of the energy content of uranium. Again, the basic effect is the production of more fuel than is consumed. The ratio goes something like this: For every four atoms of uranium used to produce electricity, a well-designed breeder will produce more than five atoms of new fuel which can then be used to fuel new reactors. You might study and become an expert on the breeder, but this equation tells its past, present and future in less than fifty words. Remember it.

So, it becomes clear that this type of reactor will insure stability in the nuclear industry. Scratch the LMFBR program and we are advised that the use of nuclear power would be limited, since the current generation of light water reactors are able to use only about 1 or 2 percent of the energy potential of uranium. Projected figures, both private and government, show that, fused in this manner, all the currently available supplies of uranium would be exhausted by the turn of this century. Nor is it likely that additional domestic high-grade uranium districts will be discovered, as there have been none since 1957, and the experts see none in the immediate or far future. What I'm trying to say is that the shelves are getting empty and the store is just about to close. And, yes, our credit has run out and out and out. . . .

Nevertheless, the breeder's attractiveness as an efficient user of uranium has yet to really catch the financial imagination of utilities. Like everyone else, they're busy counting dollars and trying to produce more than they consume. That's an honest reaction known as the American work ethic.

For example, the government's share in the Clinch River facility, located at Oak Ridge in East Tennessee, was originally 65 percent of a "modest" $699 million investment. That was 1972. Since then, the figure has soared to $1.4 billion, or 82 percent of the latest estimated cost of $1.7 billion, and the actual ground-breaking is not officially scheduled until August, 1976—if the Good Lord's willing and the creeks don't rise.

It doesn't take a financial wizard to realize that these initial cost estimates were given prior to the current inflationary spiral. The reflected increase is just at $718 million. In addition, construction and equipment costs have more than doubled due to inflation, design requirements, plus other factors. Also, the hike increase in engineering costs reflects participation of three manufacturers rather than the original one.

Stricter licensing requirements, geared toward a "rather be safe than sorry" philosophy, have also meant more analyses and inspection teams. And this extra mile of safety has not been crossed without some criticism, or perhaps, the better word is questioning. As of May, 1975, for example, the GAO (Government Accounting Office) wondered aloud about ERDA's proposal to allow participating utilities to withdraw from Clinch River—the nation's first large-scale demonstration breeder nuclear power plant—if they disagreed on significant changes in the original design. This would cover too much ground, the GAO hinted.

The Nuclear Regulatory Commission has given a few gentle nudges as well. It has suggested a number of safety features that would alter, by addition, original plans for the breeder. One specific item is a core catcher, designed to prevent a "core disruptive accident." This "extra padding" is a device located just below the reactor which, in the case of an accident in the core, would spread out the core debris to prevent its residue from melting through the bottom or from reforming into a critical mass capable of a chain reaction.

"There are very strong indications," noted the GAO report, "that utility participants are opposed to including a core catcher in the 'design' and that its inclusion would permit the utility companies to begin termination proceedings." Despite this threat, if you want to see it that way, the NRC is very likely to get its way, even if this means reshuffling the deck of cards in favor of that extra mile of safety. As one army engineer said: "Contrary to that popular joke, construction jobs for power plants—especially breeders—don't go to the lowest bidder. No, sir. We go first class in this business!" This should be a real consolation to those rural farmers of East Tennessee who have also been wondering aloud about their new neighbor-to-be. . . .

First conceived by Dr. Enrico Fermi and his associates in the early forties, the breeder's basic design, as stated, is to simultaneously produce power and breed new fuel. Each time a neutron causes an atom to fission (split apart), two or three neutrons are ejected, only one of which is required to continue the chain reaction, that is, to cause another fission. Almost every time one of the extra one or two neutrons hits and is absorbed by a fertile atom, the fertile atom is changed into a fissionable atom. Plutonium is burned with fertile material—uranium-238—to produce more plutonium, and uranium-233 is burned with thorium 232 to produce more uranium-233. Thus, you can tell from this simplified cycle that the breeder makes fuel (fissionable material) by consuming fertile material. It's nice and clean all the way.

Some still may have difficulty in understanding just why we don't "throw our all" into standard nuclear plants rather than venturing too far into what they see as the deep, dark breeder waters. After all, these are fairly safe, as long as we go by the rule book, and we know pretty well just what we're dealing with . . . so why stretch things with a breeder?

The best and easiest answer to this comes from a scientist at the Hanford Engineering Development Laboratory (HEDL), at Richland in southeastern Washington, which is operated for ERDA by the Westinghouse Hanford Company, a subsidiary of Westinghouse Electric Corporation. This is the key engineering development laboratory for the nation's Liquid Metal Fast Breeder Reactor Program. A primary facility there is the Fast Flux Test Facility (FFTF), which will be used for fuel development and component testing.

The man told me (there are, I discovered, very few female scientists in the United States) that the breeder's winning combination—conserving fuel and reducing requirements for enriching facilities—is impossible to beat right now. "These are two things that a nuclear 'power' plant is simply 'powerless' to do. So, let's face facts and proceed accordingly," he shrugged.

Also, the dwindling uranium supply hinted at earlier is just that: a dwindling supply. The best estimates say we have maybe twenty years' supply of the uranium required to fuel the non-breeder or "burner" nuclear plants now on production lines. And although there is a great deal of uranium in the world, only about seven atoms per 1000 are in fissionable form (U-235), while virtually all the remaining 993 are fertile U-238. To add another stack to the pile, to be truly economical, the uranium must be found in relatively high grade ore in accessible places. Most of the existing ores are not in the low-cost, bargain basement, category. Breeders then, should remedy this situation and soon, unless a rash of moratoriums squelch the flames.

Design-wise, an absence of moderators* is the principal difference between the fast reactor and thermal ones like the Light Water Reactor (LWR). For example, since water is a moderator, it cannot be used as a coolant in a fast reactor. There are two alternatives in the form of metals and gases, with gases being quite suitable because of their low density. Pursuing this line of thought, Gulf General Atomic Corporation is developing its own Gas-Cooled Fast Reactor (GCFR).

Relatively heavy nuclei prevent a liquid metal coolant from acting as a moderator, and this factor differentiates between the design of a fast reactor (i.e., a reactor minus a moderator where neutrons are not appreciably slowed down) and thermal reactors such as the LWR (Light Water Reactors) which utilize moderators. Since water is a moderator, it is not acceptable as a coolant in a fast reactor. There are two possible substitutes and these are liquid metals and gases. Gases are good because of their low density, i.e., not enough atoms are present to do much moderating. The most suitable liquid metal coolant is sodium (Na) but it carries problems of its own.

There are two very ticklish things about using sodium. One, it can become highly radioactive from certain key reactions. Two, violent chemical reactions can take place when sodium comes in contact with water or steam. How might this happen? By a leak in the heat exchanger in which the hot (and it *is* hot) sodium produces steam to drive the turbine. But can it happen? Such an accident is virtually impossible, because an extra sodium heat transfer loop has been thoughtfully inserted so that the radioactive sodium from the reactor never comes near water or steam, and the sodium which converts to water or steam is not radioactive.

Sodium's real selling point is its unique ability to remain liquid at high temperatures, which means an efficiency not attainable with water-cooled reactors. It also offers excellent heat transfer properties. So, you have to weigh these against its highly corrosive tendency and violent reaction on contact with water or air. Nothing, it seems, comes

* Material inserted in a thermal reactor to slow down the neutrons.

easy. Stated another way: Every cloud with a silver lining also has a dark underside. . . .

There are risks involved in using sodium as a coolant in the LMFBR, but the breeder also has safety mechanisms designed to make certain these risks remain just that—risks. Table 9, for example, illustrates the actual time and planning required on a breeder nuclear power plant. It doesn't pop up overnight, and there are a series of checks and re-checks to make certain every nut and bolt fits exactly where it's supposed to fit.

Various facilities at Hanford enable testing of reactor components, such as refueling equipment, control rod drives, core restraint devices, etc., and in such technological areas as design of reactor internals, heat transport system layout, and sodium chemistry. Other facilities are used to analyze nuclear fuel, examine irradiated fuel, and improve the processes involved in fuel fabrication. Principal existing units at HEDL include the High Temperature Sodium Facility, which incorporates the composite reactor component test activity, and the Hydraulic Core Mockup, used to verify the design of the reactor vessel and core cooling system. There are also a variety of supporting laboratory facilities that include technical shops and fabrication facilities, advanced electron microscopes, a welding facility, metal testing and development operations and computers. In addition, HEDL has pilot-scale plants for controlling effluents and processing waste.

The above paragraph doesn't mean jobs just to keep people busy and off the streets. The emphasis is on safety. Hanford exists so that when the "go" switch is pushed at places like Clinch River (tentatively mid-1982), it will go until something better comes along and energy moves up another peg on the scale of life. Personally, I like such extra padding and I'm quite willing to pay the tab.

The term "enriching facilities" may pose a question. In order to use nuclear fuels more efficiently—and this is the name of our game—the ratio of fissionable to fertile uranium (i.e., U-235 to U-238) must be increased above that found naturally. Officially, this is called enriching the fissionable content of uranium. It is accomplished at huge and expensive gaseous diffusion plants found at Oak Ridge and a place in Kentucky called Paducah. These little jewels are so expensive that only a few nations can afford the price tag. So, if fast breeders don't provide the fissionable materials, it will mean that additional plants, perhaps using other methods of enrichment, will be required to supply the fuel needed for "burners." This spells money.

Still not convinced? Maybe the term "fast" has caused some alarm. Relax. It does not necessarily mean the breeder is a wild lady who's difficult to control. "Fast" is simply technical jargon referring to the speed of neutrons in the reactor. In essence, it means that the better neutron economy which results from the predominance of such fast neutrons will lead to more efficient or rapid breeding. Therefore, while breeding is technically feasible in certain other reactors, fast breeders certainly seem to hold a valuable monopoly.

Nor is the Liquid Metal (Sodium) Cooled Fast Breeder Reactor the only pebble on the beach. Study Table 10 and you can readily get a glimpse of the world view. There is additional information in the section called "Kernenergie" (Chapter 14) which delves into the international aspects of nuclear power. At present, the United States seems to be somewhat behind the USSR and Western European countries in breeder research and output. The helium gas-cooled breeder reactor (heat produced in the reactor is carried away by helium gas, which is then used to boil water in a steam generator) is being studied both in the U.S. and Western Europe as a possible alternative or even backup to the pioneer LMFBR. It has numerous attractive advantages unique to its gas-cooled design. Since gas is not dense enough to serve as an effective moderator, carbon is mixed with the fuel to slow down neutrons. One clear advantage of this "machine" is that gas can sustain higher temperatures at lower pressures than water (gas emerges at 1,400° Fahrenheit and produces 1,000° Fahrenheit steam) which means an increase in overall efficiency to 39 percent.

Ready for the bad news? There are three basic burrs that color the gas-cooled reactor less than bright. One, gases are far less efficient at carrying off heat than water. (An example of this is that only small cars are able to tolerate air-cooled engines while larger models utilize water). Burr Number Two is a sticky one which should make all potential bomb makers and would-be terrorists downright ecstatic: Due to its insufficient moderator, gas-cooled reactor requires highly enriched U-235, which makes nice bombs if diverted to the wrong sources. And, finally, while fuel handling costs are lower, this type reactor costs more overall than its cousin, the light water reactor, or the molten salt breeder reactor (MSBR) which operates, interestingly enough, with slow neutrons. In the MSBR, U-233 and thorium as fluorides are in the form of molten salts, which are pumped as a liquid between the reactor vessel and the heat exchanger. The advantage of the MSBR is that U-233 is not as dangerous a material as plutonium. The cards still stack in favor of LMFBRs, however, because thermal breeders simply have a much lower breeding gain.

Actually, interest in the fast breeder concept dates back not only to Fermi's early work at a self-sustaining reactor, but also to the wartime Manhattan Project and the first nuclear chain reaction at the University of Chicago. Since this period, much of the preliminary effort on the LMFBR has been conducted at national laboratories now under ERDA's watchful eye. An experimental reactor called "Clementine" at Los Alamos was used from 1946 to 1953 to demonstrate the feasibility of operating a reactor with fast, unmoderated neutrons, plutonium fuel, and a liquid metal coolant.

The Experimental Breeder Reactor I (coded EBR-I and located in Idaho) was historic in that it produced the first nuclear-generated electric power. The EBR-I was built and operated by Argonne National Laboratory from August, 1951, through December, 1963, to confirm the breeding principle in a fast neutron reactor and to establish the engineering feasibility of using a liquid metal coolant. This experience, along with other early liquid metal-cooled experimental reactors, contributed much vital information to reactor design and safety. Specifically, other reactors included the Sodium Reactor Experiment, the Hallam Nuclear Power Facility, and the Submarine Intermediate Reactors Mark A and B.

In the mid-1950s, construction began in the United States on two fast reactor electric generating facilities: the 625 MW(th) Experimental Breeder Reactor II, and the 200 MW(th) Fermi reactor. The first has operated successfully for more than ten years, producing more than 5×10^8 kWh of electricity and clearly confirming the feasibility of breeder power plants.

The Fermi reactor, now decommissioned, achieved criticality in 1963 and produced 32×10^6 kWh of electricity before the project was officially terminated in 1972. This pilot reactor provided valuable technological information as well as operating and maintenance experience which will be useful in future LMFBR power plants.

In the mid-1960s, construction was initiated on the 20 MW(th) Southwest Experimental Fast Oxide Reactor (SEFOR) facility, a joint U.S./Federal Republic of Germany effort directed at providing a facility to conduct research on the safety of LMFBRs. This project was concluded with a warm *auf Wiedersehen* in early 1972 after successfully completing its mission of confirming the safety character of mixed oxide fuels for LMFBRs.

A report submitted by the Project Steering Committee of the Clinch River Breeder Reactor Project, narrates how, in July of 1969, the U.S. Congress (through Public Law 91-44) authorized the AEC to begin a two-phase approach for the first large-scale LMFBR demonstration plant in the United States. A summary of this endeavor was provided by W. B. Behnke, executive vice president of Commonwealth Edison Company of Chicago, J. E. Watson, power manager for TVA, and Thomas A. Nemzek, director of ERDA's Reactor R&D. The overall purpose of this first phase was simply to define the technical and economic characteristics of the plant. Participating in this effort were Atomics International, General Electric, Westinghouse, and several American utilities.

Shortly after the called "planning session," Public Law 91-273 was enacted by Congress in June, 1970. It authorized the old AEC to undertake a second phase: design, construction and operation of a demonstration plant.

It soon became apparent, in fact as early as 1971, that the logical next step should be an actual demonstration of the concept in a working, large electric generating facility. The feasibility would be further tested by operating it as an integral part of an electric utility system. The breeder was just about to graduate to OJT, that is, On-The-Job-Training.

For the present, the Clinch River facility is the focal point, because this will be the first commercial prototype. It is designed as a 380 MW(e) sodium-cooled, fast-neutron reactor that is fueled with a mixture of plutonium and uranium oxides. As with any first-of-a-kind plant, the cost will be greater than for a commercial generating station of the same size. Its developers maintain, however, that viewed in the context of the total program, the cost of Clinch River will not be overly excessive and will, in the long run, save valuable dollars.

The plant will be an integrated electric generation facility, designed to operate as a part of the vast Tennessee Valley Authority distribution network. The major components will include: a reactor and steam supply system; steam-turbine-driven electric generation system; cooling towers; electrical switch-yard, and related auxiliaries and support facilities.

Clinch River is designed for a thirty-year plant life. It will:

- Be engineered to meet utility requirements for reliability, inspectability and maintenance.
- Have an initial breeding ratio of approximately 1.2 with the prospect of increasing this to 1.3 later on.
- Have minimum environmental impact, including minimal planned releases of radioactivity from routine plant operation and a closed-cycle evaporative cooling system.
- Provide complete built-in safety features for protecting the public's health and safety.

As a further breakdown in the division of labor, it is interesting to note the principal parties involved in the Clinch River Project and their respective roles:

- Project Management Corporation (PMC) is a non-profit corporation organized just to manage the design, construction and testing of the demonstration plant. PMC is responsible for ensuring that project resources are effectively utilized. Its Board of Directors is composed of five members: two Commonwealth Edison representatives, two TVA representatives and one Breeder Reactor Corporation representative. A three-man Project Steering Committee oversees the project on behalf of PMC.
- Breeder Reactor Corporation (BRC) is a second non-profit corporation created in 1972 that provides senior utility counsel, disseminates information on project progress to the utilities and the general public, arranges for participation of utility employers in the demonstration phase of the program, and provides a continuing industry review of PMC's operations. BRC also coordinates the electric utility industry's financial contribution.
- Commonwealth Edison Company and the Tennessee Valley Authority jointly submitted the basic proposal of cooperative arrangements involving industry and government. These two utilities are providing the project with management and years of operating experience. Specifically, Commonwealth Edison provides its engineering management and procurement service to PMC, while TVA provides management to oversee the construction and quality assurance activities for the plant. Both organizations repre-

sent the utility community's best interests by ensuring that the design and operating features of the plant properly figure in the pertinent technical and economic requirements for operating a fast breeder generating plant as part of a utility system.

According to TVA's power manager, J.E. Watson, the Tennessee Valley Authority will initially own and operate the complex. Later, at the conclusion of the demonstration period, he said, TVA will have the option of either retaining its "lease-purchase" type ownership at a price based on its value to the utility, or putting a For Sale sign in the front yard. . . .

Occupying a strategic role in this chain of command is the ERDA, whose responsibility it is to direct that gremlin called the Environmental Impact Statement. The EIS is regarded as about as much fun to prepare as playing a rousing game of tiddily-winks in a closed container with the non-human lead in *Jaws*. Its preparation—often stretching into years and dollars—involves a lot of blood, sweat and tears.

By mid-1975, a proposed final Environmental Impact Statement had been issued on the Liquid Metal Fast Breeder Reactor Program. It involved a nice piece of paperwork: a seven-volume statement consisting of more than 5,000 pages and representing more than fifty man years of effort by the government and its research labs.

In addition to this total LMFBR effort, ERDA develops and manages related base program research and development efforts. For a demonstration project, ERDA would provide major funding support and has the lead role responsibility for the design and construction of the nuclear steam supply system. The government also makes available to the project its staff, laboratories and contractors, in addition to providing technical supervision and administration.

Summarized, you might say that ERDA is a nice lady to have around the house, especially when things go topsy-turvy and you need a soothing hand, or even occasionally, somebody to blame.

Key contracts in the design process come under three specific listings: First, the primary contract with ERDA, Commonwealth Edison, TVA, and the PMC (which specifies principal agreements concerning the design construction, testing, and operation of the Clinch River plant); and second, separate agreements of the Breeder Reactor Corporation which represents the electric utility industry with PMC and ERDA. Third, the PMC has also entered into prime contracts for project design services with Westinghouse Electric Corporation, as lead reactor manufacturer;

and with Burns & Roe, Inc., for architect-engineering services. Holmes and Narver also provide services as a sub-contractor to Burns & Roe.

Not to bore you with detail, but the Clinch River facility offers a truly comprehensive look at the tremendous amount of detail and organization that goes into a nuclear complex. You've heard the expression "too many chiefs and not enough Indians." Well, here is a case where you have scores of Indians and only five or six chiefs: a ratio offering a division of responsibility which ensures—like a check and balance—that each component functions in its capacity to the best of its ability.

Table 11 elaborates even further. For example, as the lead reactor manufacturer, Westinghouse is developing, according to latest (1975) plans, the overall design and will provide the nuclear steam supply system for the nuclear portion of the breeder plant. How would you like to be the salesman who got that commission? As sub-contractors to Westinghouse, GE and Atomic International Division of Rockwell International Corporation have also contracted to provide major assistance on this project.

A postscript to this would go under the heading "Jobs" and in a big way. Clinch River is nothing short of economic adrenalin that shoots manpower on all levels—laborer to executive—on a long-term basis. It's something to think about. Another fact bears repeating: Ax the nuclear industry and you *immediately* cut *six million jobs,* with others going like dominoes in an irregular pattern. That's something else to think about if we really want to get into this question on a realistic basis. As I said earlier, no giant hand is going to reach down and pluck us up and over. It's a nice thought, but that's all.

When you talk about the Liquid Metal Fast Breeder Reactor there are perhaps four or five primary issues of public concern that surface immediately. These are: environmental effects, plutonium usage, uranium, safety and cost.

Nuclear abolitionists are busy telling East Tennesseans that establishment of the Clinch River installation is going to make the Cumberland Plateau a garbage dump for the world. This is bad talk to those folks who live in this lush hill country where creeks run and dart 'neath moss-covered trails. It pains the heart, as one old Appalachian farmer said, shaking a boney finger. "Just ain't right," he sighed.

The farmer is right. It ain't right—'cuz it ain't right.

Point Number One: Breeders are superior to most present nuclear power plant reactors in minimizing environmental impact. They are designed to operate with high efficiency (about 40 percent) which lowers the thermal discharges, compared to the current water-cooled (PWR and BWR) nuclear power plants (about 33 percent efficiency) and roughly equals thermal discharges from the current High Temperature Gas-Cooled (HTGR) nuclear power plants.

Point Number Two: Estimates of recoverable uranium resources currently contain a high degree of uncertainty, but a thorough new resource assessment is now in progress, according to ERDA planners. Technology and price—as with other resources—determine recovery estimates and the current rate (1975-76) assumes a recovery cost of $30 per pound or less of uranium concentrate. However, and this is a big however, the effective size of the uranium resource base in energy terms depends more on the technology for using uranium than on the amount of uranium in the ground. Thus, the size of the uranium resource, if used in current light water reactors, becomes nearly as large as that for the remaining domestic oil and gas resources. (Note: The extraction of this amount of energy assumes the use of plutonium recycle in reactors and assumes an assay of 0.2 percent unrecovered uranium-235 isotope in the fuel enrichment process.)

Success, then, with the breeder reactor concept could turn the domestic uranium supply into a nearly unlimited resource. Even more important, the quantities of uranium and thorium ores that must be mined will be reduced by a factor of 1000 or more, since all the uranium and thorium mined will be utilized instead of the less than 2 percent which is currently the case with a "burner" reactor. This also means a lessening of the extent and environmental impact of mining.

As Dr. James H. Wright of the Westinghouse Electric Corporation has noted, design of the Clinch River Plant is in direct line with both the letter and spirit of the National Environmental Policy Act of 1969. The plant is being constantly assessed to ensure that its operation will be safe and reliable and have no significant adverse effect on the environment.

The public is likewise protected against the health hazard of plutonium by a series of physical barriers in exactly the same way as in the water-cooled reactors and by the same waste disposal methods. The only difference—and this is made to order for critics—is that the quantities of plutonium involved are greater in the breeder fuel cycle.

The safety mechanisms, however, are appropriately stiffer and more involved.

One final point concerns the Emergency Core Cooling System. LMFBRs do require one . . . and its purpose is to protect against a loss of coolant accident. There, any similarity between the "burners" and a breeder comes to an abrupt halt. As far as the water-cooled reactors are concerned, it is a parting of the way. The main reason is that sodium coolant (which is in a class by itself) does not operate at high pressure like water and thus will not flash to a vapor-like steam. The gas-cooled fast breeder reactor (GCFB), like the high temperature gas-cooled reactor (HTGR), cannot suffer a complete loss-of-coolant accident, since air at least will always be present. Provisions are made, however, to assure that coolant circulation is always possible.

As a further reference, Table 12 is based on material provided by Westinghouse's Dr. L. E. Strawbridge, at the American Nuclear Society Topical Meeting on Fast Reactor Safety in Los Angeles. This covers three levels of design and is well-worth poring over. Again, it's a matter of asking the specialists for advice rather than relying on rumor or second-hand informaion that just cannot be substantiated.

Let's conclude this LMFBR trek by saying that after some thirty years of research and development, it is time for the bird to leave its nest and do some work. The winds have tested right, and the total concept seems ready to go in terms of size and performance as a major step toward the ultimate goal of commercially competitive breeder reactor power plants.

Table 9: The Clinch River Breeder Reactor Project (CRBRP)—the nation's first large-scale demonstration breeder nuclear power plant—is being built at a Tennessee Valley Authority site near Oak Ridge. It represents the combined efforts of government, utilities, and private industries working jointly in a national commercial breeder program.

Key Accomplishments to Date and Projections

Westinghouse authorized to begin design assisted by General Electric, Atomics International, and Burns and Roe	Feb 1973
Raw Material Procurement Authorized	Feb 1974
Plant Concept Selected	Mar 1974
Reference Design Report Completed	Jun 1974
Plant Cost Estimate Submitted	Aug 1974
Environmental Report Issued	Oct 1974
Preliminary Safety Analysis Report Submitted	Jan 1975
Limited Work Authorization (actual start of earth-moving)	Sep 1975
Receive Construction Permit	Aug 1976
Start of Concrete Placement	Jan 1977
Complete Concrete Containment to Operating Floor	Jan 1979
Set Reactor Pressure Vessel	Jul 1979
Submit Final Safety Analysis Report	Feb 1980
Start Sodium Load	Nov 1980
Receive Operating License	Apr 1982
Initial Criticality	Jul 1982

Table 10: Nuclear Breeder Power Plant Development Timetable

	Demonstration Plant (250 to 500 MWe)	Large Plant (600 to 1000 MWe)
USSR	1972*	1976*
FRANCE	1973*	1980
UK	1973*	1980
GERMANY	1978*	1984
U.S.	1980*	1986
JAPAN	1980	1987

* Plants under construction or committed.
Figures available through the American Nuclear Society

Table 11: Responsibilities For The Clinch River "Nuclear Island"—A study in the divisic of labor between government, utility, and private industry working toward a common goa

WESTINGHOUSE Overall integration Reactor and reactor enclosure systems Primary sodium heat transport system Containment structure Reactor support buildings Related control & instrumentation systems Overall plant control system	ATOMICS INTERNATIONAL Reactor Service Building Sodium purification system Auxiliary liquid metal handling and servic Radioactive wastes disposal system Containment inerting and fire protection Liquid metal leak detection system Fuel handling and reactor refueling systen Related control and instrumentation syster
GENERAL ELECTRIC Intermediate sodium heat transport system Steam Generator Building Steam Generator System SG Auxiliary Heat Removal System Related control and instrumentation systems Piping & equipment electrical heating and control system	BURNS AND ROE Building electrical power and emergency power systems Plant service building Heating, ventilation, cooling and air conditioning Related control and instrumentation syste Radiation monitoring

An artist concept of the Clinch River Breeder Reactor Plant—the nation's first large scale demonstration breeder nuclear power plant—to be built in East Tennessee near Oak Ridge.

Table 12: Three Levels of Design

evel	Definition	Application
1	Provision of simple, reliable functional design free of defects, with inherent safe performance, fabrication and operation to the highest proven standards.	Fuel of proven performance through life Core design with negative power co-efficient Adequate Doppler constant Multiply redundant heat transport systems Redundant power supplies Redundant control system Low pressure coolant system with wide margin to boiling Natural convection capability Maximum use of proven technology and hands-on maintenance
	Provision of protection systems to provide adequate response of system in the event of all identified transients.	Doubly redundant plant protective systems Decay heat removal redundancy Battery power supplies for vital services Guard vessels for leak protection Inert atmospheres in sodium cells Sodium-water reaction protection system
	Provision of extra capability to cope with extremely unlikely events which are never expected to occur, and with additional design requirements to provide prudent margin for unforeseen events.	Containment building Site requirements Containment isolation system Capability to accept extra thermal loads in the core support structure Dynamic loads in vessel and primary system components Geometric requirements in and around the vessel

Chapter 12

GETTING IT ALL TOGETHER: FUSION

Fred Olds, senior editor of *Power Engineering,* tossed this audience question to Dr. Robert L. Hirsch, director of the Controlled Thermonuclear Research Division for the Energy Research and Development Administration. Referring to funding and costs, Olds began in a teasing manner: "Since fusion—being far away—looks good today, why not take all the R&D money applied to oil, gas and fission, and simply pour this into the fusion program? Just what would that do for you?"

Without a moment's hesitation, Dr. Hirsch glanced at Olds, shook his head and replied: "Make a lot of people hate us."

Everybody laughed. Yes, everybody enjoyed a good laugh.

Then the ERDA official threw in a bucket of cold water and things settled down. There are a lot of good things coming after the year 2000, he began, and fusion is certainly one of them. "However, the problem is to get there from here . . . if we can't do that, then we simply don't win our case."

If indeed the gods should will that this tired, old, blood-thirsty planet does make it to the year 2000, then fusion will probably be the prime energy source that keeps those home fires burning. It offers a glimpse of the good life available to all, despite those who would go back to the days of candle wax and wood-burning stoves.

The United States is spending $102 million (1975) on research to determine the feasibility of deriving electrical power production solely from fusion—the process by which the sun generates its own energy. You see, the sun is a very hot gas, so much so that the atoms in it have been ionized, which means that the negatively charged electrons have become separated from the positively charged nucleus (called an ion) of the atom. This gas of electrons and ions is called a *plasma* and has some very special properties. One of these is that the ions frequently collide with each other. The hotter the plasma, the harder they collide. If the plasma is hot enough, the ions will collide with each other with enough force to overcome their tendency to repel each other due to their positive electric charge. When this happens, these ions or nuclei combine or fuse to form new nuclei (new elements) and release energy in the process. And that's what it's all about.

What is needed, then, is a way to generate a very hot plasma, like the sun, and hold onto it long enough for many fusions to take place and release energy. A major problem facing researchers is how to generate, hold, and heat this plasma in a suitable container so that the ions, which are moving more than a million miles per hour, will not strike the walls and lose their energy.

One method of confinement is to keep the plasma inside a magnetic field called a "magnetic bottle." In the early 1950s, reactor-level densities were achieved in these "bottles." In the early 1960s, fusion temperatures above 50,000,000° C. were achieved, and since then, these temperatures have been reproduced at will. Until the late 1960s, however, magnetic bottles leaked plasmas too fast to be effective. But then experiments began to show tremendous improvement, and efficient containment was achieved, thus proving that magnetic bottles can work as theory predicted.

Results of experiments conducted in 1972-74 have shown the practicality of building larger fusion machines which can produce net fusion power. These machines will probably be built by 1980, although the Soviet Union has a model Tokamak system that went operational in mid-1975. The first total fusion power plants, producing electricity on a modest scale, will be built about 1985. If the United States is successful with these and other plants, it will probably lead to the first commercial scale fusion power plant by the late 1990s.

First generation fusion power plants will generate electricity by means of thermal energy conversion cycle, which will obtain heat energy from the fusion of deuterium and tritium, two isotopes of hydrogen. Tritium is radioactive and deuterium is not. Despite the fact that tritium is one of the least hazardous of the radioactive elements, power plants that utilize it will have to be very careful that they are shielded in a manner that will safely contain it.

Eighty percent of the energy released by this reaction will be carried by neutrons. This requires that the reaction chamber be surrounded by a region in which the neutrons can slow down and release their energy as heat to a working fluid. The fluid will, in turn, be used to generate steam to drive an electric turbine generator. Fusion engineering research and development studies have already begun and will be expanded in the future. The goal of this engineering will be to develop systems that will function properly and reliably at minimum cost to the consumer.

If fusion power comes close to expectations, it will replace most other forms of electrical power in the United States by the middle of the twenty-first century. This would free the nation, forever, from the need to use precious and dwindling supplies of fossil fuels in power plants. Coal, oil and gas would simply not be required for electrical power, which would come instead from the inexhaustible and non-polluting fuel reserves of deuterium extracted from the world's oceans.

Studies made in recent years show that fusion systems could be used directly or indirectly for the manufacture of combustible fuels, such as hydrogen, synthetic natural gas, and alcohol, from water and gases found in the air.

I thought it might be helpful to list several advantages of fusion before getting into a lengthy explanation of the actual mechanics as they were detailed to me by Dr. Hirsch and Dr. Charles Fenstermacher, group leader of the Electric Discharge Gas Laser Group at the Los Alamos Scientific Laboratory.

First, fusion fuel is in good supply and comes from sea water, which nobody has a real monopoly on—yet. As mentioned, one major material for fusion fuel is deuterium, or heavy hydrogen. It is a naturally occurring material, and one out of every 7,000 molecules of water is a heavy water molecule containing an atom of deuterium. Thus, the ocean contains sufficient quantities to last for many billions of years without seriously depleting the supply. Nor is it difficult to extract fusion fuels from ore, because the technology for obtaining these is already available. The cost for large-scale activities is estimated to be substantially less per unit of energy content than the cost of uranium. The cost of fuel today is on the order of 1/100th of a kilowatt hour. Therefore, fusion fuel is in good supply, available to all, and on a energy scale its cost is truly negligible.

Second, there are no chemical combustion products in fusion, so there is no chemical pollution problem.

Third, there is inherent safety in fusion with absolutely no possibility of runaway.

Fourth, while it is true there is radioactivity in fusion, it is relatively low, as are the inherent hazards associated with it. Okay, for those of you who are hissing "tritium" under your breath, stay calm. Vast amounts of tritium are generated in fusion machines, and it is mildly radioactive, and nobody is trying to hide or conceal this fact. Point One, systems are being built utilizing shielding by which tritium can be contained, not only in routine operations but also under fault conditions. Points Two and Three, people have had a lot of experience in dealing with tritium (i.e., weapons program and the fusion reactor), so a great deal is known about it. For one thing, it has a low biological hazard, and because it is hydrogen, it tends to disperse very quickly.

One reason that we know this for certain is because a tailor-made accident occurred at California's Livermore Laboratory several years ago. Some 10,000 curies of tritium escaped and everybody went scampering with their counters, trying to measure it—except that they couldn't because it had already dispersed. The tritium had gone as quickly and quietly as it had come . . . and nobody died or even came down with cancer in the process. And to this date there is no record of any reported carcinogenesis related to the incident.

Now that this matter has been dispersed, let's continue with a few more advantages of fusion:

- No emergency core cooling problem. And all you would-be terrorists be forewarned: there is no weapons grade material, which means absolutely no possibility of diversion for clandestine purposes. No, not even for your "Flash Gordon" ray guns.
- A final and extremely attractive point is that because of the safety and environmental characteristics involved in the use of fusion, it will probably be possible to locate plants close to heavy load centers—such as urban sites where the energy is vitally needed.

Initially—if you are not a Ph.D. in physics—it might be helpful to distinguish between the terms fission and fusion. The most basic definition possible is simply this: Fission means to split apart and fusion means to bring together.

To talk about fusion power research and development is another matter. It involves very complex physics and engineering, which is another reason why there is greater interest in the nuclear community about nuclear activity than there is out among the general population. My experience has been that the vast majority of people—those concerned with eight-to-five jobs and running households—believe that nuclear energy is complicated (and it is) and there is just no way to untangle its intricacies without years of study. So, instead of digging in and possibly making some kind of beachhead, the tendency is to stay with headline reading and not go beyond.

It was a learned man named Albert Einstein ($E=mc^2$) who made a profound observation way back in 1946. "To the village square," he said, "we must carry the facts of atomic energy. From there must come America's voice." The village square, however, has been strangely silent. Perhaps, the facts are not reaching their destination, or perhaps, the babble is too loud for the people to hear.

While researching this book, I found that women generally avoided asking me too much about the subject, and men—intrigued by the unusual—would invariably ask, "Honey, whatcha writing?" Most limited me, a member of the female species, to four acceptable categories of interest; babies, cooking (preferably home canning), gardening (organic) and sewing. A fifth unacceptable but highly welcome category would be anything remotely smacking of raw sex. That is the way the cards are stacked, and believe me, I'm not knocking any of those highly relevant and interesting subjects. . . .

One government contractor who spent most of his time ogling an airline stewardess on our flight from Albuquerque, finally sobered to the point of inquiring about my topic. When I replied "nuclear power," he reacted with a glazed look and then ventured: "You must have a master's in physics?"

"No," I answered sweetly. "Home canning."

My seatmate laughed and proceeded to order us a scotch and soda. Then came not a well-worded proposition, but a horde of questions on all aspects of nuclear power. Like everyone else, this business executive wanted to know what the facts were from someone who didn't threaten his ego, and I was approachable. Like him, my expertise was, in a sense, my lack of expertise. In short, I was just an ordinary citizen who had decided to switch-off the television and go out in the fresh air after some answers.

Another point. Most of us simply don't like to admit that we don't know something. That's a fact of life. It's a part of the human condition and should be recognized as such.

Listen, friends, I got hoof-in-the-mouth disease so much from writing this book that my jaw has a permanent ache. But it's been worth the whole trip because I learned something by, first, reading solid facts, and, second, asking questions from the specialists, who, it turns out, were very kind and very tolerant. And among the scientists, there were no exceptions.

I still remember Dr. Bibb (Chief, Research & Development), personally driving me through the massive Oak Ridge complex which is operated by Union Carbide for ERDA. Nearly a year of research was just beginning, and it was obvious that the amount of research and knowledge involved was literally astounding. I can compare this, my first dabble in the waters of nuclear technology, as something like window-shopping at a country five-and-dime and then, suddenly, being plucked up and thrown into the midst of Gimbel's on Christmas Eve.

Also, during those early days, my pride was at issue, and I worried that Dr. Bibb would find out that I really didn't know what he was talking about.

So, I continued to smile sweetly, trying to look cool and make impressive notes in shorthand as he went through items like High Flux Isotope Reactors and fission do's and don't's. Finally, as a long day drew to a close, honesty won out and I looked at my patient guide and remarked, "You must think that I'm very stupid. . . ."

Granted, if he had said "yes" I would have been crushed. . . .

However, Dr. Bibb grinned in my direction and answered, "Why? because you haven't understood a word I've said all day? I know, but you will. It takes time and that's why you're here—to learn and see the facts firsthand. And, no, I don't think you are stupid."

As I said, scientists are beautiful people.

The next day brought my encounter with a youngish Dr. Clarke who heads the Thermonuclear Division at Oak Ridge National Laboratory. It was his "job" to explain the Tokamak system to me and, so, there I stood, frantically trying to decide if he was talking about a fission reaction or a fusion reaction. Yes, it was that bad. . . .

This Ormak device, as it's called at Oak Ridge, is proving to be an important step toward the goal of controlling and using the energy forces that power the sun and stars. It's cheap; it's there and all we need to do is tap it. As stated earlier, the eventual goal at Oak Ridge and related experiments in several other designated U.S. laboratories is the development—by about the turn of the century—of steam power plants that will generate electricity from the heat of the energy produced by fusing the nuclei of light atoms: namely, the hydrogen isotopes called deuterium and tritium.

Dr. Clarke explained that the Ormak device is one of a class of Tokamak experiments. The original concept was invented by the late Soviet Academician Lev Artsimovich and pioneered by various Russian scientists and engineers during the sixties. In the late sixties, the advantages of the Tokamak became obvious, and several experiments were initiated in the United States by 1970.

Such background was easy to understand and I smiled knowingly. Then Dr. Clarke began using the term "plasma" and, having never taken a physics course, my mind immediately defined it as "blood." He explained that controlled fusion has to work without an atomic trigger and within some kind of container. The problem is that the "plasma" in which the fusion reactions take place cannot touch the walls of the container, and not because it would peel the paint. . . . The difficulty with containment is that if the "plasma" does touch something solid, then it would immediately "cool off," which is far from the desired effect. Like a fool, I carelessly said, "Yeah, it would be messy."

The very learned head of the Thermonuclear Division glanced at me curiously and then—probably fearing the worst—went ahead and asked it anyway: "Messy?"

"The blood—all over everything," I sputtered, immediately wishing I could fission off somewhere. Nearby, several technicians achieved a sort of critical mass, and I glared at one and said, "Don't laugh, in the next incarnation you'll come back as a woman!"

My cover was blown and we all knew I had made a very elementary blunder, on the kindergarten level. It was the first major symptom of the "hoof-in-mouth" disease that I was beginning to contract.

"Honey," Dr. Clarke said very gently, probably wondering how to break the news, "plasma is ionized gas and not blood."

"Oh," I gagged. I mean, what can you say?

Dr. Clarke wisely suggested that we might go back to his office and review some basic principles of physics and then possibly—in the next year or two—return to the working lab. Then, he took several hours to do just that. He explained the concepts of fusion as it is being studied at Oak Ridge and other labs across the country. So, I learned one basic fact immediately: plasma is not a bloody mess.

There are many things which are made of plasma, the fourth state of matter. You may have noticed just a few: the sun, the stars, the Northern Lights, neon tube's glow, or even that pretty but deadly fireball of a hydrogen bomb. It can be created by injecting gas into a vacuum chamber and heating it above 10,000° Centigrade. It must be held there from 0.1 to 1 second for the fluid to react. The electrons separate from the nuclei—negative electrons from positive ions—and the result is ionized gas, which has one important characteristic over ordinary gas: the ability to conduct electricity and to be shaped and directed by magnetic fields. This puts plasma in a class all by itself.

Also, thermonuclear plasma gives off heat but not in the usual sense. Rather, it produces heat as energetic particles and fast neutrons, which are then used to make "our" kind of burning heat—the type that runs turbines and generates electricity.

So, to refer back to what was implied at the start of this chapter, what you will find herein is neither a graduate nor an undergraduate course in physics, but rather, a technically accurate sketch of the basics of fusion and the progress being made toward making it a prime energy source. The American program is being accomplished through detailed planning, industrial development on a timely basis, and cooperation with the universities.

"Primarily, we are hoping to develop fusion as an electrical energy source," said Dr. Hirsch, director of this particular ERDA division since 1972 and recently the head of the U.S.-USSR Joint Fusion Power Committee. "We are well aware," he said, "of the fact that there are other potential applications of fusion energy." These possibilities include breeding uranium-233 or plutonium-239, producing synthetic nuclear gaseous fuels, possibly de-toxifying nuclear wastes, and providing energy for chemical processing.

"These are possibilities from a physical point of view," Dr. Hirsch explained. "As yet, we have not done enough analyses to learn whether these applications are desirable or even cost-effective." For the present,

ERDA's interest is on the feasibility of using fusion for electrical power, and the sooner, the better.

Again, what most critics of nuclear power don't seem to realize is that the government has no intention of using it as our prime energy source forever and ever and ever. Rather, *government sees nuclear power as a temporary stepping stone to something far better and safer.* But, we have to get to that distant shore by first crossing the stream and not by leaping over it.

As previously stated, the core of any fusion system is the extremely hot ionized gas of fusion plasma. When the nuclei of these two special isotopes of hydrogen are fused, the resulting particle has less mass than the sum of the masses of the original particles. This "mass discrepancy" appears as energy, in accordance with Einstein's famous equation: $E=mc^2$ (where E represents energy, m represents mass, and c represents acceleration). This nuclear reaction produces *several million times* the energy of a chemical reaction between two molecules.

Atomic nuclei are all positively charged and normally repel one another. Because of this repulsion, they ordinarily do not get close enough to fuse. However, when they are forced to travel very fast, they can overcome this repulsion. Shorter range but more powerful forces take over to cause this reaction; and it can only happen when the temperature of the reacting particles is near or above 100 million degrees Centigrade, which is ten times hotter than the surface of the sun. Think about that for a minute. The temperature of the reacting particles must be near or above 100 million degrees Centigrade, which is ten times hotter than the surface of the sun. Needless to say, achieving such fantastic temperatures is just one of three difficult requirements for making practical controlled fusion a reality.

Neutrons carry 20 percent of the energy and are sent out from the system's center to interact with the outer region. This area then thermalizes or slows down the neutrons which are heated up. A coolant passed through this region is then heated and proceeds to go out and produce steam. It seems a lot easier to put a kettle on the stove, doesn't it? Still, you have to produce the heat, to heat the heat . . . which sometimes does get complicated.

It is necessary to put energy in to create this very hot gas. This means that an energy investment is required on the same order as making an initial deposit in a savings account aimed at drawing interest. Once that temperature is attained or confined for a suitable period of time, the deuterium and tritium atoms not only collide but sometimes fuse and become helium ions. In this process of joining, a portion of their mass is converted into energy and the amount which is released is relatively large: this means that it is possible to get more energy out than the original investment, which is not a bad transaction.

Magnetic fields are the key to holding a very hot ionized fusion gas. You might recall the junior high school science demonstration where iron filings are held by the invisible lines of a magnetic field. The particles in a fusion gas can be held in a similar way, and that is the primary approach to fusion power today. At this point, any resemblance to simplicity comes to an abrupt halt. It is difficult to hold fusion plasma in a magnetic field; so difficult, in fact, that it can be likened to holding a wiggly glob of jello between rubber bands.

So, the meanest problem in fusion research, since its inception, has been to develop a means to keep fusion plasma stable within its magnetic lines. It has been necessary to come to grips with the basics of a field called plasma physics, which has taken a big bit of doing.

By the mid-sixties, better and better confinement was achieved with magnetic bottles, and by the latter part of that decade, a variety of different machines were spawned as contenders in what became known as the fusion sweepstakes. The theoretical limits were slowly beginning to be challenged and we went on to bigger things. As a partial result, ERDA is "nesting" on four primary approaches to fusion power, and three of these are tied to the magnetic confinement system. These are: (1) Tokamak, (2) Magnetic Mirror, (3) Theta Pinch, and (4) Laser Fusion.

The Tokamak is a doughnut-shaped system with magnets on the outside and a vacuum vessel with hot plasma down inside. The regime in which it operates depends a great deal on the impurities present. Therefore, to keep the plasma as pure as possible, the plasma region is surrounded by a thin but vacuum-tight stainless steel liner. As an extra measure of cleanliness, the steel liner is gold-plated.

To produce the plasma in a Tokamak system, the inside of the stainless steel liner is filled with hydrogen or helium. Then electric current from a capacitor bank, supplemented by hundreds of truck batteries, is applied through a transformer. This provides sufficient energy to break down the atoms of gas into a plasma of ions and electrons and to resistively heat the plasma to 3.5 million degrees Centigrade.

Tokamak's toroidal magnetic field is obtained by fifty-six copper coils operated by liquid-nitrogen temperature (about –320°F), powered by four generators. A cooled aluminum shell inside these coils provides steady current stabilization of plasma and supports the toroidal winding which induces more than 100,000 amperes of plasma current. Other toroidal conductors, inside the aluminum shell, produce a vertical field necessary for plasma equilibrium. These latter windings are supplied by storage batteries.

The Theta Pinch is somewhat similar in that it is also circular-shaped, but its aspect ratio is more like that of a bicycle tire. It is also a pulse system, so that in a reactor form, it would operate like an automobile engine in a repeated pulse situation. Its purpose is to heat and confine plasma simultaneously, by rapidly squeezing (hence, the term: pinch) it with an enormous magnetic pulse flowing directly from a depository of specially-designed condensers. If this technique ever becomes operational, the Theta Pinch offers an immense advantage of producing electricity directly without the need of the usual time-consuming and complicated heat cycles: i.e., energy from a fusion reaction heats liquid metal and then, in turn, the liquid metal heats water to produce steam to turn generators.

As far as I'm concerned, the most physically striking of all the fusion experiments is the Scyllac at Los Alamos, under the direction of Dr. Fred Ribe. It sits like a multi-coiled serpent with a giant trunk or nerve center from which hundreds of white, tentacle-like cables run out in bundles from coils around its torus. And big. It is housed in a structure that is twice the size of a standard basketball court.

The view is especially striking from a steel-laced catwalk that runs neatly around the Scyllac. Clutching my battered Nikon, I climbed precariously out on this lofty perch and began to feel something like Alice in Wonderland. I have a bad thing about height and that medusa-looking fiend below didn't seem worth dying for. In fact, as I inched further and further from the small escape hatch, it became obvious that one slip and my body would probably never be recovered from the coiled mass below. Los Alamos does things like that to the human mind. . . .

The type of reactor utilizing a Scyllac system would produce a sort of breathe-in-breathe-out operation in which the plasma would be compressed magnetically to produce fusion. The energy thus released would be pushed back against the magnetic field, inducing the currents directly into the system that initially created the magnetic field.

The Magnetic Mirror is basically a straight linear operation with strong fields at both ends to hold the plasma in its place. While California's Livermore Lab has centered a lot of its efforts on laser fusion, it has also put a great deal of work into such mirror configurations, which are referred to as its 2XII Experiment. This involves a magnetic well concept to achieve plasma stability. While these may not become first generation reactors, they are aimed at the principle of eliminating the intervening heat cycle and turning fusion energy directly into electricity. Maybe not tomorrow, but possibly, the day after. . . .

ERDA began pursuing a fourth concept known as Laser Fusion in the late sixties. Basically, the idea is to use the laser technique to deliver large amounts of energy in very short periods of time. Theoretically, if this energy can be directed toward a pellet of fusionable material, then a small micro-explosion can be created wherein more energy can be produced than the original laser energy which initiated the reaction. This is something like a billiard-scale hydrogen bomb.

Sixty percent of ERDA's emphasis is on Tokamak and 20 percent on Magnetic Mirrors and Theta Pinches, according to Dr. Hirsch. The $109 million program for fiscal year 1975 had about 10 percent going to industry, he said, and this figure is expected to increase considerably in the future. There are approximately 12 percent of the funds allotted for some forty universities such as the University of California and Princeton.

These, then, are some of the confinement experiments that are operational today: (1) The Tokamak, called Ormak at Oak Ridge, showed that it was possible to scale from previously smaller-sized systems to a larger one. Experimental results were generally in accordance with theory, and this machine was also used to test and reflect heating techniques required to reach very high temperatures. (2) Symmetric Tokamak-Princeton was a "small" machine costing about $3 million which successfully tested the technique of using compression as a heating device for a Tokamak. It is also providing other information as to exact details of plasma physics. (3) Theta Pinch basically refers to the Scyllac at Los Alamos and tests plasma stabilization in a certain geometry. (4) What is probably the world's largest mirror experiment is at the Lawrence Livermore Laboratories in California. The sides contain beam heaters which not only heat the plasma but also repel it.

Such so-called small machines carry a big price tag, costing between $3 million and $10 million apiece. Consequently, the general approach to fusion research in the United States has been to do what you can in small, rather low-cost research. When the results justify it, then you can go on to bigger things.

For example, the largest fusion-research laboratory in the United States is currently the university-sponsored and ERDA-funded lab at Princeton. Until recently, its principal research machine was the Model ST (Symmetric Tokamak) that was converted from the Model C Stellarator in 1969. Now, however, it is completing a machine three times the ST's size, to be called the PLT for Princeton Large Torus. It will be about four feet high and ten feet across, with coils of pure copper wrapped around it that are even more monstrous. This gem costs $13 million and plugs into a power supply that has a value in the order of $30 million to $40 million.

Scientists are not devoid of human emotion, and there is a great deal of competition in the so-called fusion sweepstakes. Probably more than a few tears were shed when this chunk of federal money went off to Princeton, rather than Oak Ridge, Livermore or even LASL's Scyllac, which sits shivering in the cold, mountain air of New Mexico.

Princeton's machine was expected to go operational in late 1975. If larger magnetic fields are scaled, it will prove that transfer is possible from smaller to larger models. Now, theoreticians are just able to safely project themselves—based on previous experimental data—to a magnitude of "Three" without losing their footing. Slipping, maybe, but not totally losing their grasp.

In terms of size and money, probably the most notable project to date is the Tokamak Fusion Test Reactor (TFTR), the first U.S. fusion experiment to simultaneously produce reactor-level conditions of density, temperature and confinement time. It will operate on fusion fuel and a repetitive pulse basis. It will be much larger than the Princeton Large Torus (PLT), which actually represents a prototype of many engineering aspects of the TFTR. And, because it will have the shielding necessary for a fusion power system plus the mechanics for a reactor operation, the TFTR will cost $215 million. It should be operational by early 1980, according to ERDA, and is to be built—primarily by industry—at the Princeton Plasma Physics Lab. One aspect Dr. Hirsch tossed out to potential bidders: since it is research and development, proposals are now being accepted for all industrial engineering aspects as well as the system's assembly work.

After many sleepless nights, the crew at ERDA have found what the AEC learned: developing any new prime energy source is not easy—nor is it cheap. Performed correctly, it is a very complicated task that requires much engineering expertise and patience. Power systems don't go operational overnight. ERDA is willing to project two experimental power reactors plus a demonstration plant for the mid- to late-1990s. The latter would show the economics and reliability of fusion power so that residents of the twenty-first century could safely build and utilize this system. A gift, you might say, for future generations, when you care enough to leave the very best. . . .

There is a Tokamak variant called the "Doublet," being vigorously pursued by General Atomic in California, which provides a good glimpse of what a fusion power reactor might look like toward the end of this century.

Located at General Atomic's spacious, park-like complex that rests a short distance from the Pacific, the "Doublet" is a labor of Dr. Tihiro Ohkawa, who is regarded by many as a kind of Japanese Einstein. Fusion has been his speciality for some fifteen years and on a budget that has been bare, bare, bare. He had no giant motor generator sets and couldn't afford their price tag, so Dr. Ohkawa, undaunted, went shopping and came back with a rainy day special: 600 U.S. Navy submarine batteries at a bargain basement price. He then proceeded to design a complicated set of loading switches to direct the power of these batteries to his machines. It worked.

The unique feature of Ohkawa's Doublet is that it has non-circular cross-sections and looks like a giant, shelled Planter's Peanut. The shape is designed to counteract a tendency of plasmas to be cranky by kinking or buckling, even jerking and snapping, through their magnetic confinement. This tendency is referred to as instability, and, as temperatures have risen, so have new sets of such instabilities. You might liken it to a balloon or sausage where, if you squeeze one end, another juts out. In fact, it's quite possible for a lively particle to zip like an ant all the way around the torus—missing, by some strange act of nature, all the other particles and ending up exactly where it started, where it then proceeds to make problems. This is referred to in some circles as "biting its own tail." The Doublet deals with this instability by making a particle's path as long in cross-section as it is the long way around.

Yes, a circular Tokamak *can* eliminate this effect, but only by applying much stronger magnetic fields which, in turn, are more costly and time-consuming.

Laser Fusion, the fourth approach mentioned, is being conducted by ERDA's Division of Military Application, because, quite honestly, lasers have potential applications on the military side as well as in the field of electrical power. A laser, based on the principles of quantum optics, is a device that emits a very intense beam of light. Igniting the deuterium-tritium reaction with a beam offers a totally different approach to releasing thermonuclear energy. It works something like this: A pellet of solid D-T (isotopes of hydrogen) would be radiated uniformly on all sides by laser beams. This would cause the outside surface to be heated and then blown off. An effect known as the "Implosion Rocket Effect" would then drive a compression wave that would produce not only heating but also very high densities in the system's center. Such high densities and temperatures proceed to initiate a fusion reaction which could liberate more energy than was involved to produce the implosion and heating—all in terms of a billionth of a second. That's the amount of time it would require for this compressed pellet to blow apart. That is then referred to as "Inertia Confinement," because it is the inertia of the mass that determines the length of time that this pellet will hold together.

One of the questions associated with Laser Fusion is how to determine just how efficiently laser (light) energy could be absorbed into the solid material. Initially, this proved to be a "toughie," but progress has been made and it now appears that this absorption process can be fairly efficient. There is a problem, however, because when this light is absorbed, it causes instabilities and gives rise to special kinds of particles that can inhibit the implosion.

With the recognition of something called "Lawson's Criteria" for Laser Fusion and "Inertia Confinement," the next question was: just how can matter be compressed? Achieving a loss of criteria, then, is just a breakeven point. What are its limits? Because in Laser Fusion it is necessary to compress matter to densities on the order of 10,000 times.

These ideas are not classified—now. They are available in open literature, granted not the kind you might find in a local porno shop, but they are on the open market. The technique works basically this way: If one has a pellet of normal density (the density of a solid drop of hydrogen at normal temperature) how does one—on the allowable time scale—compress it? If one brings the laser onto the pellet and proceeds to heat off this outer surface, the result is that it gets hot and expands. In expanding, according to Newton's Third Law, Reaction Equals Reaction, or even some sort of ablative process, the particles escape. In so doing, however, they are literally going in both directions at once—both outward and inward.

Theoretical studies have found that if one could deposit something like 100,000 to 1,000,000 joules (a watt second) in a time scale on the order of a nanosecond, then it might be possible to achieve the "Loss of Criteria" and produce more energy, which is what it's all about.

LASL's Dr. Fenstermacher told me that Laser Fusion really comes in these two parts: (1) How do you make lasers that big? and (2) If this is accomplished, how do you measure what's going out? And how is it studied? Up to now, a bulk of this thinking-work has been done through computers, because such experiments don't scale easily, i.e., you don't build a table-top model and then presto, convert it upward to a working reality. When you talk about lasers you are essentially talking about big machines.

The relative newness of this program (started in the late sixties) is one reason why physics questions like these are just beginning to surface. Livermore is constructing a laser now which is quite large and aimed at producing on the order of 10,000 joules.

A milestone in the Laser Fusion program will be considered reached when researchers are able to produce pellet implosion as they want it, that is, to get significant burn and then to break even so that energy output equals energy input.

In 1975, the total program in this country was funded at $65 million, and the Soviet Union is applying equal efforts in its corner of the world. In the U.S., research is centered primarily at three ERDA labs: (1) Livermore, (2) Los Alamos, and (3) Sandia. Also, in fiscal year 1976, industry and universities will receive significant funding in this particular area.

Laser Fusion at Los Alamos is large and secretive, mainly because, I think, they enjoy the aura of mystery that seemingly shuts them off from the real world. Another, more official reason is that the lab has always had an above-average interest in military weaponry applications. The Manhattan Project has left its mark. A third reason for their "secretive" nature is what one federal agent termed "security problems in recent years." He refused to elaborate.

Before the alarm goes off, however, let's scale those sixteen-foot fences for just a minute and imagine, if you will, a reactor that consists of a pressure vessel, swirling with liquid lithium in a manner that produces a vortex similar to a sink basin as the water goes out. A minute drop of frozen deuterium-tritium is injected into this swirling mass. The whole concoction is then zapped by an enormously powerful laser beam that seems to come from nowhere. Ping! And it's gone.

The drop—not knowing what hit it—begins to implode, to be squeezed to great density. The heat and pressure of that squeezing produce fusion reactions that, in turn, produce high energy neutrons that are captured in the lithium and heat it enough to make steam. This is a peacetime application of laser technology. A wartime operation LASL is researching, involves using lasers to produce a form of "death ray" that burns through matter in seconds and might well be the ultimate weapon of the future. In fact, it might well be the weapon which eliminates our future. The trick is to scale it down, not up; otherwise a workable model is expected to be operational just in time for World War III. . . .

It also makes good sense to speculate that Los Alamos, perched high and remote in the hills of New Mexico, shares more than a lot of input with a sister facility, some seventy-five miles away through the shrub brush and flatland near Albuquerque. The place I am referring to is Sandia Laboratories, which consists of two major research and engineering facilities: the headquarters laboratories at Albuquerque and smaller ones at Livermore. The labs are operated by the Bell-System for ERDA through a contract with Western Electric. On another scale —also for ERDA—Sandia operates a test range that blows things up at Tonopah, Nevada.

You might recall that Dr. Hirsch mentioned that fiscal year funding for the Laser Fusion program in 1975 went primarily to three ERDA labs; and Sandia was one of them. Yet, it is far from being a household word, and I would venture a guess that most citizens are totally unaware of its presence or function. In fact, just to be smart and on the impulse, I telephoned ten college-degreed individuals and asked if they had ever heard of Sandia. Nine out of ten replied: "Sandia? Who's she? Do you drink it or take it out?" Oh, yes, Number 10 said he already had a date for Saturday night. . . .

Sandia is in a class all by herself with a fiscal year budget for 1973 of $225 million that topped even prestigious LASL, who only got $108 million in federal monies (an additional $12 million was spent on work done for other federal agencies).

From Sandia's FY 1973 budget, a total of $196 million was committed to nuclear weapons programs, and an additional $1 million was spent on other unspecified "government" work. Its main responsibility is research and development of nuclear ordnance, which means the arming, fusing and firing systems used in nuclear bombs and warheads. Components used in these systems include power supplies and timing mechanisms. If this isn't enough, Sandia also designs bomb casings for the weapons which are dropped at random by bombers in search of "the enemy" (during wartime).

At the start of 1974, the authorized investment in plant and equipment at all Sandia locations was $331 million. The Livermore facility has operations for systems development and some component work, applied research, environmental testing, and model and material evaluation. In relation to the fusion program, a tritium gas handling facility has been built.

A wide range of facilities have been developed at Albuquerque to simulate the harsh environments specified for Sandia weapons designs. Located at the main installations are transonic and hypersonic wind tunnels for aerodynamic investigations; a plasma research facility which can produce temperatures of 20,000° Fahrenheit and air velocities up to 20,000 feet per second; a Van de Graaff accelerator and a Cobat-60, 1,000 curie source for generation of intense gamma radiation; and last, but not least, a Cockcroft-Watson accelerator which produces ions with energies from 30,000 to 250,000 electron volts.

In another test area, adjacent to the environmental testing facilities, two pulse reactors and a flash X-ray machine are used to study the effects of radiation on materials and components. In terms of neutron exposure rate, the Annular Core Pulse Reactor and the Sandia Pulsed Reactor II are the most powerful of their types in the world. An extremely powerful neodymium-glass laser is used for advanced physics research.

Anyway, I just wanted you-all to know that Sandia is more than just a pretty Mexican name or a gentle Spanish wine . . . a lot more.

Research, in general, has interesting spin-offs that are not limited to intellectual benefits for the scientific community. Eighteen years ago, for example, people were exhilarated if they could measure below a millionth of a second. Now, these same people are measuring to essentially a pico-second (one trillionth).

One offspring of Laser Fusion involves research related to a process of nature called photosynthesis. Mother Nature's feat of converting light to stored energy or chemical energy in plants is highly regarded but has yet to be duplicated in labs by humans. You may remember the basics from high school biology: When sunlight hits a leaf, the molecules in that plant must, first, absorb it, and then transfer it into the cell where the chemical processes of photosynthesis take place. The entire reaction takes place in a fraction of a second. You might say, in the twinkle of an eye.

Dr. Fenstermacher noted that scientists at LASL, aided by Laser Fusion technology, are busy studying this transfer to chlorophyl, the synthesis of chemical energy from sunlight. "This indeed could be one of the most important spin-offs of the Laser Fusion effort," he said. A lot of farmers would be mighty happy if someone does figure out how to duplicate photosynthesis in a laboratory, since it would shed a totally different light on their growing cycles.

My dazed reaction, after listening to experts from throughout the world, continues to be sheer wonderment at the complexity of a mind that dabbles in fusion and then comes up with a working model. Or, for that matter, how does one "invent" a Liquid Metal Fast Breeder Reactor? It's clearly not something you purchase in a hobby shop as a basement project to relieve sexual tension. . . .

Take an unassuming person like Charles Fenstermacher, who has a Ph.D. from Yale University. He's been at LASL since 1957 and is presently group leader for the Electric Discharge Gas Laser Section. Officially, this can be defined as research and development of high-energy, short-pulse carbon dioxide laser systems for laser fusion investigation. Unofficially, it means that the doctor is very bright. Yet, he took several hours from a busy schedule to brief me—a novice—on laser technology.

This was the attitude I encountered from other researchers who granted interviews and often on short notice. I found top-quality people who are not on any ego trip because, for them, there's no need to be. As in fusion, they've got it all together. Now, it's left for us to do the same.

Dr. John F. Clarke, director, Thermonuclear Division, Oak Ridge National Laboratory. (0212-74)

above: Fusion power, achieved by harnessing the energy released by the merging of the nuclei of light-weight elements, continues to be regarded as an ultimate energy source. Technicians here check the vertical magnetic field coils that surround ORMAK which, along with TOMAK, are the principal experimental devices for fusion research at the Oak Ridge National Laboratory, in East Tennessee. (4059-75)

right: Dr. William R. Bibb, chief of the Research and Development Branch at ERDA's Oak Ridge operations, and Dr. John F. Clarke, director of the Thermonuclear Division at Oak Ridge National Laboratory which is operated by the Union Carbide Corporation, Nuclear Division, for ERDA. The ORMAK injector is a section of the principal experimental device for research at Oak Ridge to harness the nuclear fusion reaction as a source of electrical energy. (62509)

The major toroidal theta-pinch facility—called Scyllac—at Los Alamos Scientific Laboratory has been completed and is in its initial experimental phase. Previously, the linear theta pinches demonstrated ignition temperature and classical radial diffusion, but the confinement of the hot plasma was severely limited by plasma flowing out the ends.

Technicians give the Scyllac device at Los Alamos Scientific Lab, high in the hills of New Mexico, a recheck following runs to obtain the proper plasma conditions for wall stabilization of motion destructive to plasma confinement essential for the fusion process.

left and below: The Tokamak Fusion Test Reactor Concept was the subject of intense study during 1974 by two groups, one at Oak Ridge and the other at Princeton Plasma Physics Laboratory, with assistance from the staff of the Westinghouse Electric Corporation. In this photograph, ORMAK employs the doughnut-shaped plasma region as a means of confinement and these coils—as shown in the superimposed photograph—are surrounded by an aluminum shell. The Princeton Large Torus (Photo 2) makes use of impurities which result from the sputtering of particles from the walls of the confinement device and which disrupt the fusion process.

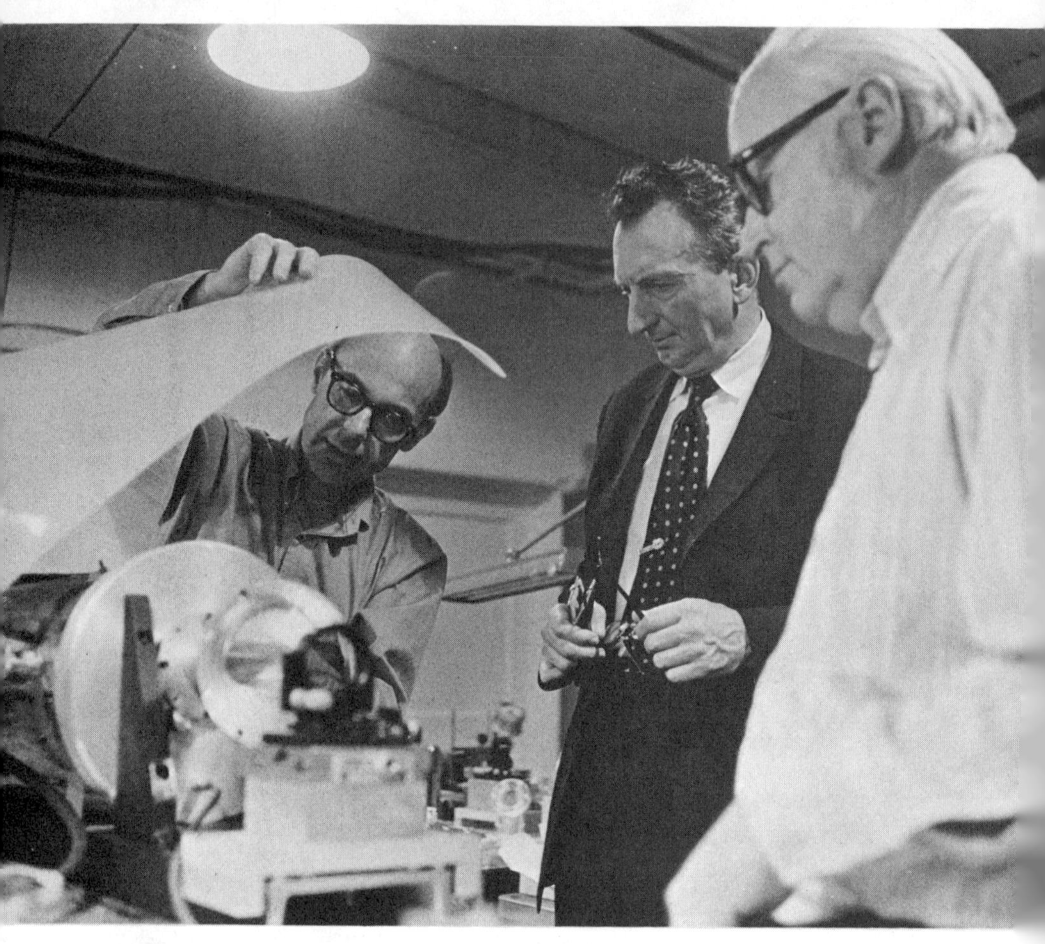

above: Dr. Charles Fenstermacher (left), group leader for the Electric Discharge Gas Laser Group at the Los Alamos Scientific Laboratory, shares with two fellow scientists and Nobel Prize Winners, Dr. Alexander M. Prochorov, a Soviet, and Dr. Keith Boyer, research and development aspects of the high energy, short pulsed, carbon dioxide laser systems.

above opposite: A 2500 Joule—two beam—carbon dioxide laser system for laser fusion under study at the Los Alamos Scientific Laboratory in New Mexico.

below opposite: The Adiabatic Toroidal Compressor at the Princeton Plasma Physics Laboratory has demonstrated the concept of heating a tokamak plasma by compression in major and minor radius. It successfully demonstrated the effectiveness of neutral beam heating at modest levels (50% ion temperature increases) and has achieved the densities needed for a practical tokamak reactor.

16
15
14

		LIFE EXPECTANCY OF KNOWN RESERVES (YEARS) At .170	At 2.80	LIFE EXPECTANCY OF POTENTIAL RESERVES (YEARS) At .170	At 2.80	LIFE EXPECTANCY OF TOTAL RESERVES (YEARS) At .170	At 2.80
los alamos scientific laboratory of the University of California FINITE ENERGY SOURCES	FOSSIL FUELS (COAL, OIL, GAS)	132	8	2,700	165	2,832	173
	MORE ACCESSIBLE FISSION FUELS (URANIUM AT $5 TO $30/LB OF U_3O_8 BURNED AT 1.5% EFFICIENCY)	66	4	66	4	132	8
	LESS ACCESSIBLE FISSION FUELS (URANIUM AT $30 TO $500/LB OF U_3O_8 BURNED AT 1.5% EFFICIENCY)	43,000	2,600	129,000	7,800	172,000	10,400
"INFINITE" NATURAL ENERGY SOURCES	WATER, TIDAL, GEO-THERMAL & WIND POWER	Insufficient		Insufficient		Insufficient	
	SOLAR RADIATION	10×10^9	10×10^9			10×10^9	10×10^9
	FUSION FUELS (DEUTERIUM FROM OCEAN)	45×10^9	2.7×10^9			45×10^9	2.7×10^9
"INFINITE" ARTIFICIAL ENERGY SOURCES (ELEMENTS TRANSMUTED FROM OTHER ELEMENTS BY NEUTRON BOMBARDMENT)	FISSION FUELS (PLUTONIUM 239 FROM URANIUM 238; URANIUM 233 FROM THORIUM 232)	8.8×10^6	536,000	21×10^6	1.3×10^6	30×10^6	1.8×10^6
	FUSION FUELS (TRITIUM FROM LITHIUM a) ON LAND b) IN OCEAN	48,000 120×10^6	2,900 7.3×10^6	UNKNOWN		48,000+ 120×10^6	2,900+ 7.3×10^6

A partial illustration of the full range of energy reserves and their known life expectancy. In regard to Laser Fusion and Thermonuclear Fusion, it details the abundance and cheapness in terms of raw materials and the amount of energy that can be produced.

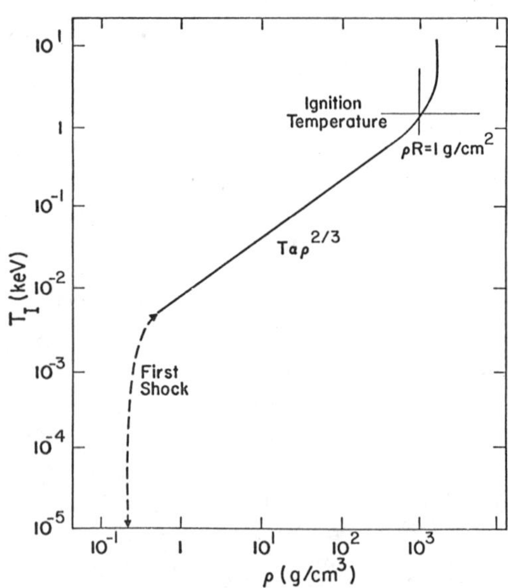

An interesting illustration of the approximate adiabatic compression of Deuterium-Tritium fuel by an increasing pressure, to a point satisfying ignition and burn criteria.

Lawson Criteria – Laser Fusion

$$n\tau = 10^{14}\ \text{cm}^{-3}\ \text{sec}$$

becomes, in other, more convenient units

$$n \rightarrow \rho\ \text{g/cm}^3$$

$$\tau \rightarrow \frac{R}{v_{shock}}\ \text{(hydrodynamic disassembly time)}$$

or $n\tau = \text{constant}$

becomes $\rho R = \text{constant}\ (\sim 0.3\ \text{g/cm}^2)$

The "Lawson Criterion" is expressed here as the product n τ where n is the plasma density (in particles per cubic centimeter) and τ is time (in seconds) for which the plasma of that density can be confined by the magnetic field. Note: For a practical fusion reactor, n τ must exceed about 10^{14} for the Deuterium-Tritium Reaction and about 10^{16} for the Deuterium-Deuterium Reactions. Since the "Lawson Criterion" is a product of particle density and confinement time, there is a range of conditions over which it can be satisfied. For example, if the density is at the lower practical limit of 10^{14} particles per cubic centimeter, the confinement time for the charged particles in a deuterium-tritium system must exceed one second. However, in the vicinity of the upper limit of about 10^{16} particles per cubic centimeter, the confinement time would have only to be longer than 0.01 (i.e., one hundredth part) of a second.

Lawson Criteria

Energy Produced = Energy to heat + Energy losses + Useful Energy

leads to $n\tau \sim 10^{14}\ \text{cm}^{-3}\ \text{sec}$

In addition to the specifications of temperature and plasma particle density, a fusion reactor must satisfy another condition. This is known as the "Lawson Criterion" because it was first pointed out by J. D. Lawson in England in 1957. It is based on the requirement that in a self-sustaining system, the reacting nuclei must be confined long enough to produce sufficient recoverable energy by fusion to compensate for the amount supplied initially to heat the plasma.

An illustration of laser requirements for generating energy to heat a mass of Deuterium-Tritium. The major and immediate problem tied with laser fusion is concerned with the development of lasers capable of repeatedly delivering large amounts of energy in very short and properly shaped pulses. A solution does seem to lie within the realm of technical possibility. Neutrons produced in a Deuterium-Tritium Reaction carry some 80 per cent of the fusion energy and are taken up by lithium. From this point on, the basic scientific points are the same as for a fusion reactor with toroidal magnetic confinement.

Laser Requirements

los alamos
scientific laboratory
of the University of California

Energy to heat a mass of D-T

$$E_L = C_p M \Delta T$$

and $$M = \frac{4\pi}{3}\rho R^3$$

so that $E_L \propto \rho R^3$

and under the constraint

$$\rho R = \text{constant}$$

$$E_L \propto \frac{(\rho R)^3}{\rho^2} = \frac{\text{constant}}{\rho^2}$$

at $\rho = \rho_0$ (normal liquid density)

$$E_{L_0} \sim 10^9 \text{ joules}$$

$$\tau_0 \lesssim 10^{-9} \text{ sec (one nanosecond)}$$

$$\text{Power} \sim 10^{18} \text{ watts}$$

if $$\frac{\rho}{\rho_0} = 10^2$$

$$\frac{E_L}{E_{L_0}} \sim 10^5 \text{ joules}$$

below opposite: This is a concept—not an engineering design—of how Laser Fusion might be applied to power generation. The pellet is inserted into a reaction chamber and the laser then fired in. In the subsequent reaction, it is necessary to capture the energy produced, energy that is all in the form of neutrons. Thus far, the best way of catching these fast neutrons is by utilizing lithium which converts neutron energy into heat. This lithium flows through a heat exchanger which goes into steam, and the power is taken out as these other cycles are ended. The lithium might be introduced into the reaction chamber as a film of liquid covering the interior surface or as a swirl of droplets (or even both). The energy of the neutrons would be left in the lithium and at the same time tritium would be generated for subsequent use in the fusion reaction.

In this process the laser pings and evaporates the outside of a pellet, causing the particles to expand and thus escape. In so doing they go in opposite directions: gas escapes out one way and gas escapes inward, causing compression. This Figure illustrates compression of a fuel by the reaction and thermal pressure of hot plasma generated by the absorption of laser light.

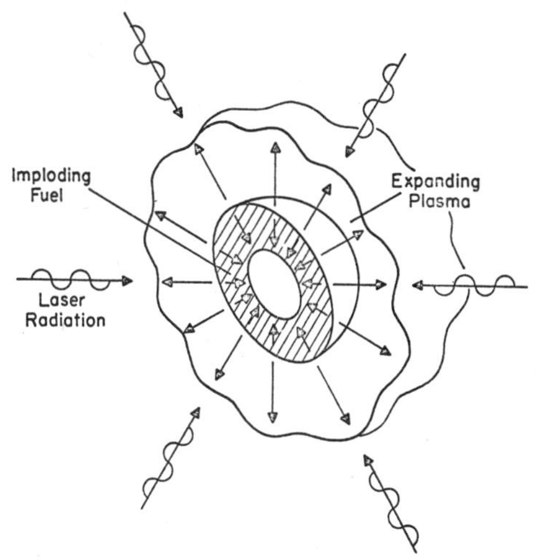

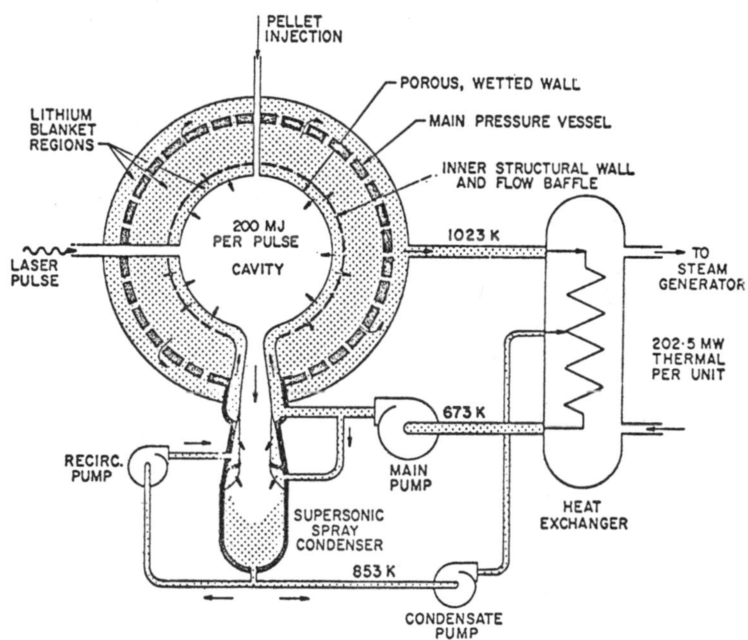

Lasers are a priority item and can be utilized in varied ways. This illustration details "Laser Amplifier Essentials" based on a two-level Quantum System that may be either atomic or molecular in structure.

Laser Amplifier Essentials

Consider a two-level Quantum System (atomic or molecular.)

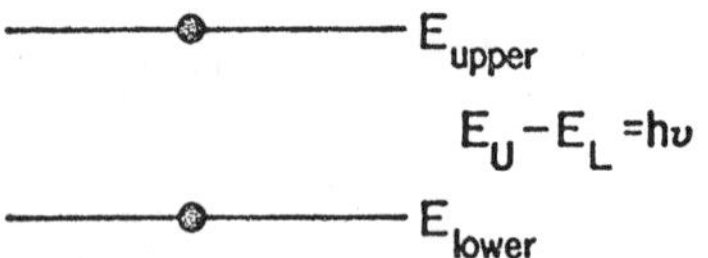

With optical transitions allowed, there are two possible processes under irradiation by photons of frequency ν:

I. $E_L + h\nu \longrightarrow E_U$ (absorption)

II. $E_U + h\nu \longrightarrow E_L + 2h\nu$ (stimulated emission)

In a laser amplifier,

1.) the upper state is prepared by pumping—optically, electrically, or chemically,—to produce a population inversion, i.e. $N_U > N_L$

2.) an incoming photon pulse is amplified through process II

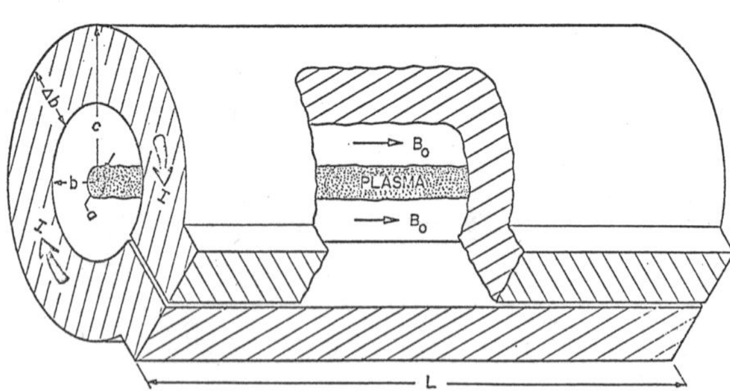

Linear Theta Pinch Geometry as utilized at Los Alamos Scientific Laboratory and dealing with Fusion Reactor Applications of the High Density Linear Theta Pinch.

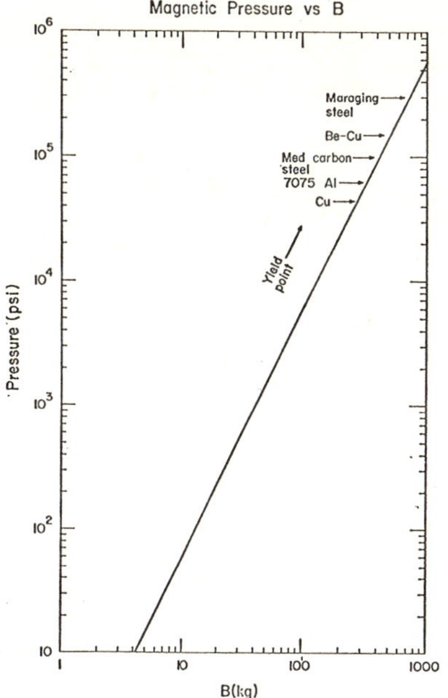

Magnetic Pressure vs. Magnetic Field Strength in regard to High Density Linear Theta Pinch at Los Alamos.

Arrangement of the dual energy supply for the staged Theta Pinch Research Experiment at LASL. These concepts were presented at the IV National School on Plasma Physics (1974) at Novosibirsk, USSR.

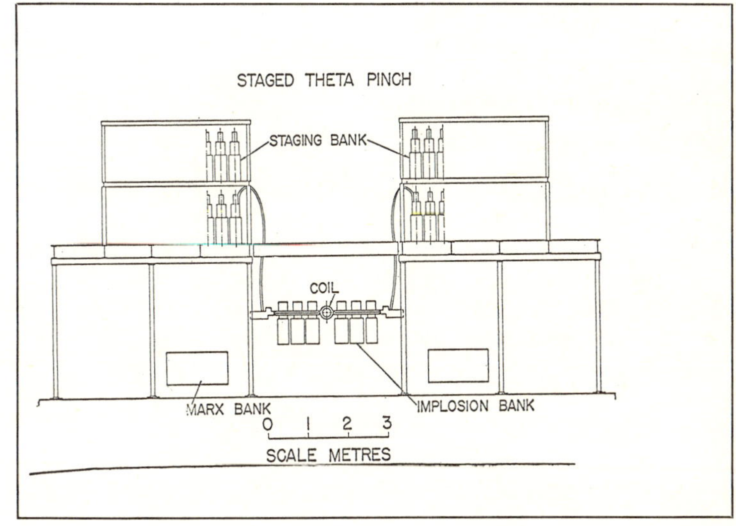

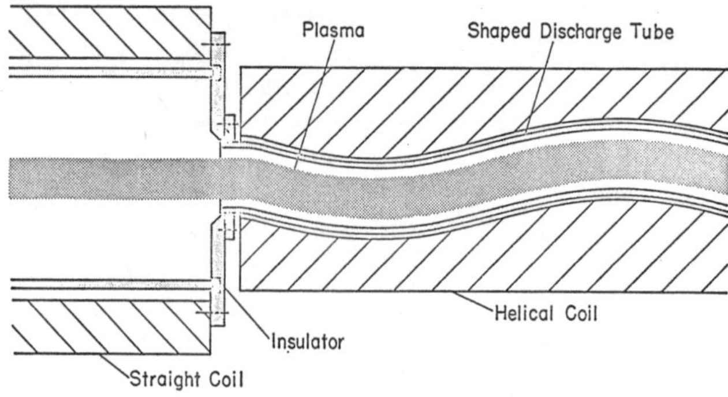

Arrangement for the dual energy supply for the staged Theta-Pinch Experiment at Los Alamos.

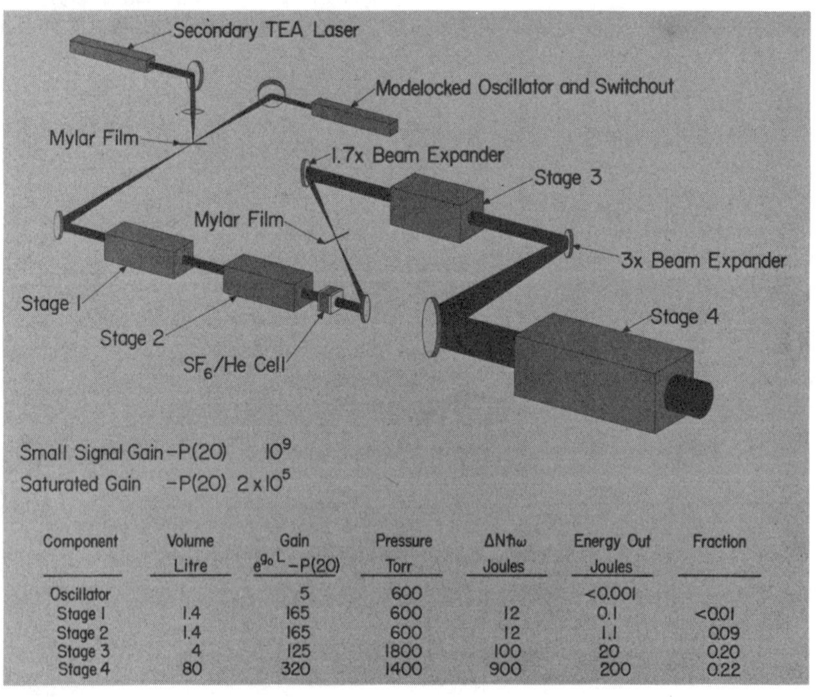

Component	Volume Litre	Gain $e^{g_0 L}$ – P(20)	Pressure Torr	$\Delta N \hbar \omega$ Joules	Energy Out Joules	Fraction
Oscillator		5	600		<0.001	
Stage I	1.4	165	600	12	0.1	<0.01
Stage 2	1.4	165	600	12	1.1	0.09
Stage 3	4	125	1800	100	20	0.20
Stage 4	80	320	1400	900	200	0.22

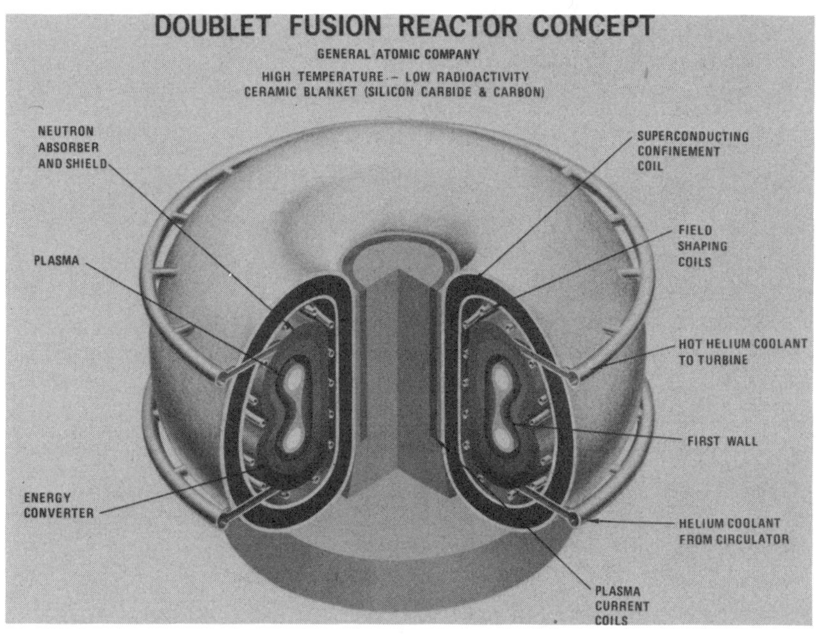

1. EFFECTIVELY INFINITE FUEL SUPPLY AT LOW COST (<<1 mill/kwhr)
2. INHERENT SAFETY, NO RUNAWAY
3. NO CHEMICAL COMBUSTION PRODUCTS
4. RELATIVELY LOW RADIOACTIVITY AND ATTENDANT HAZARDS
5. NO EMERGENCY CORE COOLING PROBLEM
6. NO USE OF WEAPONS GRADE MATERIALS SO NO DIVERSION POSSIBILITY

Advantage of a fusion reactor

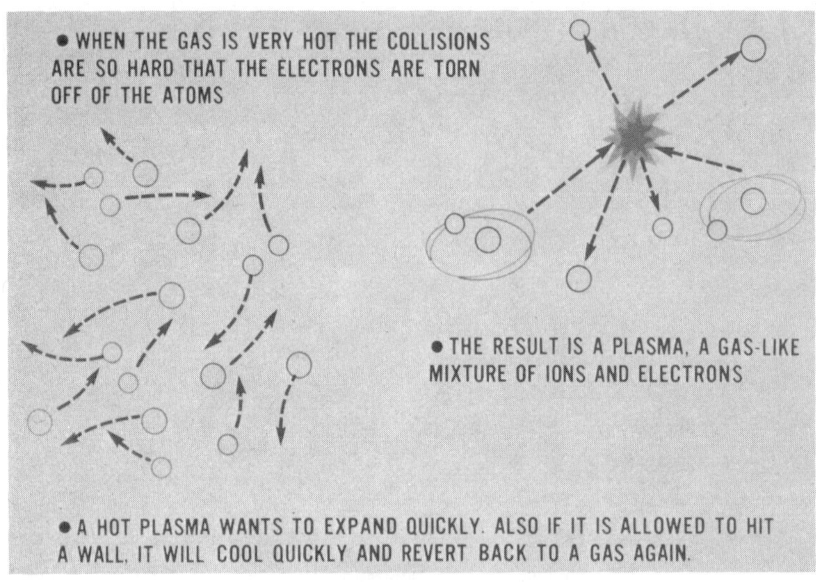
• WHEN THE GAS IS VERY HOT THE COLLISIONS ARE SO HARD THAT THE ELECTRONS ARE TORN OFF OF THE ATOMS
• THE RESULT IS A PLASMA, A GAS-LIKE MIXTURE OF IONS AND ELECTRONS
• A HOT PLASMA WANTS TO EXPAND QUICKLY. ALSO IF IT IS ALLOWED TO HIT A WALL, IT WILL COOL QUICKLY AND REVERT BACK TO A GAS AGAIN.

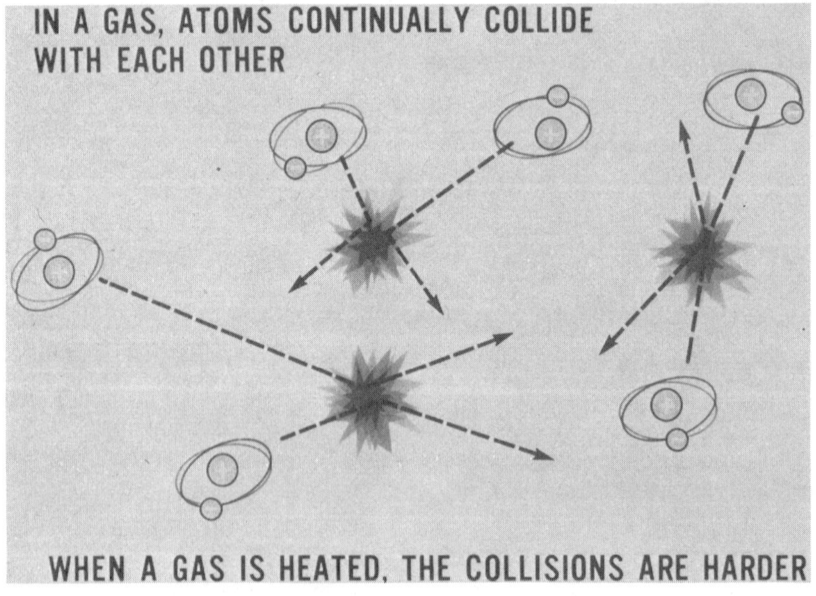
IN A GAS, ATOMS CONTINUALLY COLLIDE WITH EACH OTHER
WHEN A GAS IS HEATED, THE COLLISIONS ARE HARDER

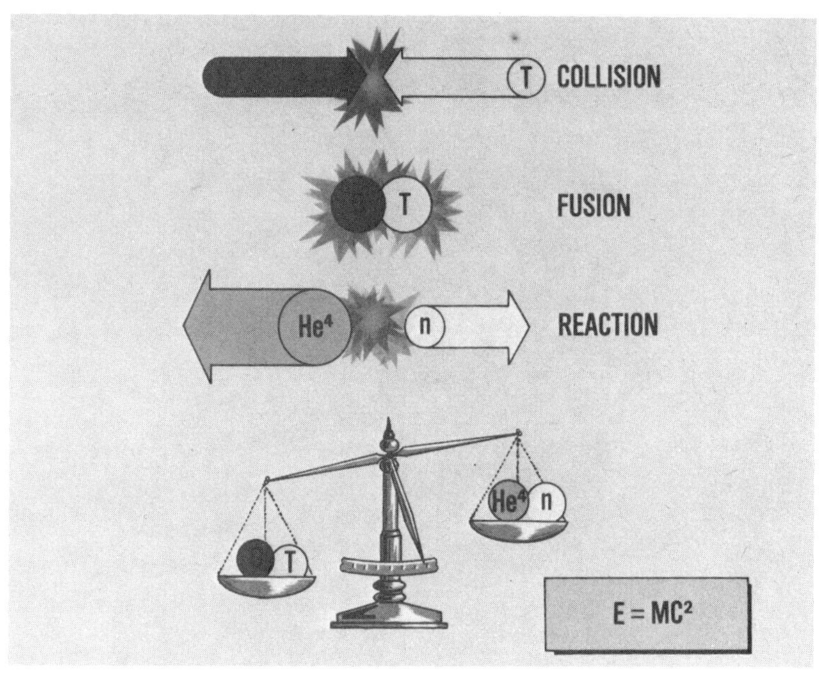
T
COLLISION
T
FUSION
He⁴
n
REACTION
He⁴
n
T
E = MC²

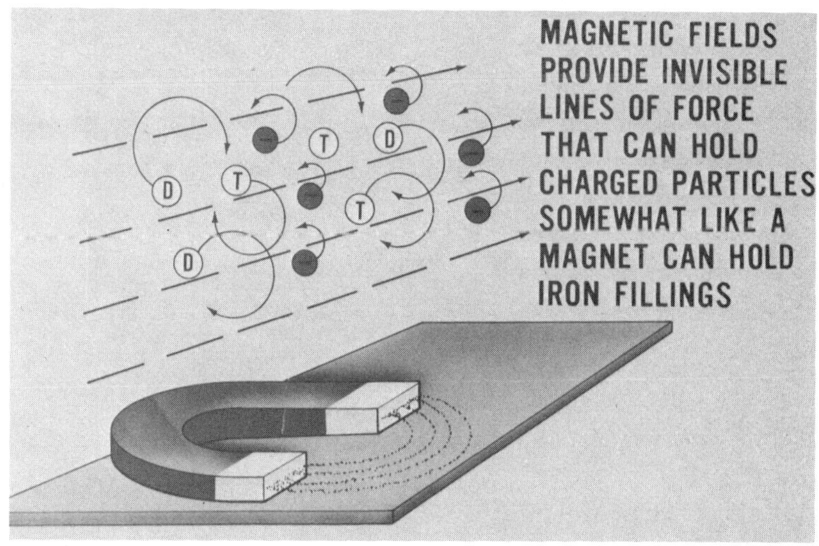
MAGNETIC FIELDS
PROVIDE INVISIBLE
LINES OF FORCE
THAT CAN HOLD
CHARGED PARTICLES
SOMEWHAT LIKE A
MAGNET CAN HOLD
IRON FILLINGS
T
D
D
T
T
D

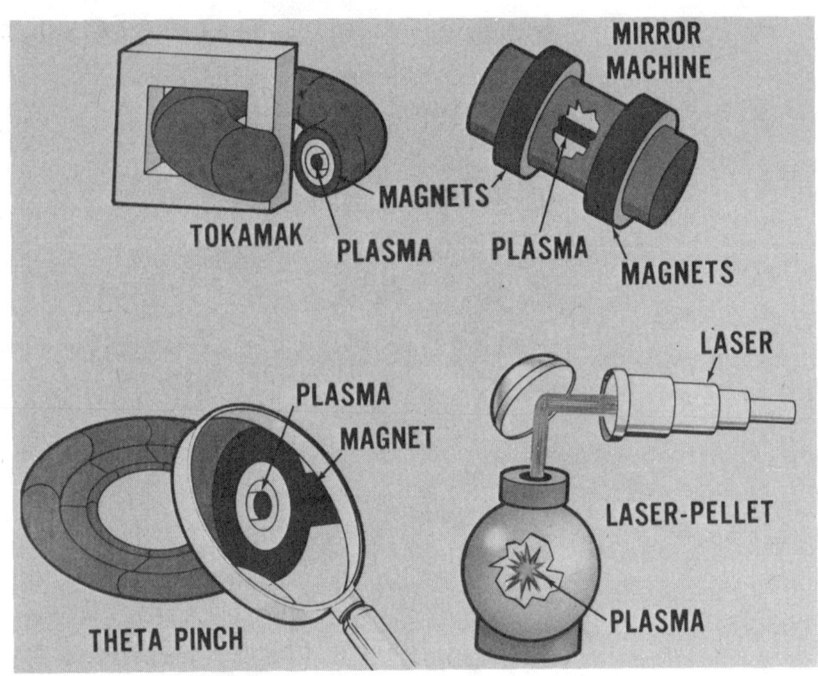

Research of various types of fusion devices

		DATE ACHIEVED
DENSITY:	GREATER THAN 10^{13} cm^{-3}	1953
TEMPERATURE:	GREATER THAN 5 keV $(50{,}000{,}000^{0})$	1962
CONFINEMENT TIME:	GREATER THAN $\left(\frac{10^{13}}{\text{DENSITY}}\right)$	1969 FOR NEAR CLASSICAL CONFINEMENT

GOAL: ALL THREE TOGETHER

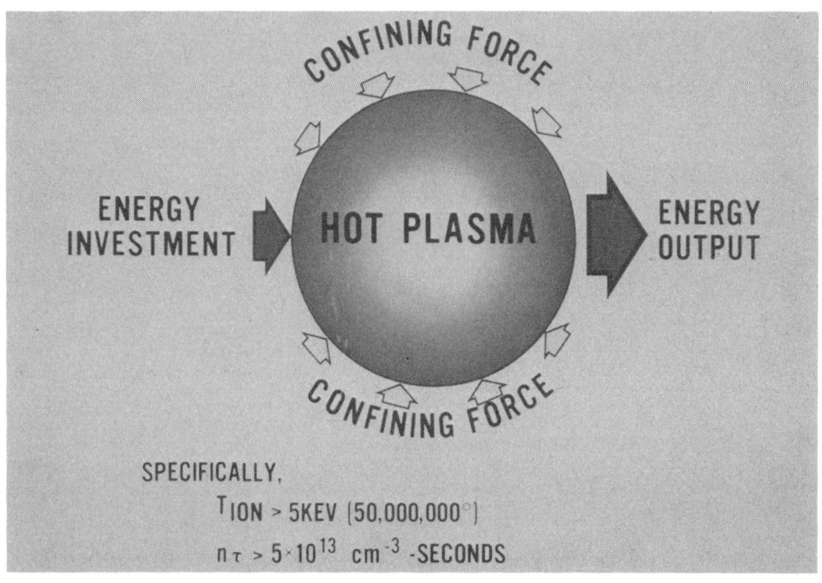
CONFINING FORCE
ENERGY INVESTMENT
HOT PLASMA
ENERGY OUTPUT
CONFINING FORCE
SPECIFICALLY,
TION > 5KEV (50,000,000°)
nτ > 5×10^13 cm^-3 -SECONDS

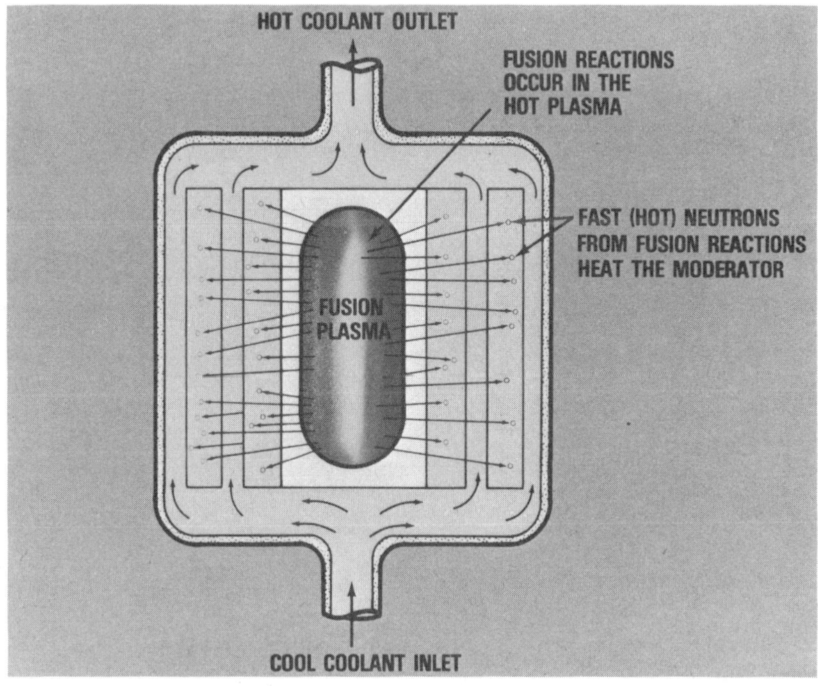
HOT COOLANT OUTLET
FUSION REACTIONS OCCUR IN THE HOT PLASMA
FAST (HOT) NEUTRONS FROM FUSION REACTIONS HEAT THE MODERATOR
FUSION PLASMA
COOL COOLANT INLET

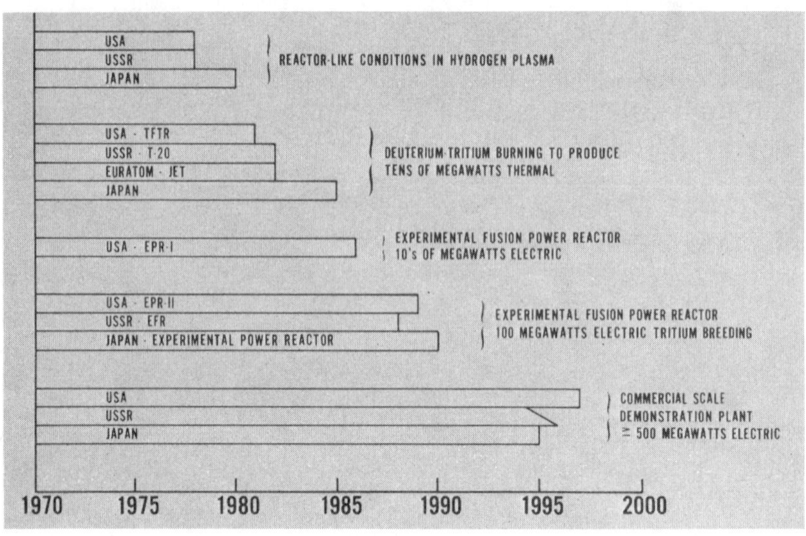
USA
USSR
JAPAN
REACTOR-LIKE CONDITIONS IN HYDROGEN PLASMA
USA - TFTR
USSR - T-20
EURATOM - JET
JAPAN
DEUTERIUM-TRITIUM BURNING TO PRODUCE
TENS OF MEGAWATTS THERMAL
USA - EPR-I
EXPERIMENTAL FUSION POWER REACTOR
10's OF MEGAWATTS ELECTRIC
USA - EPR-II
USSR - EFR
JAPAN - EXPERIMENTAL POWER REACTOR
EXPERIMENTAL FUSION POWER REACTOR
100 MEGAWATTS ELECTRIC TRITIUM BREEDING
USA
USSR
JAPAN
COMMERCIAL SCALE
DEMONSTRATION PLANT
≥ 500 MEGAWATTS ELECTRIC
1970
1975
1980
1985
1990
1995
2000

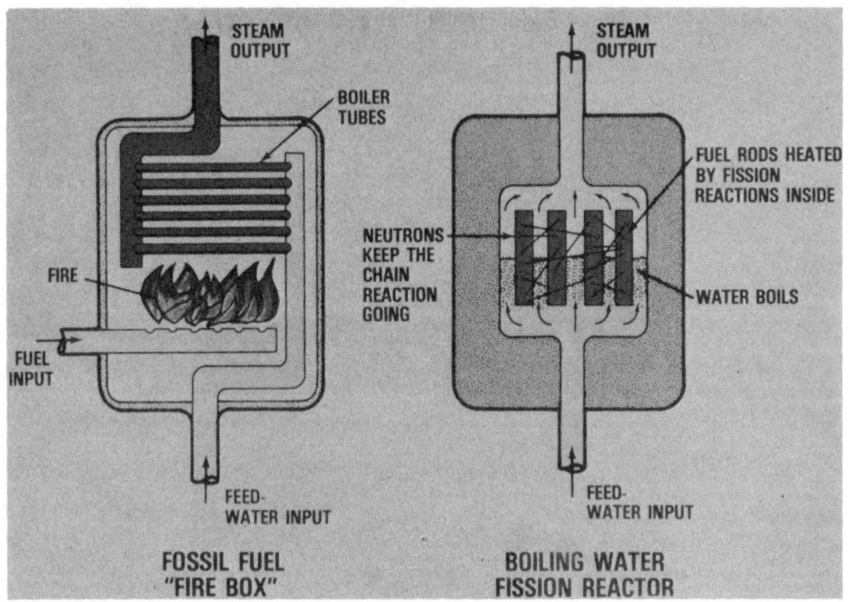
STEAM
OUTPUT
BOILER
TUBES
FIRE
FUEL
INPUT
FEED-
WATER INPUT
FOSSIL FUEL
"FIRE BOX"
STEAM
OUTPUT
FUEL RODS HEATED
BY FISSION
REACTIONS INSIDE
NEUTRONS
KEEP THE
CHAIN
REACTION
GOING
WATER BOILS
FEED-
WATER INPUT
BOILING WATER
FISSION REACTOR

High Energy Gas Laser Facility
los alamos
scientific laboratory

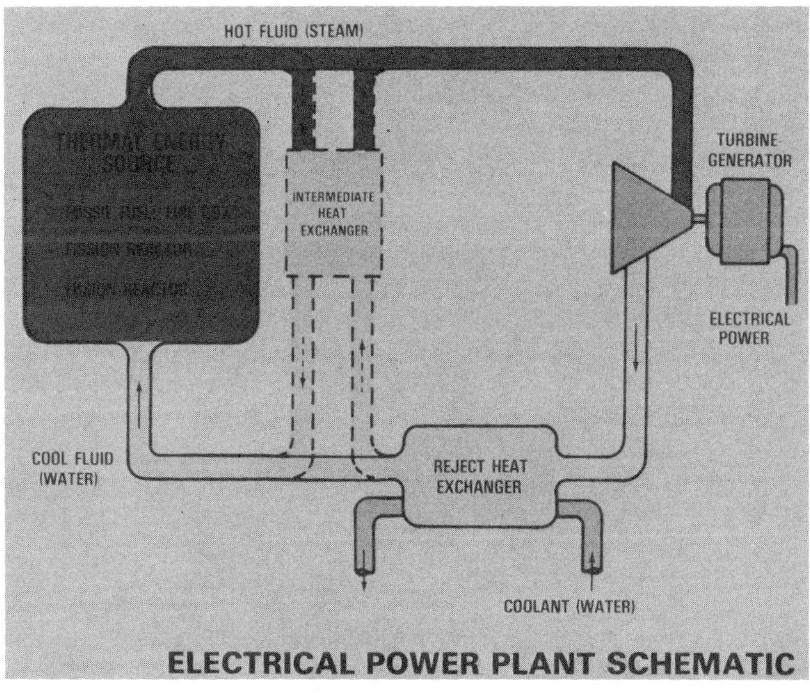
HOT FLUID (STEAM)
THERMAL ENERGY SOURCE
INTERMEDIATE HEAT EXCHANGER
TURBINE-GENERATOR
ELECTRICAL POWER
COOL FLUID (WATER)
REJECT HEAT EXCHANGER
COOLANT (WATER)
ELECTRICAL POWER PLANT SCHEMATIC

Charged particles repel each other with an invisible force that acts like a spring to try to keep them apart.

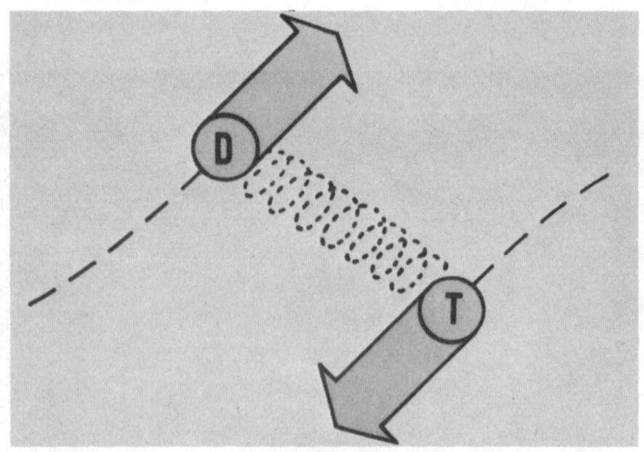

Charged particles with high energy (= Velocity = Temperature) can overcome the repulsion and fuse.

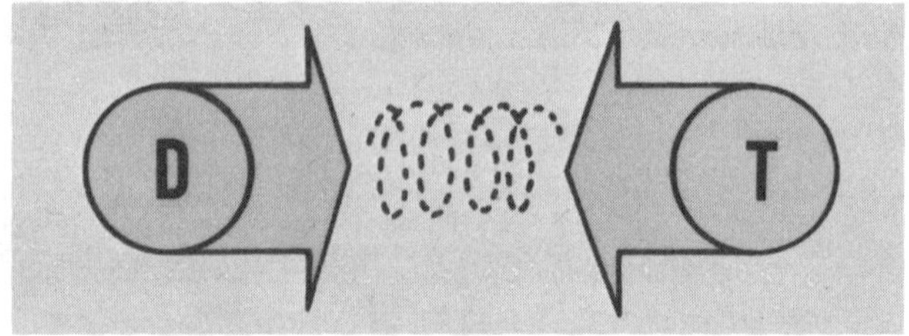

The Tokamak method holds the biggest slice of the fusion research pie

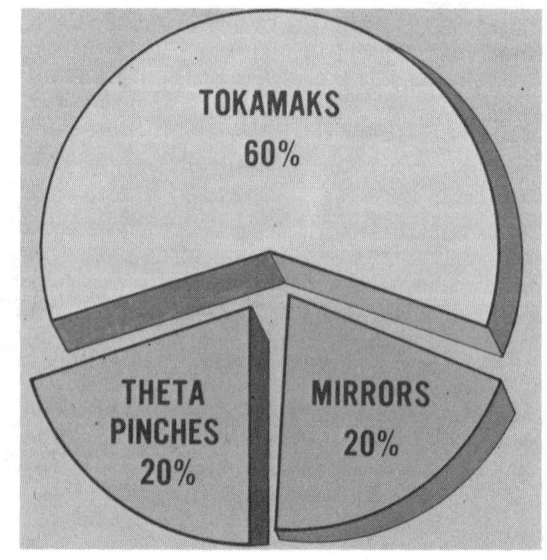

Chapter 13

"SILKWOOD, KAREN GAY: FORMER KERR-McGEE EMPLOYEE"

Not long ago, on national television broadcast live from Oklahoma City, comedian Bob Hope quipped that in this part of America, you can tell what a fella does by looking at his boots. . . .

The man from Hollywood meant oil, and the audience, most of whom owed their life to this industry, caught the tempo and screamed their approval. However, there's another empire here that is held in the same reverence, except that nobody makes fun of Kerr-McGee, because the Kerr-McGee Nuclear Corporation seemingly owns Oklahoma, lock, stock, and barrel.

Granted, that's an observation by one outsider. But you do get that distinct impression when driving through all that flat land, dotted with oil wells and nice blue and white signs that proclaim the K-M trademark. Then, there's that dazzling, square-block building in downtown Oklahoma City which says pretty clearly—thirty stories high—that here are folks doing something awfully right, or awfully wrong in the right way. . . .

No criticism intended. There's nothing wrong in making a little money for yourself, and most Okies state right-off that the state is mighty grateful to Kerr-McGee for its economic contribution. Even the Arabs have yet to make Kerr-McGee an offer they can't refuse.

In 1946, the Atomic Energy Commission (AEC) was created to

supervise mankind's dream of "harnassing the atom for peaceful purposes," and this took the direct form of building reactors to generate electricity. Soon, a multi-billion dollar industry began to take shape. There was wealth to be had from uranium mining and processing, not to mention reactor manufacturing with its myriad of components, ranging from special wiring to 750-ton containment vessels.

Leading the team was Kerr-McGee, an Oklahoma-based company which had been founded in 1929 to drill for oil. The company was a pleasant blend of two gifted men, the late Senator Robert S. Kerr, "the uncrowned king of the Senate," and Dean A. McGee, often referred to as one of the most technically brilliant and visionary men in the energy industry. Here was a rare combination of political clout and look-aheadedness.

Testing the winds of change, K-M moved very naturally into the exploration, mining and processing of uranium, first for the nuclear weapons program and then for reactors. It seemed the gold of the future: an unpretentious metal buried mostly in isolated pockets under Western deserts, and the oil companies were in on the ground floor. Man, what a time to be alive, especially if you were a Kerr or a McGee.

Uranium. This ninety-second element on the Periodic Chart of Elements accounts for a high percentage of the business conducted by K-M Nuclear Corporation at both its Sequoyah facility near Gore, Oklahoma, and Cimarron near Crescent. In addition, the firm is actively mining uranium near Grants, New Mexico, and in the Powder River area of Wyoming.

As uranium occurs in nature, it has three isotopes, that is, atoms of different atomic weights. The most common isotope is U-238, which accounts for 99.274 percent. The other two isotopes, U-235 and U-234, make up .72 percent and .006 percent respectively. For the nuclear process, the U-235 is the isotope needed, since it is fissionable (can be split). In order to increase the amount of U-235 in the uranium, the yellow cake substance must be sent to an enrichment plant and the federal government is the only agency that has these—there are three in the United States. There, the U-235 isotope gets the royal treatment and promptly, with a little urging, increases to between 2 and 3 percent for use in reactors.

The enriched uranium is then shipped from these government plants in the form of uranium hexafluoride, a gas, in two-and-one-half ton cylindrical containers. These containers are then encased in what are termed "bird cages" which are insulated and designed to withstand tremendous forces. Once the cylinders arrive at their destination, they are checked and weighed and then placed into steam chests in order to vaporize the uranium hexafluoride. The gas will solidify at below room temperatures. As this gas is pumped, it reacts with ammonium hydroxide in the form of ammonia water to form a precipitate of ammonium diurinate. The ammonium water containing ammonium diurinate is then sent through a centrifuge, where the solid formed in the reaction is separated from the solution. This solid is then dried in a twenty-foot long tunnel and comes out as a yellow powder. The powder is fed into a calciner where it is mixed with steam and ammonia to form a fine uranium oxide powder.

This uranium oxide is stored in safe-geometry containers—in this case, cylinders with a diameter of ten inches (the cylinder can be any length as long as the diameter remains the same). In such an enriched state, this material has the potential of reaching critical mass (capable of nuclear reaction) unless such adequate safety measures are taken. A lot of uranium is composed of thirty-two of these cylinders which measure about two feet long. The cans are then placed on a large machine in groups of eight, where the uranium oxide can be mixed so that it has about the same makeup in all eight cans. The powder is then stored in "donuts" or large concrete dollars to provide protection between the cans so they will not heat. In addition to this process, each lot is carefully checked for metallic and non-metallic impurities.

Once the uranium oxide is mixed in the blender, the powder then goes to the pelleting operation. This means that it is placed in a mill and then put through various feeders to three hungry presses. These compress the powder into a pellet that is ⅜-inch in diameter and ¾-inch long and weighs between 10 and 13 grams. The press compresses the powder with 40 tons per square inch. As a safety precaution, each press is enclosed to prevent dust from escaping, but the pellets, safe enough to handle in their new state, are open to the atmosphere as they come off the press.

These tiny granules are then placed in "furnace boats" which are actually molydenum pans. These "boats" are put through a furnace at 1,700° Centigrade—a temperature too hot for steel, which would melt. This furnace provides what is called the high density sintering process. The pellets have a density of 6 grams per cubic centimeter before the process and about 10.5 grams per cubic centimeter after it.

The furnace is about thirty feet long and fully enclosed. It contains an atmosphere of ammonia with some hydrogen.

After completing this sintering process, the pellets are then ground to their exact dimensions. The contract does not require measuring every pellet, but quality control personnel check sample pellets from each lot for size, chips in them or other impurities.

After this grinding operation, the pellets are placed on a miniature aluminum I-beam that is twelve feet long, the same as the fuel rod which they are made to serve. The pellets and I-beam are weighed to make certain that each group of pellets weighs between 2,505 and 2,551 grams. They are then encapsulated between two layers of plastic on a special machine that has been patented by Kerr-McGee for just this purpose. The sheets are then placed in a box that is less than four inches thick and is another geometrically safe container. The boxes are laid beside each other and are never stacked one on top of the other. The pellets or powder could be placed in a slab four inches thick and as long and wide as you wanted without affecting the uranium oxide, but they are particular little critters and do not like to be stacked on top of each other. This you don't do, otherwise, it could get very messy because a reaction occurs.

Each box weighs approximately 25 kilograms and contains ten sheets of pellets. The boxes are then placed in drums that look like two, 55-gallon drums welded together. Each drum contains four boxes of pellets and one trailer load has seventy drums on it. These are well insulated to protect the pellets from a fire and are tested to insure they won't rupture after being dropped from a height of say, thirty feet. At this stage, the uranium can be handled with the human hand, although workers at the plant are required to wear gloves. Face masks are required to be worn only at the blender where large amounts of airborne uranium oxide are likely.

Until these uranium oxide pellets are actually placed in a reactor and begin the process of rapid nuclear decay, they will not harm a person. All of the ammonia water used in the process is recycled and goes through an ion-exchange process to remove excess uranium still in the solution. The ammonia water then goes through a "still" for reclamation of the ammonia for re-use in the cycle. None of the water is ever discharged into outside bodies of water.

Security measures in the uranium process plant differ from those at the plutonium plant which requires stricter control. In order for the uranium to be worth stealing, it would need to be enriched beyond 20 percent of the U-235 isotope, and the Cimarron facility, for example, doesn't use that kind of uranium oxide. In addition, someone would have to steal a great deal of the pellets before he or she would have enough fissionable material for a bomb, and then they would be faced with the problem of having an amount too unwieldly to be of much use.

The entire process in the uranium plant is monitored through a central control room. If there is a failure in one of the systems, an alarm automatically sounds. The chances of a fire in the concrete and metal building are slight, and precautions have been taken to prevent any such occurrence. In addition, there is a physical health team which monitors all working conditions, and all personnel are not knowingly exposed to more than the maximum permissible concentrations of uranium.

In the 1940s, plutonium was first discovered among the waste products of fissioned uranium, and, because it barely exists in nature, our present supply of plutonium is almost entirely man-made.

When planning began in the late sixties on the Liquid Metal Fast Breeder Reactor, Kerr-McGee contracted to produce plutonium fuel pins for the experimental forerunner of the LMFBR, the Fast Flux Test Facility near Richland, Washington.

In 1970, K-M put up a plutonium processing plant alongside its uranium factory near Crescent, Oklahoma, a nice thirty mile drive from Oklahoma City. The two completely separate installations are within five miles of ninety-two gas and oil wells, two popular resort lakes, and the churning Cimarron River from which the facility takes its name. There is only a chain-link fence and a handful of armed guards, yawning in the Western sun, to protect the glaring barnlike structures.

Commenting on this, a federal agent speculated that a small amphibious assault force just might be able to take the complex if they came in from the river. "I've studied it. There's no other way," he muttered, and then smiled. "Still, it's like the dog chasing a car. What the hell does he do if he catches one!"

Since Cimarron opened, Kerr-McGee has managed to mine and mill tons of yellow-cake uranium, in addition to acquiring 800,000 acres of uranium leases—a definite corner on the market. With assets approaching a billion dollars, it is now the nation's largest uranium producer. Dun & Bradstreet recently named it among the five best-managed corporations in the country. And it was this same company, apparently on good terms with the AEC since Kerr's old congressional days, which was awarded a $1.4 million contract to process plutonium into pellets and pour them into fuel rods.

Nevertheless, it took one small, frightened girl, working at this plutonium processing plant, to toss a pebble that has forever—right or wrong—smeared the white cloak of Kerr-McGee a very ugly red. The blemish has been scoured, but the world still wonders what really happened there during the late fall of 1974.

While nuclear industry officials are not too happy about it, Karen Gay Silkwood, who was a $4-an-hour technician, has probably created more waves by her strange death than during the entire twenty-eight years she spent on this planet. The girl who cried "wolf" has become the "anti-nuclear forces' first tailor-made martyr."

The Silkwood story smacks of melodrama and sensationalism. It's an editor's dream because there are so many unresolved and quite possibly, unresolvable questions about her life and death: such tiny "radioactive" morsels like her charges of unsafe working conditions and falsified records at the processing plant where she worked until that dark night in November, when her tiny Honda Civic crashed off a two-lane road into a drainage ditch.

Adding to this mystery, Karen Gay Silkwood died enroute to Oklahoma City to meet a reporter from the *New York Times* and an official from the Oil, Chemical, and Atomic Workers Union (OCAW). Her interview, anxiously awaited, was to officially document alleged health and safety problems at the Cimarron Facility, charges that had triggered an AEC investigation. The large manilla envelope, supposedly containing documents to verify these claims, was never recovered from the crash scene. . . . If indeed, the folder ever existed.

Karen, the gnat, was beginning to buzz until someone decided it was time to swat her permanently. If her accident was engineered, the question is by whom? There appear to be three distinct possibilities.

Kerr-McGee has been an obvious and immediate choice to many. And that's the problem: it's too obvious, almost intentionally colored in that direction. You can't discount the fact that this company—if they are as smart as their profits show them to be—would have better sense than to kill a pesky technician who was beginning to draw flies. Let her blow out naturally, they might have reasoned, rather than blow her out and thereby create a massive boomerang that would go straight back to the home office. That's not very sound business judgment. Besides, just what could Karen Silkwood, alive, really do? Shut down Kerr-McGee? Dream on, or better still, drive through Oklahoma and see how powerful this company really is.

Death, however, can be a great equalizer of impossible situations. Hypothetically, it is possible that the union might have realized that it stood to gain leverage with its very own martyr, "killed" by sinister Establishment forces. Especially since the Eleventh Hour was at hand and the union had simply not been able to deliver the promised goods that would put the screw to Kerr-McGee. Was a sacrifice then in order and Karen made the unknowing victim? It is doubtful that this girl would have known about her scheduled execution. Perhaps, she was told she would merely be forced off the road, and that's how she should report that the folder was snatched from her.

Then, there is the final possibility that her accident really *was* an accident. Maybe, as the Oklahoma Highway Patrol states, Karen fell asleep at the wheel and drove off the road. Blood tests did reveal traces of alcohol and therapeutic levels of methaqualone, a sedative. Pathologists, however, still dispute whether the amount involved would have caused her to doze, especially in view of where she was headed.

Only hours after the wreck, a radio report immediately called her the "unwitting catalyst in a raging controversy that challenges the ability of both industry and the government to safely manage plutonium."

A *New Times* magazine article described her as "the patron saint of the anti-nuclear movement."

Ms. magazine declared that her death "raises the specter of murder and a terrifying technological reality," while *Rolling Stone* rapped the nuclear industry's terrible power and how it silenced a critical opponent.

In addition to a spate of articles and broadcast reports in the United States, at least two British papers, the *Manchester Guardian* and London's *Observer,* have featured investigations of her death.

A prestigious Paris magazine carried a wide spread with a lead that translated: "The Lady of Plutonium—that is the title of a gloomy sequence. But it's a true FBI investigation. A woman dead on a U.S. road. Was she killed by a gang of plutonium thieves who wanted to make a bomb? Did she fall into a trap? The mystery remains entire. A mystery which frightens America. . . ." The article then went into fanciful detail about the investigations of an FBI Special Agent named Steve Olson who begins to sound like Steve Austin, the Six-Million-Dollar Man. Alas, it concluded, even the tall, handsome federal agent is baffled by the strange case and the "mystery remains entire."

It's a situation that has disturbed Kerr-McGee executives and others connected with the nuclear power industry. Their feelings were perhaps articulated by Oklahoma Republican Senator Dewey Bartlett when he charged, in a Senate speech, that a "nationwide hostile publicity campaign by some press and media" is being "fomented" by the Atomic Worker's Union. He expressed the fear that over-sensationalized coverage of the Silkwood affair could further slow down development of nuclear power, which he feels is essential to solve the energy crisis.

However, the journalists involved consistently argue that the safety issues raised by Ms. Silkwood are crucial to consider as the nation moves toward greater reliance on nuclear power. There is legitimate argument here.

Even further down the line, some connected with the industry believe that a large part of the problem in how the accident has been portrayed is Kerr-McGee's failure to communicate. From the beginning, the company has been forced to adopt a policy of making little or no comment. Company officials have explained this by saying that they were acting on the orders of government investigators. And since Kerr-McGee hasn't been able to do much talking, reporters have helplessly realized there is an important element lacking in their copy—the other side of the issue.

Karen Gay Silkwood was born February 19, 1946, in Longview, Texas, and grew up in Nederland, a town midway between Port Arthur and Beaumont. This area is the heart of Texas' petrochemical region, and the night sky is lit by floodlights and tall torches of the refinery stacks burning off gaseous wastes.

For some reason, it's difficult to refer to her as a Ms., Miss, Mrs., or even, to use cold newspaper jargon, Silkwood. She just seems, even in death, like Karen. It's a name that seems to denote gentleness, passion for a cause, and a quiet kind of courage: a dangerous combination in a tricky world that has a habit of slaughtering innocents who stray off the known path into those ever-waiting brambles.

No, I'm not going to make a heroine out of Karen. The best I can provide is a brief psychological profile which is after the fact. It is fairly clear just what happened. It has happened before in other ways. She was not a chaste Joan of Arc, but on the other hand, she wasn't, as one investigator implied, an accident waiting to happen. Long before Crescent, she was a girl who started sending out signals that no one apparently took the time to pick up.

Karen was a sensitive human being who somehow found it difficult to adjust to our often insensitive society. When the thorns pierced, she bled and bled deeply. She died, as she had lived the latter part of her years, alone on a dark road. . . .

It's a pity. Just how many creative people do we manage to lose through death or, somehow, silence until their death? The System somehow seems to have chewed her up and then spit out the pieces in ugly chunks.

A 35-cent pocket guide to astrology lists the characteristics of a person born under Aquarius as ". . . unquestionably, the natural reformer of the Zodiac." A later section reads, not encouragingly, ". . . you are most bound to step in where others fear to tread . . . you are probably a real humanitarian, a willing and hard worker, a true friend to many, an intellectually-inclined idealist with so much spirit that you can well-afford to spread it around, among your friends. And those very friends may sometimes murmur that you are awfully unpredictable, a rebel, a bit of an eccentric."

Strange, how personality traits often seem to follow a pattern tied to the infinite universe. This same '74 fall forecast advised all Aquarians that the first part of November should be fairly uneventful but warns those born under this star sign to take care and be prudent during the latter part of the month. For Karen, this may have proven deadly prophetic.

As a high school student, she played flute in the band, served on the volleyball team, and belonged to the Future Homemakers of America. She turned down a place in the majorette corps on the advice of her band teacher, who said it was better to stick with music rather than kicking up her legs. Her record as a student was impressive. She was a member of the National Honor Society and one of twenty-two honor graduates in the Class of 1964.

Excelling in science, she was especially strong in chemistry and, on her own time, decided to enroll in a six-week course on radiation. In her senior year, she was accepted into an all-male advanced chemistry class which she passed with honors.

Karen was a serious sort of person who shunned local teen spots for library reading and hospital volunteer work. She had only one really irritating characteristic that an acquaintance recalled: she had the habit of talking back to her teachers, often correcting them with an uncanny firmness when they slipped up, say, on the atomic weight of tritium.

In the early sixties, such behavior on the part of a student was not yet acceptable. "She was," according to one teacher, "a very nice person who always wanted to be right about everything."

Much, much later, Karen would scream, not so much in rage as frustration, at a supervisor of the Kerr-McGee plant: "Goddamnit, I am right and you are wrong. If you want to tell me what to do, then you oughta learn how to do the job right."

Another childhood friend remembered her as the kind of person who, if something was wrong, would not simply stand by and ignore the matter. "She was not afraid to stick her neck out. When she went into anything, she put everything into it and she stayed with it. She was intensely loyal. She'd stick up for her friends. And she had a great capacity for having a good time, for laughing. Oh yes, I think that's the one thing most people from school would remember."

Much later, encouraged by the Union's interest in her claims of safety violations by Kerr-McGee, Karen stubbornly told a colleague: "We're really gonna get those [condoners of safety violations] this time." And, as he recalled, she was determined to do just that.

After high school graduation, she went to Lamar College in Beaumont to study medical technology. Her expenses were paid with a scholarship from the Business & Professional Women's Club, an organization as respectable as the DAR.

After a year of college, Karen made a fatal mistake. She took a short trip that was to change the scholastic direction of her life and squelch forever the dream of any scientific career. She went to Kilgore, Texas, to spend the summer with her grandmother. There she met Bill Meadows, who was a pipeline supervisor at Mobil Oil. They eloped.

The young family struggled from one town to another across Texas and Oklahoma. Seven years, three children, one bankruptcy and a divorce later, she tried to retie the shattered remains of her life.

Eventually, Karen moved north to Oklahoma City and agreed to give her former husband, who was remarrying, custody of their children. She painfully realized that there was just no way she could financially care for three small children.

At first, she found work as a hospital clerk. Any further college was just wistful thinking. Finally, on August 4, 1972, she was accepted for employment as a laboratory analyst at the Cimarron Facility of Kerr-McGee. It was the beginning of the end for the girl from Longview, Texas.

Company officials admitted in a confidential interview that initially, Karen Gay Silkwood was an excellent worker. She was, they said, cheerful, energetic and apparently looking forward to a career in the nuclear industry. After three months of testing plutonium fuel rods, however, her attitude began a slow, burning change. Shortly thereafter, she was outside the chain fence, carrying a strike placard and marching with other dissidents.

A colleague, Drew Stephens, had listened and shared Karen's concern about plant safety. As a union activist, he had gradually caught her enthusiasm about labor's struggle to retain recognition. But as her involvement grew, her mood also polarized, Stephens would later recall:

"She became more irreverent, almost giddy sometimes. She joked about the dangers in the plant and about the safety procedures. When a corner of the work room was contaminated and tape would be laid down to mark the 'hot' area, she'd laugh and say: 'See, there's the magic tape. It keeps the plutonium in the corner of the room.' But her laughter was really a defense mechanism. Behind that ironical sense of humor was real fear."

To the townspeople and to many workers at the plant, Karen Silkwood was as unknown as the plutonium that figured so strongly in their lives. Some openly labeled her a pro-union malcontent, "looking for trouble where there wasn't any." In local cafés and shops, she was sometimes referred to as a "radical," a "spy" and an "emotionally troubled divorcee seeking attention."

Others in town saw her as friendly, outgoing and dedicated to improving the plant's safety. One close friend remarked: "Some people thought Karen was bitchy, hard to get along with, too sharp with her tongue. Others thought she was very warm—you could sit down cold and be talking about personal, relevant things in a very short time." A former colleague added: "She had a temper, and strong compassion for other people. The temper was as strong as the compassion. And both were forceful."

Her work record began to reflect some negative observations, according to another Kerr-McGee official. "After a while, she was no longer viewed as dependable, but when she was here, she did work hard. There was also the problem of her sleeping on the job and taking narcotics on duty. . . ."

Numerous associates of Karen Silkwood were interviewed by official investigators and, under questioning, indicated that Karen had the reputation of using marijuana. More tangible and documented evidence of this association with drugs is as follows:

1. Sherri Ellis, Karen's roommate, told investigators that Karen and Stephens had smoked marijuana in her presence on several different occasions.
2. Marijuana and a hypodermic kit were found in Karen's apartment during decontamination.
3. Rick Fagan, Oklahoma State Highway Patrolman who investigated the accident, has stated that upon gathering the personal effects at the scene, he saw a partially filled capsule, one full capsule, and two hand-rolled cigarettes which he assumed to be marijuana, an assumption later verified by the Narcotics Division of the State Crime Bureau. A physician at the Guthrie hospital inspected the capsules and reported that they were of a barbiturate base.
4. Records of the Gilliam Prescription Shop #3, in Oklahoma City, indicate that Karen Silkwood had purchased methaqualone on a prescription issued by Dr. Clarence Shields of Oklahoma City. According to these records, purchases were made of thirty tablets each, on the following dates: August 13, August 19, September 25, October 4, October 13, and November 4, all in 1974. The prescription called for Karen to take one tablet upon retiring; this record of purchase involved a total of 180 tablets purchased for a period of 92 days.
5. An official statement provided by Evelyn Emerich of Crescent indicated that Karen spent the night of October 16, 1974 with her and her son, Carl Themmons. She testified that she had observed her friend in a café in Crescent and realized that Karen was under the influence of drugs and would be unable to drive.

Concerned for Karen's safety, she invited her to spend the night. Miss Emerich also advised investigators that Karen had told her that she had attempted suicide on two occasions: once in Duncan, Oklahoma and once in Oklahoma City, both suicide attempts made by means of drug overdose. Miss Emerich said that on the last occasion, a fellow employee by the name of Connie Edwards had responded to Karen's call for help.

6. Connie Edwards was then interviewed by investigators and said that in September, 1973, Karen had called her at her home in Orlando, Oklahoma, saying that she had tried to kill herself by taking an overdose of drugs. Miss Edwards stated that she drove to Oklahoma City to Karen's apartment, then located behind the Lady Classen Cafeteria on May Avenue, and that upon entering, found the girl in a stupor on the sofa. She managed to get Karen to her feet and then tried to take her to the hospital, but she resisted. Eventually, Karen responded and Miss Edwards took her to Orlando where she spent the night at her home.
7. Medical report indicates that an autopsy disclosed excessive amounts of methaqualone in Karen's vital organs, in addition to a slight amount of alcohol in the blood system. Lethal level is .50 milligrams. She had .35 milligrams and ordinary therapeutic levels are between .20 and .30 milligrams.
8. A handwritten note retrieved by a decontamination squad in Sherri Ellis' and Karen Silkwood's apartment indicates a purchase of $300 for drugs by Karen and a purchase of $75 for drugs by Sherri Ellis.

After a period of time, allegedly unsafe working conditions became an obsession with Karen, until finally, in September, 1974, she notified the AEC in Washington, as an official representative of the Oil, Chemical, and Atomic Workers International Union. A "Division of Inspection Report" issued by the AEC's Kenneth H. Jackson and dated December 19, 1974, stated that a composite of the total information obtained on this subject (Alleged Falsification of Quality Assurance Records) had resulted in the following allegations:

1. Falsification of photomicrographic negatives of weld test samples;
2. Improper use of quality control sample analytical data by K-M supervisors and employees;
3. Irregularity in pellet inspections; and
4. Falsification of computer input data.

A subsequent investigation by the AEC revealed some evidence to support two of the allegations: falsification of photomicrographic negatives of weld test samples and improper use of sample analytical data. The photomicrographic tests are used for verifying and recording the thickness of the fuel cladding next to the weld; they supplement metallographic examinations for determining weld quality.

On this question, the analyst involved, who is no longer employed at K-M, admitted touching up the negatives in question. In a signed statement W. Scott Dotter told the AEC and K-M that he acted alone without either the consent or knowledge of management or supervision. He also stated that he made the alterations in an effort to conceal small artifacts (scratches) that were incurred during darkroom processing. The man explained that his supervisors would reject what they considered defective negatives and that the entire sample would have had to have been rerun. He also stated that at no time did he endeavor to cover defects in the actual weld itself, nor did he have any intent to defraud or to falsify such records.

In addition to the required inspections conducted by Kerr-McGee, the report also noted that the fuel pins are subjected to stringent over-inspection after arrival at Hanford. At this final point, both destructive and non-destructive examinations are made of samples from each shipment. Of the fuel pins delivered to Hanford, about half have been so inspected, with the result that less than 2 percent have been rejected.

Kerr-McGee has a contract with the Energy Research and Development Administration (ERDA) to provide 18,500 plutonium fuel rods for the FFTF located in Hanford. As stated earlier, the FFTF is a facility for testing components in the breeder reactor program. None of the fuel has yet been used in the reactor, since it is not scheduled to begin operation until early 1978.

And, although the investigation showed some violation of quality assurance procedures, the Hanford inspections—as late as mid-winter 1975—have revealed no evidence or indication that the quality of fuel pins had been compromised in any way.

The AEC investigation reports (dated January 7, 1975) are available for public inspection at the local Public Document Room in the Guthrie, Oklahoma Public Library, at the old AEC Regional Regulatory Operations Office in Chicago, and at the old AEC Public Document Room in Washington. A full "Brief of Findings" (spanning 50 pages) which was issued by the AEC on December 19, 1974, follows and is contained in this official report which is covered at length under Table 20.

Another incident that really hit the Cimarron Plant was a still-mysterious sequence of events that occurred between November 5 and 7 when, the AEC later reported, Karen Silkwood swallowed "a small amount of plutonium." The "why" still remains unanswered. Several theories have been aired:

1. She was smuggling out plutonium, perhaps for use by terrorist bomb makers. Despite such rumors, there is no firm evidence to support this allegation; the NRC says that many pounds of plutonium could be "unaccounted for" but not really missing, since it accumulates in pipes and equipment during normal runs.

Note: Such "MUF" (material unaccounted for) figures still continue to be a center of controversy. One Oklahoma legislator claimed that he had been told there were more than 66 pounds of plutonium missing during one year of operation. Pressed further, this man attributed his information to a *New York Times* reporter but would elaborate no more. The Atomic Energy Commission and subsequently the Nuclear Regulatory Commission strongly denied the allegations. "There was no basis for it at all," said Jan Strasma, AEC spokesman. "It was absolutely untrue. There are complete inventories taken every two months," he continued, "and there was just no significant amount [of plutonium] missing. That little which was unaccounted for was lost in processing."

2. She was intentionally sloppy to embarrass Kerr-McGee. Again, however, evidence is lacking. One theory speculates that she ate some plutonium by putting it in a capsule. This would have gotten it by her lungs and esophagus and would have accounted for a high fecal reading but a relatively low lung reading. While such behavior might temporarily dramatize alleged safety hazards, a full investigation would likely expose a contrived incident, discrediting both her and the union. In addition, there was her reported fear of plutonium which would cast serious doubt upon her use of it in this deliberate manner. Again, the key word here is *deliberate* on her part.
3. She was deliberately contaminated by another person who sought to exclude her from the plant's plutonium-producing area, where she was said to be documenting even further alleged safety problems. The AEC's investigation points out quite clearly that her

> urine-sample kits were accessible to any employee before she left the plant. While this theory cannot be proven, available evidence indicates that a second party could very easily have been contaminating this girl without her knowledge.

Point Number One, however, simply won't go away. One theory—gaining increasing respectability—points to the possibility of a plutonium blackmarket directed by nuclear gunrunners who sold their merchandise of death to the highest national or international bidders. It could have been a neat operation until someone unknowingly came across documents which would have literally blown this covert operation to bits. If Silkwood did have records, it is quite possible she may have died never realizing the time bomb she was carrying.

MUF and LEMUF are two acronyms which have been added by the Nuclear Regulatory Commission to the vocabulary of nuclear materials management. MUF stands for Materials Unaccounted For. It is defined by NRC in the following formula:

$$\text{MUF} = \text{BI} + \text{A} - \text{EI} - \text{R}$$

where MUF is Materials Unaccounted For; BI is Beginning Inventory; A is Additions to the inventory; EI is Ending Inventory; and R is Removals from the inventory.

LEMUF stands for Limit of Error of Materials Unaccounted For, and the NRC has established the following limits of error for Materials Unaccounted For:

Plutonium or uranium-233 in a chemical reprocessing plant . . . . 1.0 percent
Uranium in a reprocessing plant . . . . 0.7 percent
Plutonium, uranium-233, highly enriched uranium in all other locations . . . . 0.5 percent

In all honesty, it should be emphasized that the discovery of a MUF figure when an inventory is taken and compared with the books, does not automatically mean that such materials have been stolen or lost. The MUF could be caused by errors, inaccurate measurements and analyses, unknown hold-up of materials in equipment, or unanticipated amounts of material in scrap. Until these possible causes are investigated, theft or diversion cannot be assumed unless there is some reason to expect it. This situation, for example, differs little from that of other industries which handle valuable materials such as platinum, gold, other precious metals, or certain drugs.

Most nuclear materials in the United States are held by prime contractors of the Energy Research and Development Administration. Those contractors who operate ERDA-owned facilities need no NRC license and conform to instructions from ERDA as to their nuclear materials accounting systems. However, those contractors who work in their own facilities must have an NRC license and comply with NRC requirements. Individual companies which produce nuclear products, such as fuel pellets or completed fuel elements, for commercial sale also hold and use these materials under NRC licenses and NRC regulations.

ERDA prime contractors can be directed by ERDA to take certain measures for nuclear materials accounting, and the costs of those measures are reimbursed. NRC, on the other hand, cannot issue direct order to licensees holding nuclear materials. Rather, it can inspect them for compliance with terms of NRC licenses and NRC regulations. If inspection reveals noncompliance, NRC can then act to require the licensee to bring his system into line with NRC regulations. However, this process takes substantially longer than do ERDA directions, particularly if NRC wishes to make a change that requires modification of the published regulations. As a result, the existence of two systems can lead to different standards and procedures.

The experts—a cross-section of individuals in industry, universities and government—say that a large percentage of MUF is believed to be missing only because it is imbedded in processing machinery or lost in the crude statistical methods used to keep track of such material. However, with only forty pounds of enriched uranium or twelve pounds of plutonium, there exists the capability of killing thousands.

For example, a report issued by the AEC in late 1974 concluded that because of the widespread dissemination of instructions for processing special nuclear materials and for making simple nuclear weapons, acquisition of special nuclear materials remains the only substantial problem facing groups or nations desiring nuclear weaponry status. And, the NRC claims it is too risky—and they may have a valid point—to disclose plutonium disappearances from nuclear plants—even to Congress. The agency explains that it would be too easy for an extortionist to learn how much of the stuff was gone and use the figure in some phony blackmail attempt. It is, in effect, a double-edged sword. Damned if they do and damned if they don't.

It is not out of the realm of possibility that some companies, perhaps not wishing to create trouble for themselves, may not always be eager to report that they have lost plutonium. For example, at the Nuclear Fuel Services plant in east Tennessee, investigators hinted strongly that the company was juggling its plutonium figures. Subsequently, the NRC began an investigation in 1976 and no report has yet been issued to the public. In another action, Representative John Dingell, D-Michigan, who chairs a subcommittee on Energy and Environment, has raised some strong questions as to whether the safeguard reporting system for nuclear facilities is totally adequate.

For example, physical security at many nuclear installations is less than stringent. At the Kerr-McGee Plant it was later learned that one of the guards turned out to be a convicted felon working under an alias. And, as part of my own test of security measures, I was able to leave every government lab and nuclear plant—including some especially sensitive—with the special clearance badge and papers which had been issued at the security station. The only exception was the San Onofre Plant in California, where the man held out his hand and said: "No Go" until I submitted the plastic entry visa.

And, a column distributed by Jack Anderson in June, 1976, notes that more than 200 pounds of highly enriched uranium, suitable for manufacturing bombs, disappeared from a Pennsylvania defense plant in the mid-1960s. No one, Mr. Anderson claims, knows exactly where it went. He adds that the NRC will admit only that 50 pounds of uranium disappeared but that "secret documents" reveal the loss was more than four times that amount.

Then the widely syndicated columnist drops in the tidbit that it has been whispered in nuclear circles that uranium has been smuggled out of the country to help the Chinese or Israelis detonate their first nuclear bombs. I can still remember the puzzled look on Chang Chih-Hsiang's face when I asked him about nuclear power in the People's Republic of China.

"Nuclear Power?" the Communist Chinese liaison officer replied. "No have nuclear power in China. China use OIL."

"Sure you do," I answered, realizing that that interview was going to be a real winner. . . . "Let's go through that one more time," I persisted, but it was oil, oil, and then oil. It soon became obvious that smiling Chang was not Mr. Anderson's source.

Then there were several earlier stories carried by the New York

Times News Service, usually under the by-line of veteran news writer David Burnham, that made the following allegations:

• A highly placed official in the commission (AEC) said one of its plants was unable to account for about 9,000 pounds of the highly enriched uranium it had produced since the plant began to operate. The official also said the controls at a second government uranium-processing plant were so bad that it was impossible to estimate the amount of highly enriched uranium that could not be accounted for during the plant's lifetime.

• A second scientist with many years of experience in the commission (AEC) confirmed that the government's gaseous diffusion plants have a cumulative MUF measuring in tons.

• An executive in the Kerr-McGee Corporation nuclear plant near Crescent, Oklahoma, said that in 1974 there had been times when the technicians in the plant were unable to account for up to about 60 pounds (30 kilograms) of plutonium.

Again, the AEC and later, NRC strongly deny the allegations but refuse to give a figure that tells exactly how much plutonium is missing and where—citing national security as their reason.

However, the AEC, in December, 1974—shortly after Karen Silkwood died—was considering the "smuggling theory" to be more than just a distinct possibility. It subsequently issued a confidential report on this.

"It [smuggling] is a possibility," one AEC source said privately. "I'd put it this way: that would not be a bad tact. I mean, it's not an out-of-line statement but I can't say any more right now." And, he didn't.

Then, National Public Radio reported in late 1974 that 44 to 66 pounds of plutonium and uranium were unaccounted for at the Cimarron Facility of KM—amounts, again, that the AEC protested were "erroneous." Later, however, an AEC official did admit to the Associated Press that his agency was having significant problems with "accountability" at KM but refused to reveal the amounts of nuclear materials that were involved. The AEC again asked the National Security Council for advice on whether to disclose details of atomic materials unaccounted for on a national level. The NSC said "no-no" and it was so.

One interesting aspect that I uncovered involving Karen Silkwood was a report showing that the young worker had been repeatedly contaminated with plutonium from outside the area in which she was working. An area she had no official access to.

At about 6:30 P.M. on Tuesday, November 5, while working at a glove box at Cimarron, Ms. Silkwood discovered that her left hand, right wrist, upper arm, neck, face and hair were contaminated with plutonium. (A glove box is a sealed chamber, with air-tight rubber gloves made a part of one wall, so that workers can handle radioactive materials safely from outside.) She discovered her contamination when, as plant rules required, she checked herself with a radiation detector before leaving the work area. Following standard procedure, she decontaminated herself by showering with a very strong soap, and returned to work.

As a result of this contamination incident, a health-physics technician prepared collection kits so that Ms. Silkwood's stool and urine could be tested to see if she had inhaled any plutonium. That night a technician marked these kits with Karen's name and placed them on a shelf in the hallway through which all workers must pass to enter or leave the plutonium building. Karen then picked up these kits when she left the plant at about 1:10 A.M.

Karen later told AEC investigators that she drove from work to her apartment in Edmond, Oklahoma, a small community just north of Oklahoma City. She shared an apartment with Sherri Ellis, 22, another Cimarron analyst. (Some eight months after her roommate's death, Ms. Ellis elected to storm the Kerr-McGee facility armed only with a high-powered rifle. Guards said she pleaded: "Kill me. Go ahead and kill me." She was shortly disarmed and taken into custody. An interesting footnote is that the guard who went to apprehend her unknowingly carried an unloaded gun.)

The next day, Karen Silkwood returned to the plant at 7:50 A.M., according to federal investigators, and worked for an hour on routine paper work. While preparing to leave at 8:50 A.M. to attend a union meeting, she discovered that her hands, her right forearm, and the right side of her neck and face were contaminated with plutonium. (There is a strong possibility that she could have been contaminated with plutonium before she came to work, since plant rules then required personal radiation checks of workers only as they left the plant, not as they entered.) She scrubbed herself, attended the meeting, had lunch outside the plant, and returned to the health-physics office at 4:30 P.M. At her request, the health officer checked her locker and car but found no contamination in either. She then left the plant for the evening.

At 7:50 A.M., Thursday, November 7, Ms. Silkwood brought the health-physics officer four urine samples and one fecal sample. A radiation detection scan of her body showed high contamination levels in and around her nose and lower levels. The sample kits also showed high radiation levels, an indication that somehow she had swallowed plutonium, but again there was no contamination in her locker or car. Two samples later proved to be so radioactive that they had to have been doctored, the AEC report has stated.

"By this time she was really getting scared," a colleague noted. "She was leaving work in a safe condition and coming back contaminated, and she just couldn't figure out why."

Next, Ms. Silkwood and a health physicist checked her apartment. They discovered dangerous radiation levels, especially on the toilet seat and floor mat in the bathroom and in the bathroom and kitchen sink traps. A package of cheese and bologna in the refrigerator were also highly contaminated.

On November 7, Kerr-McGee teams began to decontaminate the apartment, a four-week process. Although the radioactive count was high and easily detectable, the actual amount of plutonium found in personal and apartment contamination was quite small. The weight of plutonium involved was actually less than .0003 grams (0.3 milligrams) or about 1/100,000 of an ounce—not a massive dosage.

Finally, on November 10, Ms. Silkwood, her roommate and Drew Stephens were flown to Los Alamos Scientific Laboratory for two days of radiation counts. Dr. George L. Voelz, of LASL, who examined the trio, reported that the internal contamination was sufficiently low that it would not present a "significant health hazard from this exposure either now or in the future."

The three were in LASL from November 11, 1974 to November 12, 1974, returning that day by commercial aircraft to Oklahoma City and arriving sometime between 10:30 P.M. and 11 P.M. According to an investigator, they then went to Stephens' home and spent the night. In an official report, Sherri Ellis has testified that they had purchased some 190-proof alcohol in Albuquerque, plus other assorted liquors. She described these as "survival kits" containing four four-ounce, miniature bottles of assorted whiskeys.

Ms. Ellis testified that she and the others were drinking Bloody Marys made with this 190-proof alcohol until about 2 A.M. when she decided to retire. When asked just how much alcohol Karen Silkwood had consumed, she stated that she really had no idea since they were mixing drinks without a jigger. In other words, she further explained, they would pour the tomato juice into a glass and then pour enough alcohol by what she termed "eye-balling" it.

Ms. Ellis further stated that she had no idea when Karen had finally gone to sleep, but she did get her up at 7 o'clock on the morning of November 13, 1974, to attend union negotiations at the Cimarron Plant. Needless to say, Karen must have been tired after her ordeal of the last few days. The accident was to happen that night, shortly after 7 P.M., as she drove toward Oklahoma City to meet with the New York reporter and a top union official.

So, Wednesday morning Karen drove to work. Contract negotiations between Local 5-283 and Kerr-McGee had begun the previous week and, as a committeewoman, she was supposed to take part in the bargaining. She spent the morning in negotiations, arguing the union demands for better safety training and higher injury benefits. In the afternoon, she met for several hours with AEC inspectors who were still trying to unravel the mystery of her contamination.

At this point, another serious question of credibility arises. If, indeed, Karen Silkwood finally possessed the long-sought-after documented evidence of safety violations, why didn't she show this directly to the AEC investigators who spent three hours questioning her?

AEC Inspector Bill Fisher was interviewed by an official investigator and described Karen as appearing emotionally disturbed at times. He testified that she cried and told the interviewing team that she was worried about her health. Fisher further stated that upon checking the car for contamination on the night of the accident, they found a whiskey flask in the wreckage and that he and Wayne Norwood, a health-physicist, took samples of the contents for contamination tests. He testified that the contents appeared to be tomato juice.

Mike Wilding, a Kerr-McGee employee, advised in his statement that he had received information from another source that the sample submitted to Idaho Falls for tests was a Bloody Mary and supposedly did contain alcohol. Subsequent lab tests confirmed this.

Karen finally left the AEC investigators and drove to Crescent and the Hub Café for a supper meeting to discuss negotiation strategy with the Local. Jack Tice, who headed the negotiating team, told the assembled union members that, as expected, K-M was going to hold its hard line.

It is interesting to note that the union has never been particularly strong at the open shop Crescent Facility. In 1974, before and after the controversy, there were 45 union members out of some 270 employees.

At about 6 P.M., Karen excused herself to telephone Stephens, reminding him to pick up Steve Wodka, a union representative from Washington, D.C., and David Burnham, a *New York Times* reporter. She told him to expect her at the Holiday Inn-Northwest about 8 P.M. She then left the Hub Café and headed for Highway 74.

Frank Murch, another union negotiator, has testified in an official statement that he and Tice were very concerned about Karen's ability to drive. He stated that she appeared nervous, very upset and emotionally unstable. Murch stated that he even offered to drive her to Oklahoma City but that she refused.

QUESTION: Why did Karen elect to drive thirty miles to Oklahoma City when the plant was in Crescent and the three men could have just as easily driven there, especially in view of the negotiations that were taking place? The story was in Crescent, not in the cocktail lounge of the Holiday Inn-Northwest. So, why a clandestine meeting, some thirty miles from the center of action?

Even after Karen died, there were further attempts to maintain this aura of mystery: such things as "dead-dropping" photographs of the accident by taping them beneath the directory stand in a telephone booth for a local journalist to retrieve at a set time and date. . . . As one FBI agent noted dryly: "Stephens could just as easily have handed them to the reporter. No problem. No need to go to all that trouble."

In any case, at 8:05 P.M., a lone truck driver sitting high in his cab and rolling along the two-lane highway, spotted the white, 1,638-pound Civic Hatchback, almost hidden in the muddy culvert.

Karen Silkwood apparently died instantaneously some seven miles and ten minutes from her original starting point. The official explanation issued by the Oklahoma State Highway Patrol was that, exhausted after her Los Alamos ordeal, she had fallen asleep and drifted off the road to her death. In addition, Oklahoma City's chief forensic toxicologist, Richard W. Prouty, discovered .35 milligrams of methaqualone in her bloodstream, which, he said, was conceivably enough to lull her to sleep on the highway. Normal dosage is about .21 milligrams. Lethal dosage is .50 milligrams.

But that was not sufficient for Steve Wodka. Understandably, he found it difficult to believe that she would actually pass out on the way to one of the most important meetings of her life. And, no manilla folder, crammed with incriminating Kerr-McGee documents, was ever located at the accident scene. If indeed, such a proof of fraud actually existed.

The medical examiner listed as her personal effects the following: an ID badge, an electronic security key for the plant, two marijuana cigarettes,* two used tissues, a Bradley Mickey Mouse pocket watch, a small notebook, her clothes, $7 in bills, and $1.69 in change.

Wodka called Tony Mazzocchi at the OCAW International in New York, and they agreed that an outside expert was needed to investigate the crash. Three days later, A. O. Pipkin, an ex-cop from the Accident Reconstruction Lab of Dallas, arrived in Oklahoma City. What he initially imagined a routine investigation at $300 a day or $40 per hour, soon turned into a headache of the first degree.

Dressed in a Day-Glow orange jumpsuit, Pipkin examined the Honda and found what were, to him, two curious dents, one in the rear bumper and another in the rear fender. They seemed fresh. There was no road dirt in them and they appeared to have been made by a car bumper.

Returning to the scene, he noted that the car had crossed over the yellow lines and hit the culvert on the left side of the highway. If the girl had fallen into a stupor, he reasoned, then she would have drifted to the right. He also noted the zagged direction of tire tracks in the bloodish-red clay, which indicated to him that the car was out-of-control prior to leaving the road.

As far as A. O. Pipkin was concerned, the evidence suggested one very possible answer: Karen Silkwood's Honda had been hit from the rear by another vehicle.

Could the union have been more delighted to learn such tragic news?

---

* Lieutenant Kenneth Vanhoy, Oklahoma State Highway Patrol, said marijuana verification had been reported to the Oklahoma Department of Public Safety after documented analyses by chemists of the State Crime Bureau.

A can of worms had been opened and there stood Pipkin with his hands out. . . .

Much later, a frustrated accident investigator would relate to another potential customer that the labor union "had gotten me in a lot of trouble over this accident in Oklahoma City." He would further advise that his speciality was not investigating car accidents but heavy truck accidents. He then turned down the client, saying again that he could not get involved in another case at this time because he was "in trouble in Oklahoma City."

According to official records, Pipkin was contacted by another investigator and was, at first, hesitant to discuss the matter. Even brisk. Recontacted later, however, he apologized for his earlier curt attitude and said that he had been hounded by "a union, attorneys, and press from Oklahoma City," after having "reconstructed an accident, where he felt a girl had been run-off the road and felt foul play was involved." He remarked that "based upon all the problems he was having, in the future he would stick to civil litigation." Furthermore, he said, "he was in trouble with the State Police in Oklahoma, and planned to send them a copy of his report, against the wishes of his clients."

As for Pipkin's credentials, shortly after the accident, an official inquiry with the Secretary of State, Charters Division, Austin Texas, revealed that neither Accident Reconstruction, Inc., nor Accident Reconstruction Labs, Inc., is listed as operating in Texas or as having been formed in Texas. The 1974 Metro Edition of the City Directory for Dallas does not show a listing for Accident Reconstruction, but does list an Adolphus O. Pipkin at 5346 Mercedes Street, Dallas (the address has since changed to an unlisted one). Inquiry by an official source further ascertained that Pipkin is simply not licensed to operate in the state of Texas, even though he has an office in Dallas.

(A check of police records revealed that Pipkin does not have a criminal record in Dallas. Court records do list a divorce suit dated October 8, 1974, which he brought against his wife on grounds of incompatibility. According to the file, this suit was "amicable" and Pipkin's wife waived any action.)

An official check with a Mr. McQuarter of Austin, Texas, at the Texas State Board of P.I. and P.S.A., reveals that neither Pipkin nor his business were then listed as licensed in the state of Texas. McQuarter stated that he was familiar with Pipkin's work in New Mexico, specifically, Albuquerque, but was not aware that he was now operating in Texas. He then advised that a Dr. Tonn (Ph.D. in physics at Rice University) was the primary authority in the South, Southwest, and West in accident reconstruction, his knowledge, technique and reputation being irrefutable. Dr. Tonn, he added, was quite active in this area and could testify primarily on physical evidence from accident photographs, but for some reason this expert had not been consulted in the Silkwood case.

Public records obtained through the Professional Engineers of Texas, Registration Branch, Austin, state that Adolphus O. Pipkin was not then a registered engineer in Texas.

Also, a search of newspaper files from the *Dallas Times Herald,* and the *Dallas Morning News,* including clip files, art files and microfilm records from 1925 to November 21, 1974, revealed no information concerning this "nationally recognized investigator" or his firm. Strange, very strange. And, as to his business reputation, of twenty-six private investigation firms contacted, only one was familiar with Pipkin.

Further data on Pipkin was received by an official source from a Captain Wimberly of the New Mexico State Police Intelligence, Santa Fe. He stated that "Pipkin was employed by the Albuquerque Police Department from 1951 to 1954. He resigned in 1954 to go into business for himself," and subsequently attended a short course in Albuquerque in 1955. "This short course consisted of two weeks training and was an extension of the Northwest Traffic Institute." Captain Wimberly further advised that Pipkin was unable to pass the required examinations to obtain a license to operate as a private investigator in New Mexico and consequently moved his base of operations to Dallas in 1963.

During another official interview, Pipkin, himself, provided a background summary. He was born in Missouri but spent most of his life in Albuquerque until 1969, when he moved to Dallas. After two years of engineering school at New Mexico University, he joined the Albuquerque Police Department in 1951, where he worked until 1955. Pipkin stated that he had been assigned accident investigation in the Traffic Division while with the APD, but that other than a seminar at NMU on accident investigation, he had no formal education in this area except "twenty-three years investigating at least 100 accidents per year." He added that he was neither an engineer nor a metallurgist.

In detailing his experience and rates, Pipkin stated that he kept up with pertinent periodicals and had considerable experience in courtroom testifying. He said his primary experience and work load had been in heavy truck accidents and that his testimony dealt primarily with the "logical deduction of the physical evidence."

In another official interview, Pipkin said that on November 12, 1974, he was contacted regarding his availability to make instant accident investigation by an attorney named Frederick Barron of Dallas. He said Barron gave him the name of Steve Wodka, who was then in Washington. Barron told him that Wodka would give him the background information necessary to make the investigation.

Pipkin then telephoned Wodka and the information was given. He said that, at that time, he realized that he just might be stepping into an embarrassing situation and stressed to Wodka that there would be no "hanky-panky"; that he would make the investigation and strictly report the facts as he found them.

Following the conversation with Wodka, Pipkin went to Oklahoma City on November 16, 1974, and contacted Drew Stephens, who took him to see the damaged car, which was in a garage in the northwest part of the city. Pipkin examined the car, then proceeded to the scene of the accident, where he made a series of photographs of tread marks which he could identify as having been made by the Honda.

In an official report, he stated: *"the only thing . . . found in the accident in his investigation that would adversely affect the position of Kerr-McGee was damage to the left rear of the automobile indicating that it could have been hit by another vehicle from the rear, causing the accident."*

At this point, the official interviewer paused from his note-taking and pointed out to Pipkin that he had not seen the automobile involved until *three days after the accident.* Was it not possible that damage to the left rear could have resulted from other causes during that interval? This Pipkin admitted.

A final reading to this episode might be provided by Oklahoma State Highway Patrol Trooper Rick Fagan, who was contacted by investigators on December 9, 1974. He was told that it had been learned from the wrecker operator who moved the Silkwood car that the vehicle had been pulled out by its rear end.

Interrupting, and without further prompting, Fagan declared: "You don't have to tell me that. I stood above it, and watched the wrecker pull the Honda out by its rear end over the lower end of the south retaining wall of the culvert, *and that is where the left lower mark of the Honda was damaged."*

Official investigators attribute the other small dent to another accident Karen had, some two weeks earlier, when she swerved to avoid a cow and went off the road into a ditch.

This was a difficult chapter to write in the sense of pulling information from many sources and sifting between what was credible and what was fanciful. One clear picture has emerged: there has been a great deal of con, but virtually no pro, regarding industry's position in this matter. In many cases, outright distortions were given by journalists who seemed to prefer the sensational rather than the factual.

Deliberate distortion or carelessness, apparently there aren't many people who will take the time to read through an AEC report and accurately interpret it, or who have access to official investigations that shed a totally different light on the individuals concerned. Every bit of material in this chapter can be verified, and much has been reproduced in the form of official documents and pictures.

This chapter is sensational to the degree that it is probably the first complete and accurate dossier on the death of Karen Gay Silkwood. In fact, I can still remember my final interview with a top official of Kerr-McGee who had agreed to talk (unofficially) only at the request of a third party.

This man had very carefully retraced Karen's employment record, her initial contacts with the union, and the final events leading up to her death. He even mentioned, after some three hours of heavy talk, recent rumors of a proposed movie on the entire episode.

There are several factors, the K-M executive concluded, which throw doubt on Ms. Silkwood and her allegations:

- OCAW's Tony Mazzocchi and his questionable ties with a little-known, New York-based anti-nuclear group. Kerr-McGee wonders if Mazzocchi could remain totally impartial in a case pertaining to nuclear-related materials. "You might say," mused my contact, "that Tony could very easily 'pre-judge' the situation or even 'pre-shape' it."
- Documented drug usage by Ms. Silkwood and those who associated with her. Another previously unpublished piece of evidence is official testimony by Harold Smith, an employee of the wrecker service which handled the November 13th accident. He said one of the documents removed from the scene was a letter from

Ottawa, Canada which invited Karen to a "drug party" there. According to Smith, the letter read: "The machine will roll [rolling of marijuana cigarettes] and everything will be ready when the invitee arrives." This letter is an official part of Justice Department records.

Then it was my turn to ask a few glaring questions. Sift through all this—drug parties, emotional trauma, plutonium contamination—and one point starts ticking. Let me state it quite bluntly: Why would Kerr-McGee allow a person or persons who were demonstrating such questionable emotional stability to continue working in a sensitive plant involving the handling of nuclear-related materials?

Point Two: What about the substantiated charges of "bored" workers shooting one another with pellet guns loaded with plutonium? Others were said to have thrown handfuls of the stuff over a fence just to tease security.

Point Three: How many other Silkwoods are there in the country, male or female, who are working in classified or sensitive areas and are set and waiting to go off. Are they also in the process of becoming accidents waiting to happen . . . ?

Point Four: Shouldn't thorough personality profiles and background investigations be a pre-requisite for working with nuclear-related materials? And if union membership means an employee cannot be technically "fired" from a plant, then perhaps the nuclear industry is one area where the union has no business representing anyone at any time.

Okay, in fairness, there is a partial answer to this one. My Kerr-McGee source explained that AEC regulations—which I confirmed—did not permit full security checks on personnel until some six months prior to Karen's death. Then the company was only allowed to institute total checks on new personnel and not on those employed at the nuclear facility before the stricter regulations went into effect. Sounds incredible, but it's true.

Out of all this, Kerr-McGee has understandably beefed up its security and is paying out a lot of money to ex-employees—union members—who are no longer working at the Cimarron facility. "In one way, it's a pleasure to pay them [unemployment] just to stay away," my K-M contact mused. "And, on the other hand, it's a crime that we have to."

Nor are there any more bored workers who pass the time by firing loaded pellet guns. And the contamination detectors no longer start clicking outside safety zones. The floor has been swept and things are ship-shape once more. You might even venture to say that "Romper Room" is no longer a daily feature at the Kerr-McGee plant. Their "star pupil" has gone away and things will never be the same again. . . .

Some six months after compiling this chapter, I'm editing news copy on a video display "tube" at the *Tennessean*. The slug (tag for a news story) is PROBE and I quickly type in the computer sign-on and then the code letters identifying it as United Press International wire copy, thus transferring it over to the computer's news section. For modern newspapers, gone are the days of paper and pencil editing. It's all electronic.

Waiting, I sip a styrofoam cup of hot coffee and watch quietly as the story flashes on the screen. Incredible. A probe is being initiated by a Senate subcommittee on the death of Karen Silkwood. UPI continues: "Sen. Lee Metcalf, D-Mont., chairman of the subcommittee on government operations, said he will undertake the probe with the idea of recommending government or legislative action."

There's a brief but familiar sketch on Karen which I quickly scan through. Then there's the partial listing of sponsors of the probe, names like the National Organization of Women, United Auto Workers, Coalition of Labor Union Women, and Ralph Nader's Congress Watch. These groups, according to UPI, believe that the Justice Department's investigation was not complete. They want more.

A month later, I'm flying to Washington to meet with Senator Metcalf's staff and determine if they've uncovered any new evidence. It's a safety mechanism on my part—the book must be as accurate as possible.

The capital is cold in December but exciting as always. The FBI's Clarence Kelley is testifying on the Hill before Frank Church's committee, and the Rosenberg files have just been opened to the press. For a journalist, this city is a field day with stories breaking everywhere.

I'm late getting over to Metcalf's office and wind up parking blocks away from my destination. Finally, breathless, I'm directed to the basement of a Senate office building. It's old, with grey-haired men and pretty secretaries moving briskly down long corridors. After a short wait, I'm introduced to E. Winslow Turner, Chief Counsel. He's a colorful, interesting personality and for a few minutes we talk in generalities. Finally, Turner says, "Jacque, we think there are so many pieces to this mosaic that just don't ring true."

At that point their probe was in the limited stages, meaning basically that the investigators were investigating the accuracy of the previous investigators. From this level, a second stage can be reached which means subpoenas and testimony before Congressional hearings. "It's a little touchy," Turner added.

And then the cat ceased playing with its newly found mouse. On a cold, windy afternoon in Washington, the chief counsel of the Senate subcommittee turned and faced me directly. His eyes were steel and direct. I could imagine what it would be like to meet this man in a courtroom.

Four words shot in my direction: "What do you know?"

A few minutes later, I had apparently established some credibility. It remained for him to establish his. I later learned that Turner and my newspaper publisher, John Seigenthaler, were acquainted and that he had telephoned John at the *Tennessean* to determine if I was what I claimed to be. Surprised, John had informed Turner that if I had authored the manuscript, then "it is unimpeachable. . . ." That was nice to know.

The clock ticked on and the Washington sky darkened. Turner then informed me that he wanted my files on the Silkwood case. Probably, it should have made me feel important, but it didn't. I wanted to say, "No, do your own research just like I had to do."

Instead, I thought for a few brief seconds and then replied bluntly, "No way. I'm not after publicity. I've spent a lot of time on this—"

Turner sensed what I was thinking. "They'll be kept confidential," he assured me.

"In Washington there is nothing confidential," I replied.

"I'll subpoena them," he shot back.

Subpoena? That sounded serious. . . .

"After all," he continued, "you came in here prepared to offer your help for this investigation."

Now I was really angry. "Correction. I came to Washington to double check my findings and be absolutely certain that your investigators hadn't uncovered a new development to this case."

Turner nodded and then made a telephone call. I fully expected to be arrested. "You are not taking the popular stand," he nodded in my direction.

"I realize that. I've got no ax to grind either way. My concern is the truth."

"It's refreshing to hear a journalist talk like this, but you're going to have a lot of people mad at you. Watch out for Bella Azburg and *now*. Don't talk to the unions. God, are they going to be mad. You're going to need some friends."

Turner wasn't trying to threaten me. Rather, he honestly seemed to be stating the case as he saw it. The man seemed concerned.

"I'm not afraid," I said. But I was. Regardless, I always keep in mind something a nun once told me: "They can destroy the body," she had said softly, "but they can never kill the human spirit."

Turner then informed me that the Justice Department had officially closed their files on the Silkwood case. This had the effect of denying outside investigators access to FBI files involving Silkwood and Kerr-McGee, a problem I had not encountered as Turner was well aware.

The hour was late. Outside the darkness pressed like thick wool. The chief counsel and I shook hands. Neither of us had allowed the other much yardage. I got the distinct impression that the Senate probers were lacking substantial material and I was equally certain that they hadn't uncovered any new evidence. That aspect of my trip had been successful. A few hours later and I was back in Nashville—tired but reasonably satisfied. In my mind, it was all over, but not as far as Turner was concerned. . . .

The telephone rang the next afternoon and it was E. Winslow Turner, polite, congenial and pressing. He was preparing to send down one of his investigators to question me further and see my material. At first, I balked and then back-pedaled some. For some reason, I feared that the government—or some other equally powerful party—might try to stop publication of the book. Eventually, I agreed to be interviewed, and a Senate investigator named Peter Stockton arrived in Nashville from Washington. We talked and then talked some more. He looked at some of my material and requested permission to copy it.

I looked at Stockton. He was the clean-cut, all-American boy type, but after some eight hours of intensive questioning, his motive loomed clear—the Xerox machine and then hop, skip back to Washington. "You may copy two or three pages of these reports but that's all," I replied, pointing to a stack of pages, some stamped "Evidence—FBI". His interest was understandable. I explained that once the book was published, only a matter of months, he would be welcome to copy all the reports that he wanted and without charge or buying the book. "I just want my ideas intact until publication," I grinned in his direction.

Stockton nodded. It seemed clear in his mind that the field mouse had become a reasonably intelligent rat who wasn't going to make it easy for investigators on either side. We shook hands and he returned to Washington, promising to make it all a part of official record. The government man was doing his job and I was just trying to do mine. Sometimes, it works like that.

Little did I realize on that cold, windy day in Washington that less than six months later my life would be turned into a nightmarish roller-coaster of wild publicity and strange threats. My eventual testimony before a congressional subcommittee was to rip Washington and cost me my copy editor's job at the newspaper. And, as the storm heightened, I would even be forced to flee—with my children—to safety where we lived tense days and nights praying that no one would find us.

You see, Chief Counsel E. Winslow Turner was strangely prophetic when he said ". . . but you are going to have a lot of people mad at you." I did.

The two-lane road on which plutonium worker Karen Silkwood was to take her last ride—Highway 74 which runs north and south from Crescent, Oklahoma, to Oklahoma City. It was on this lonely stretch of asphalt that the Kerr-McGee analyst died in an auto accident enroute to deliver documents she said showed safety violations at the Crescent plant. These records were never found and have been a major point of controversy.

The now-defunct Cimarron Facility of Kerr-McGee's Nuclear Division at Crescent, Oklahoma. A federal agent observed that the most likely strategy a band of terrorists might use to "take the plant" would be an assault from the river. The death of Karen Silkwood, however, was to be the lever which eventually halted KM's plutonium processing and what was possibly a lucrative blackmarket for nuclear gun runners.

The plutonium plant at Kerr-McGee's Crescent, Oklahoma nuclear division—no longer operational—stressed "safety first," but found itself embroiled in controversy over allegations of falsification of quality assurance records and other related irregularities pertaining to the quality of fuel pins being manufactured for the Westinghouse Hanford, Washington company. It was outside this chain-linked fence that members of the Oil, Chemical and Atomic Workers International Union marched in their demands for better safety conditions for employees.

KM's now-defunct processing plant was an economic mainstay for many residents of Oklahoma, who now find themselves unemployed and on welfare.

Entrance to the Cimarron Facility of Kerr-McGee's Nuclear Division—uranium and plutonium processing—which had been awarded a $1.4 million contract to process plutonium into pellets and pour these into fuel rods.

Bits of broken glass and a ripped taillight from the Honda Civic lie scattered around the south retaining wall of the culvert where Karen Silkwood's car plunged. The Oklahoma red clay shows tread marks caused by the wrecker and impact of the Honda being lifted from the basin.

The south retaining walls of the culvert which crushed Karen Silkwood's Honda Civic and killed the young nuclear worker as she was enroute to Oklahoma City and a meeting to deliver documents—never recovered—of alleged safety violations at Kerr-McGee's Crescent operations.

was this culvert to which Silkwood ove a 1973 Honda er her car left the dway on the east e and traveled proximately 270 t along a rough, ping shoulder ore impact occred.

An official picture of the lower end of the culvert's south retaining wall—located off Highway 74—taken shortly after the accident, shows smudges and marks caused by pulling the wrecked Silkwood car out by its rear end.

ACCIDENT RECONSTRUCTION LAB
1710 Boll Street
Dallas, Texas 75201
214-826-2100

December 15, 1974

TO: OIL, CHEMICAL AND ATOMIC WORKERS
INTERNATIONAL UNION

RE: KAREN SILKWOOD
D/A: 11-13-74

CASE SUMMARY

Case: D-401 Car/Fixed Object Collision

SYNOPSIS

On Wednesday, November 13, 1974 at approximately 7:30 p.m., Ms. Karen Silkwood was fatally injured as a result of a vehicle accident. This accident occurred on Oklahoma State Highway 74 just south of Crescent, Oklahoma. Ms. Silkwood was driving a 1973 Honda and going south on Oklahoma State Highway 74 when her car left the roadway on the east side and traveled approximately 270 feet before colliding with a culvert wing wall. My investigation into this accident and the subsequent analysis and reconstruction was to determine whether the car left the roadway due to natural causes such as the driver going to sleep or whether the car left the roadway due to either an unknown vehicle striking the car, causing it to go out of control, or a combination of driver over-reaction and the unknown vehicle striking the car.

IDENTIFICATION

Location Data:	Oklahoma State Highway 74 approximately 7.3 miles south of Crescent, Oklahoma.
Location Type:	Open highway
Type of Area:	Rural
Date:	Wednesday, November 13, 1974 at approximately 7:30 p.m.
Type of Accident:	Car and fixed object with possibility of prior collision with unknown car.
Severity:	Fatal

AMBIENCE

Light:	Night, dark
Weather:	Light clouds
Road Condition:	Dry

ROADWAY

	Oklahoma State Highway 74
Type of Roadway:	Rural highway
Width:	34 feet, shoulder to shoulder
Number of Lanes:	One northbound, one southbound
Surface Type:	Asphalt
Coefficient of Friction:	0.65 dry (est.)
Road Edge:	Improved asphalt shoulder
Configuration:	Straight
Grade:	Approximately 2-1/2% downgrade as you travel south
Compass Direction:	North-South

TRAFFIC CONTROL

Speed Limit:	55 miles per hour
Pavement Markings:	Interrupted lane lines.

VEHICLES

	Vehicle Number 1
Type:	1973 Honda Civic - Hatchback
Specifications:	
Suspension:	(front and rear) Independent MacPherson strut, coil spring
Steering:	Rack-and-pinion type, 31 feet turning circle.
Braking System:	(rear) Power assisted leading-trailing shoe drum type
Tires:	600S x 12
Curb Weight:	(Hatchback) 1638 lbs.
Seating Capacity:	Four
Fuel Tank Capacity:	10 gallons
Damages:	
Impact:	Left rear
Second Collision:	Frontal
Vehicle Deformation:	Substantial

	Vehicle Number 2
Type:	Unknown

DRIVERS

Vehicle Number:	One
Name:	Karen Gay Silkwood
Age:	28 years
DOB:	2-19-46
Sex:	Female
Occupation:	Chemical worker
Vehicle Number:	Two
Name:	Unknown

INVESTIGATORY AGENCY

Agency:	Oklahoma State Police
Personnel:	Rick Fagan, Trooper

SUPPLEMENTAL DATA

Police Report

Photographs

METHOD OF ANALYSIS

Study of Accident Scene.

Study of vehicle.

Study of Police Report.

Reconstruction of scene from personal inspection and photographs

Scale diagram of pre-crash and crash positions.

Reconstruction of vehicle number one from personal inspection and photographs.

DESCRIPTION

Pre-Crash: V-1 was traveling south on Oklahoma State Highway 74 at approximately 50 to 55 miles per hour as it approached the intersection of Oklahoma State Highway 74 and an unnamed county road. Shortly after passing the intersection, V-1 went out of control and left the roadway on the east side, traveling in a general southerly direction with the left front leading to where the right rear of V-1 would be directly behind the left front. This is more graphically illustrated in one of the diagrams included with this report. V-1 travelled approximately 240 feet in a slight arc to the drivers right before first coming in contact with the north wingwall of a culvert on the east side of the roadway.

Crash: The right front of V-1 passed over the lower section of the north wingwall and as you view the car from the rear it turned in a counter clockwise direction due to the contour of the land and the wingwall. V-1 then became airborne for approximately 20 feet and just before it struck the south wingwall it landed and skidded across a portion of the mud covering the area in front of the south wingwall and then collided with a front, head-on collision against the south wingwall. The attack position that V-1 was in during its collision with both the north and south wingwalls are illustrated in diagrams included with this report. Following the final collision with the south wingwall V-1 came to rest on its left side.

INVESTIGATION

My initial investigation was conducted on November 16, 1974 at the request of Mr. Fred Baron, an attorney in Dallas, Texas, and Mr. Steve Wodka, a representative of the Oil, Chemical and Atomic Workers Union.

Mr. Drew Stevens met me in Oklahoma City on November 16, 1974 and we drove to the accident scene. Mr. Stevens identified the accident scene and assisted me in making my inspection and measurements.

There were two things that called my attention to this accident. The first thing was the fact that the car went off the lefthand side of the roadway, rather than the righthand side. In most one-vehicle accidents where the driver has gone to sleep, or because of impaired abilities the vehicle has always gone off to the right because of the contour of the road, namely the crown.

The second thing was the configuration of the tracks through the grass as the vehicle left the paved shoulder on the east side of the roadway. There were three distinct tracks through the grass, rather than two or four. This indicates that the car was tracking in the attitude shown on the diagram with the left front leading and the right rear tracking behind it. This showed that the car was out of control before it ever left the paved portion of the road. The only way that this vehicle could achieve this attitude was for the car to have been out of control before it left the roadway. The only way that this car could have been put in that attitude was either an impact by an unknown vehicle or a combination of an impact by an unknown vehicle and and driver over-reaction and subsequent loss of control.

The tracks through the grass were identified as belonging to V-1 by identifying the tread print in certain areas.

Extensive measurements were taken and are reflected in the large accident site diagram included with this report.

After inspecting and measuring the accident scene, V-1 was inspected and photographed at a residence in Oklahoma City. There was considerable damage to the front of the vehicle. The rear of the vehicle was inspected and evidence was found of a prior collision which was related to me by Mr. Stevens. It was later learned that this prior accident occurred on October 31, 1974 on Oklahoma State Highway 37 near Guthrie, Oklahoma. Mr. Stevens related to me that he had made the repair estimate on this car and this damage on the right rear was unrelated to the accident which occurred on November 13, 1974.

Investigation Continued

The rear bumper was covered with road film in the form of dust. On the left rear corner of the bumper there was a fresh dent approximately 2 inches in length. Initial inspection showed this dent was made by an object moving from rear to front. Careful study was made to rule out any possibility of this dent being made by stones or the concrete on the wingwalls. On the left rear fender of V-1 there was a large dent which was also made by an object moving from rear to front and subsequent study was made to rule out any damage being done by rocks or the concrete wingwalls. This vehicle was photographed extensively and these photographs are included with this report.

After having made an initial analysis of the accident, consideration was given to employing Dr. B. J. Harris to render an opinion in this investigation. Dr. Harris is a consulting engineer in Dallas, Texas. He was a former professor at the University of Oklahoma and is well versed in the analysis and reconstruction of accidents and structural failures.

On December 8, 1974 Dr. Harris and I went to Oklahoma City for the purpose of his investigation. Dr. Harris' Addendum to this report, based upon his inspection and analysis, is included.

Addendum by Dr. B. J. Harris

On December 8, 1974, Dr. B. J. Harris, consulting engineer from Dallas, Texas, inspected the subject vehicle garaged in Oklahoma City and also the scene of the accident. It was understood that the vehicle inspected was driven by Miss Karen Silkwood and that she was killed in the accident. This accident scene was on Highway 74 just north of Oklahoma City. It was further understood that the accident occurred near 7 p.m. on November 13, 1974.

The 1973 Honda Civic was travelling southbound on a long straight stretch of asphalt concrete pavement which was in excellent condition. The highway had a visibly well-defined crown sloping each way away from the centerline. The vehicle apparently crossed over to the opposite side of the centerline, travelled down grade into a shallow ditch, and crashed into the south wingwall of a concrete box culvert.

Mr. A. O. Pipkin made several observations and deductions which Dr. Harris was called on to study. These were understood to be (1) the fact that the vehicle crossed over to the opposite side of the highway from what one would normally expect under the given accident conditions, (2) the vehicle appeared out of control as it left the highway, and (3) a dent in the left side of the rear bumper appeared fresh and unrelated to the accident. These items were studied and the report herein addresses each.

A vehicle travelling on a straight highway which has a noticeable center crown, will, in the opinion of Dr. Harris, tend to gradually pull off to the right due to the component of gravity force acting to the right. This would only be counteracted by crosswinds or vehicle mechanical problems such as misalignment. Therefore, it would be reasonable to expect that a person falling asleep while driving on a stretch of highway, such as in the case of Miss Silkwood, would have travelled off the highway to the right.

A vehicle gradually travelling off the highway would, undoubtedly, track very closely. That is, the front and rear tire marks would closely coincide. Mr. Pipkin inspected the scene of the subject accident a few days after it happened and reported that tire marks in the grass were clearly visible and defined the vehicle path up to the culvert on which it impacted. He reported the tire marks showed the vehicle was yawing (or rotating about a vertical axis) as well as translating forward. The photographs taken by Mr. Pipkin do clearly show tire marks which could be interpreted as defining the yawing motion of a vehicle. At the time Dr. Harris visited the scene, the tracks were not sufficiently clear to define the vehicle motion. In addition to the passage of several weeks of time, large tractor tire prints were numerous in the same location. A tractor had worked, it appeared, in spreading a small amount of dirt on the north side of the north wingwall of the culvert. Assuming the tire marks referenced by Mr. Pipkin were those of the vehicle in question, it would appear highly likely that the vehicle was out of control when it left the pavement.

The vehicle was inspected very closely, and in particular the dent on the left side of the rear bumper. A powerful magnifying instrument was used to closely check the bumper surface condition. The dent in question was located on the corner of the bumper where it wraps around the side of the body. The dent was approximately 3/4 inches wide vertically and 2 inches in length. The dent was not covered by road film and appeared to be fresh. It could not be determined what material caused the dent but in careful inspection of the dent it could be seen in the scratch patterns that whatever force caused the dent did so along a line of action approximately 30 degrees to the horizontal. The smearing of the black paint on one side of the dent clearly defined that the force acting on the bumper travelled, relative to the bumper, clockwise when viewed from above the bumper.

The first reaction was to consider possible impact on something during the accident. It was observed that caked red clay at the front left corner of the roof had many pieces of broken glass embedded in it. A measurement of 76 inches was made horizontally from the left front roof corner to the dent located on the left rear bumper. At the scene of the accident, a cluster of broken glass was found which was undoubtedly the location of the left front corner of the roof. An arc of 76 inches was swung and the ground was closely inspected. Nothing was found which would have caused the dent. There were red rocks in the stream bed nearby but they would surely have left stains in the scratches. The direction of the scratches in the dent definitely ruled out their being made by the bumper hitting the concrete wingwalls or something on the ground during impact since any such blow would have made scratches in the opposite direction to those noted. The possibility of banging the vehicle during extraction from the ditch was considered carefully. There does not appear to be any way that the direction of the scratches in the dent could have been caused by handling during removal.

A dent in the left rear fender below the bumper first appears to be related to the dent in the bumper since they are in close proximity. The crinkle paint and scratches in the fender dent are horizontal in direction and appear to have been made by something moving rear to front, relative to the vehicle.

The dynamics of the vehicle are difficult to ascertain with the limited information available, however, from terrain observations and clear impact imprints on the culvert wingwall, an estimation is attempted. It appears that the vehicle slowed while skidding on the grass and then ramped up slightly on the right side, due to the embankment slope, just before it arrived at the north wingwall. This movement induced a counter-clockwise rolling action, when viewed from the rear, which is substantiated by the impact marks on the south wingwall. The vehicle's right side probably passed over the lower east end of the north wingwall and then became airborne until it impacted the south wingwall. It appears that the vehicle was airborne approximately 19 to 20 feet forward and 3 to 4 feet downward. This would produce an impact velocity of approximately 30 miles per hour.

The hypothetical question is addressed as follows on what motion might be expected from the vehicle involved in the subject accident if some force impacted it on the left corner of the rear bumper at the location of the previously discussed bumper dent. The line of action of impact has to be known accurately in this case since curved surfaces are involved. If the impact force normal to the bumper is the primary force it would induce a rotation of the vehicle in a direction controlled by which side of the center of gravity (c.g.) the line of impact would act. If the line of impact extended to the right of the c.g. then the spin would tend to be counter clockwise when viewed from above. Clockwise if line is to be the left. If friction during impact was significant then it would impart a strong force component which would contribute to a clockwise rotation. If the force made a bumper dent the size of that noted, it would have been a glancing blow of limited magnitude and likely would not have directly changed the linear direction of travel or velocity. However, it could have induced significant rotation motion which might have resulted in driver over-reaction and resulting complete loss of control.

B J Harris

Dr. B. J. Harris, P.E.
Consulting Engineer

PHOTOGRAPHS

I am including with this report 20 color photographs of the accident scene and V-1. A description is attached to each photograph.

SCALE DIAGRAMS

There are several scale diagrams included with this report and they are described as follows:

1. The accident site plan showing the path of the car as it left the roadway, its attitude and its attack position on the north wingwall.

2. A combined plan view and profile view of the culvert wingwall.

3. Attitude diagram as V-1 leaves the roadway.

4. A plan view of the culvert wingwall and the attack position of V-1 on north wingwall.

5. Profile of south wingwall and the attack position of V-1 on south wingwall.

OBSERVATIONS AND CONCLUSIONS

From all of the evidence presented and the analysis, I can draw the following observations and conclusions:

1. V-1 was proceeding south at approximately 50 - 55 mph.

2. V-1 left the east side of the roadway at an attitude which is unnatural to ordinary driving conditions and the mechanical capacity of the vehicle.

3. V-1 travelled in this attitude for approximately 240 feet on the grass area east of the shoulder and arcing slightly to the west before striking the north wingwall.

4. V-1 struck the north wingwall as indicated on the scale diagram.

5. Because of the contour of the land, the space of the north wingwall and its own velocity, V-1 rotated in a counter clockwise motion as viewed from the rear and became airborne.

6. V-1 was airborne for approximately 20 feet, struck the mud just north of the south wingwall and then impacted head-on with the south wingwall.

7. The attack position on the south wingwall is as indicated on one of the scale diagrams.

8. The impact speed at the south wingwall was approximately 30 miles per hour.

9. The dents on the left corner of the rear bumper and the left rear fender were not made due to the impact with either the north or south wingwall or the subsequent wrecker handling of V-1 after impact and during its removal to the garage.

10. Based on all of the evidence present, along with the conclusions drawn by Dr. Harris, it is my opinion that there is enough circumstantial evidence present to indicate that V-1 was struck from the rear by an unknown vehicle, causing it to go out of control, due to either the initial impact or the combined impact and driver over-reaction.

Respectfully submitted,

ACCIDENT RECONSTRUCTION LAB

A. O. Pipkin, Jr.

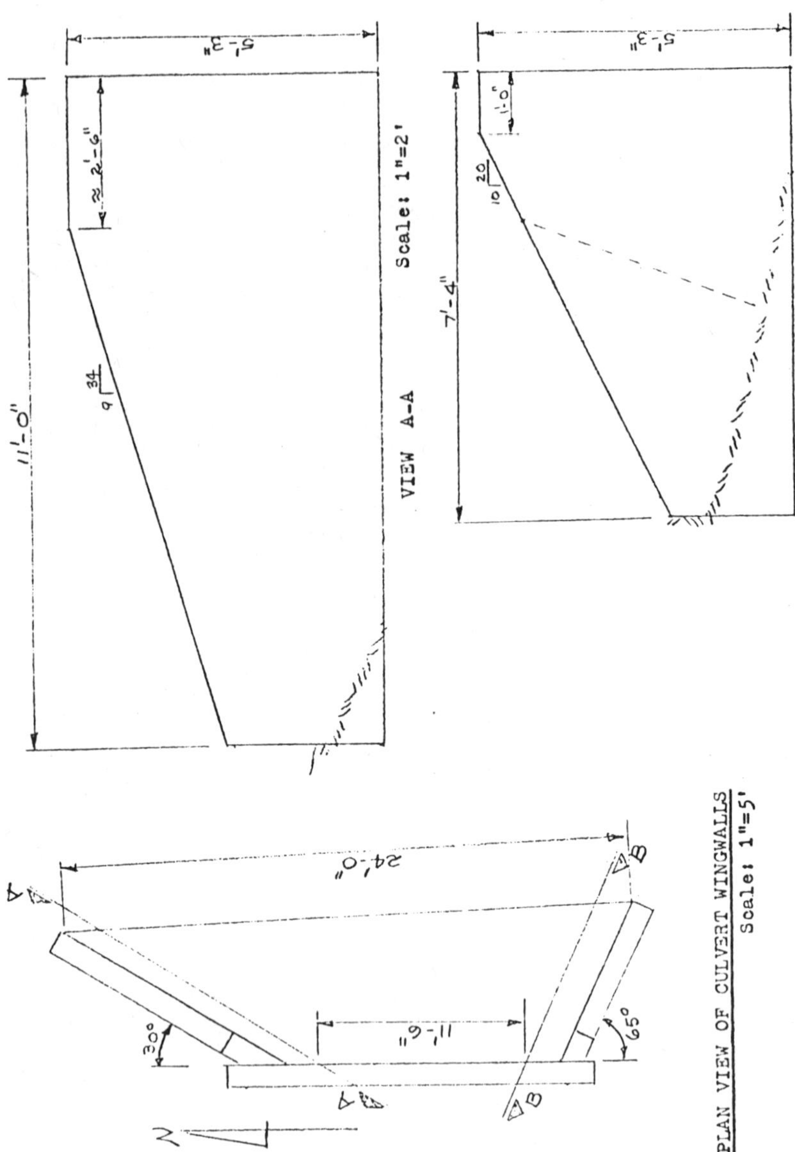
5'-3"
≈ 2'-6"
11'-0"
Scale: 1"=2'
VIEW A-A
5'-3"
1'-0"
7'-4"
24'-0"
11'-6"
30°
65°
PLAN VIEW OF CULVERT WINGWALLS
Scale: 1"=5'

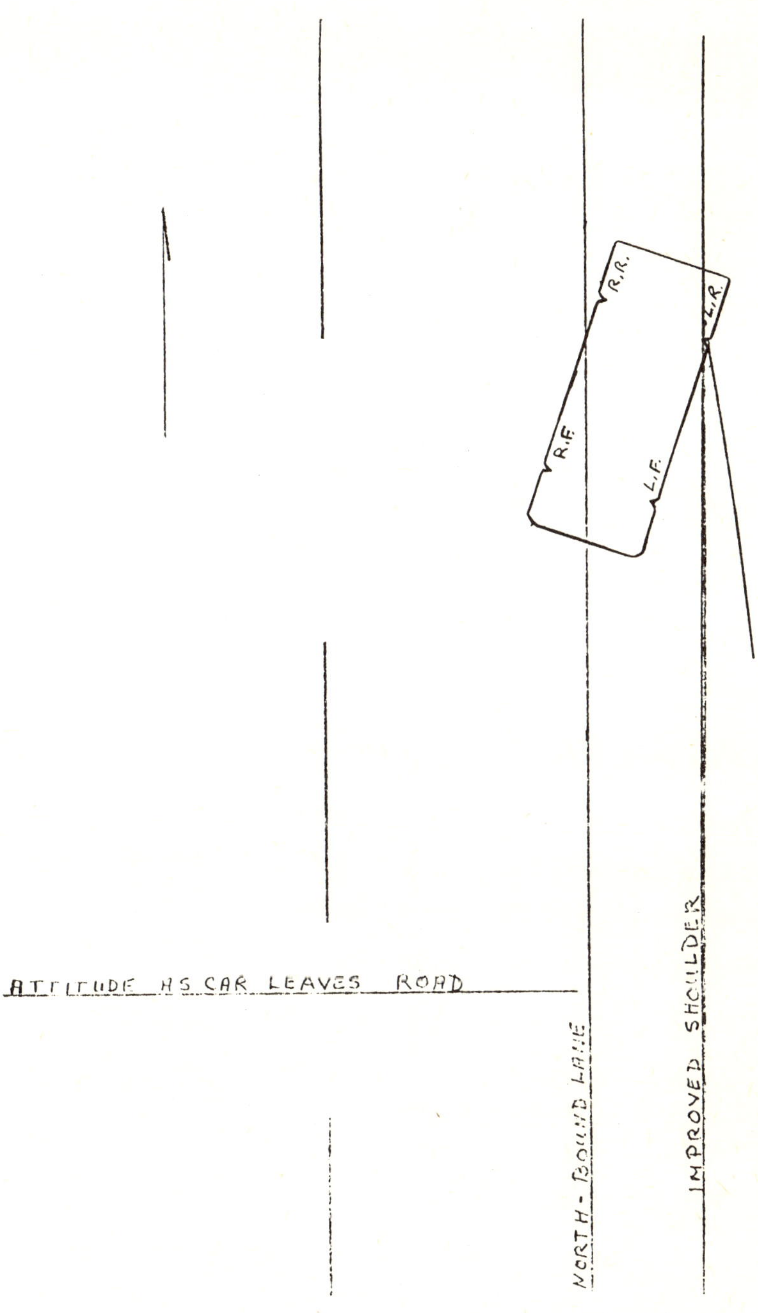
R.R.
L.R.
R.F.
L.F.
ATTITUDE AS CAR LEAVES ROAD
NORTH-BOUND LANE
IMPROVED SHOULDER

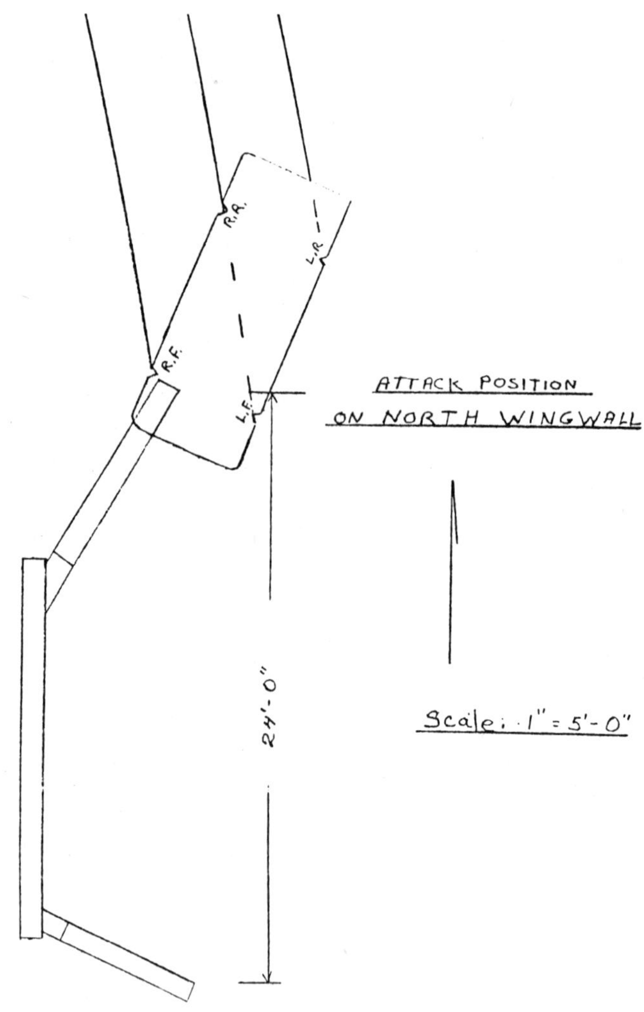

PLAN VIEW OF CULVERT WINGWALLS

ATTACK POSITION ON SOUTH WING WALL

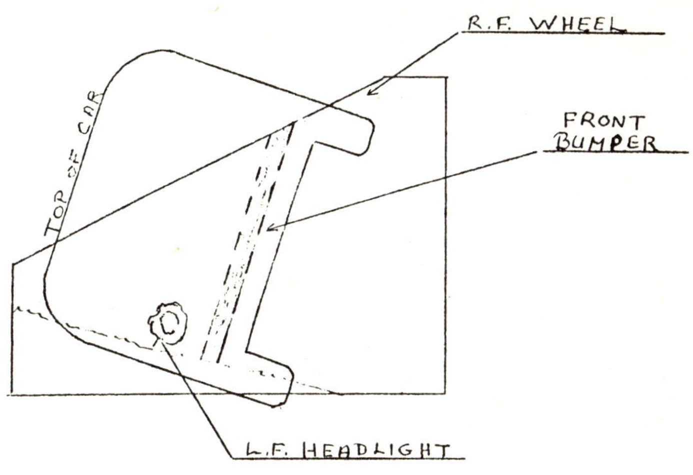

PROFILE OF SOUTH WING WALL

SCALE : 1" = 2'-0"

REPORT OF ACCIDENT OF KAREN GAY SILKWOOD OBTAINED FROM THE OKLAHOMA DEPARTMENT OF PUBLIC SAFETY

PARTMENT OF PUBLIC SAFE'

"Don't say it - Write it"

TO Lt. Darrell W. Wiemers #32 DATE November 26, 1974

FROM Lt. Ronnie Johnson #92

SUBJECT Injury Accident Involving Karen Gay Silkwood

Vehicle One, driven by Karen Gay Silkwood, went out of control, backed off a 13-foot embankment, and struck a fence post approximately 40 feet north of the north edge of the roadway and approximately 110 feet northwest of control post number 42-06-10.00. This information was obtained at the scene after contacting Mr. Martin of Martin's Wrecker Service. Mr. Martin, Trooper Adams #293, and myself went to 3.8 mile west of Guthrie on SH 33 and Mr. Martin told me that Miss Silkwood called his service and asked him to meet her at this location to pull her vehicle out of the ditch at approximately 1:30 p.m. He met the subject, pulled her vehicle out of the ditch, and she paid him $13 cash and drove her vehicle to work.

Dr. Clarence Shield, Jr. was contacted by phone on November 23, 1974, at 2200 hours. I asked Dr. Shields if he treated a Miss Silkwood on October 31, 1974, for injuries resulting from an accident. Dr. Shields stated that he did treat Miss Silkwood for an accident. Dr. Shields stated that he did treat Miss Silkwood for whiplash which she said was from a collision, but she was treated November 1, at approximately 5:00 p.m., and also on November 4. Dr. Shields said that he saw Miss Silkwood in October and she was complaining about not being able to sleep in the day time (while working nights) and he prescribed Methaqualone. Dr. Shields advised his records would be turned over to Dr. Chapman, if any further information is needed.

Respectfully,

Lt Ronnie Johnson

Ronnie Johnson, Lieutenant
Supervisor, District One

1. Go ahead
2. Turn left
3. Turn right
4. Make "U" turn
5. Stop
6. Slow for cause
7. Start from park
8. Change lanes
9. Overtake or pass
10. Back
11. Start in traffic lane
12. Remove stopped parked
other

1. Went ahead
2. Turned left
3. Turned right
4. Entered "U" turn
5. Stopped
6. Slowed
7. Started from park
8. Entered other lane
9. Overtaking
10. Backed
11. Started forward
12. Removed stopped parked
other

1. Oneway road
2. Alley
3. Two lanes
4. Three lanes
5. Four or more divided
6. Four or more not divided
7. Driveway
8. Turn bay
9. On ramp
10. Off ramp
other

1. Stop sign
2. Traffic signal
3. Flashing signal
4. Yield sign
5. Warning sign
6. RR gates, signals
7. No passing zone
8. Officer
9. No control
10. Abnormal control

1. Level
2. Straight-upgrade
3. Straight-downgrade
4. Straight-hillcrest
5. Curve-level
6. Curve-upgrade
7. Curve-downgrade
8. Curve-hillcrest
9. Sharp curve (add to above if applicable)
other

1. Apparently normal
2. Drinking-ability impaired
3. Odor of alcoholic beverage
4. Very tired
5. Sleepy DRUGS
6. Sick
7. Condition not known

Body defects

other

OBJECT STRUCK BY VEHICLE OR LOAD ON FIRST CONTACT

1. Street light pole
2. Other utility pole
3. Guard rail
4. Guard post
5. Culvert
6. Traffic signal
7. Barrier
8. Curb
9. Island
10. Traffic control sign
11. Ditch
12. Embankment
13. Tree
14. Dividing Strip
15. Retaining wall
Bridge (u.v., abutment, etc.)
Other highway struct.
other

POINT OF FIRST CONTACT ON VEHICLES

1. Front-center
2. Front-right
3. Front-left
4. Rear-center
5. Rear-right
6. Rear-left
7. Right-side-center
8. Right-side-forward
9. Right-side-aft
10. Left-side-center
11. Left-side-forward
12. Left-side-aft

ROAD CONDITION: 1. Dry 2. Wet 3. Ice/Snow 4. Muddy other

ROAD SURFACE: 1. Concrete 2. Asphalt 3. Gravel 4. Dirt other

LOCALITY: 1. Residential 2. Business 3. Industrial 4. School 5. Not built-up other

LIGHT: 1. Daylight 2. Darkness 3. Lighted 4. Dawn 5. Dusk other

WEATHER: 1. Clear 2. Partly cloudy 3. Overcast 4. Raining 5. Snowing other

VEHICLE CONDITION: 1. Apparently normal 2. Brakes 3. Steering 4. Headlights 5. Rearlights 6. Tires other (Abnormal Condition)

TIRE CHECK: 8 8 8 8

WHAT PEDESTRIAN WAS DOING

1. Crossing at intersection
2. Crossing not at intersection
3. Crossing at other than ...
4. Getting on-off vehicle
5. Walking with traffic
6. Walking against traffic
7. Push/work on vehicle
8. Playing
9. Other working
other

Indicate North by Arrow

DIRECTION OF TRAVEL
Veh. 1 N☐ S☐ E☐ W☐
Veh. 2 N☐ S☐ E☐ W☐

COLLISION DIAGRAM

SEE PAGE TWO

visibility obscured by

Location of FIRST Damage or Injury Producing Event on Travel Portion of Trafficway?
Yes☐ No☐

Defect in Road

REMARKS: (COMMENTS THAT WILL CLARIFY REPORT — refer to vehicles by number)

Veh. #1 south bound on SH 74. Ran off east side of roadway. Veh. travelled approx. 255 ft. in east bar ditch. Veh. struck north retaining wall approx. 3 ft. from face of bridge. Veh. airborn approx. 24 ft. and struck the south retaining wall approx. 3 ft. from face of bridge and approx. 3 ft. above ground level. Veh. landed on left side. No skid marks. Witnesses interviewed stated that they had advised the driver was in no physical condition to operate a vehicle.

UNSAFE, UNLAWFUL, OR OTHER ACTION (this section — primarily for general statistics and administrative purposes)

Unit 1 2	Describe	Unit 1 2	Describe
1. Failed to Yield		10. Improper Overtaking	
2. Followed too Closely		11. Improper Parking	UNDER INFLUENCE OF
3. Unsafe Speed		12. Inattention	
4. Made Improper Turn		13. Wrong way on –	DRUGS
5. Changed Lanes Unsafely		14. Improper Start from –	
6. Stopped in Traffic Lane		X 15. Other Improper Act or Movement	
7. Failed to Stop		16. Not Known – or – No Improper Action	
8. Unsafe Vehicle		17. Other Action – not directly related to collision	
9. Left of Center		18. Pedestrian Action	

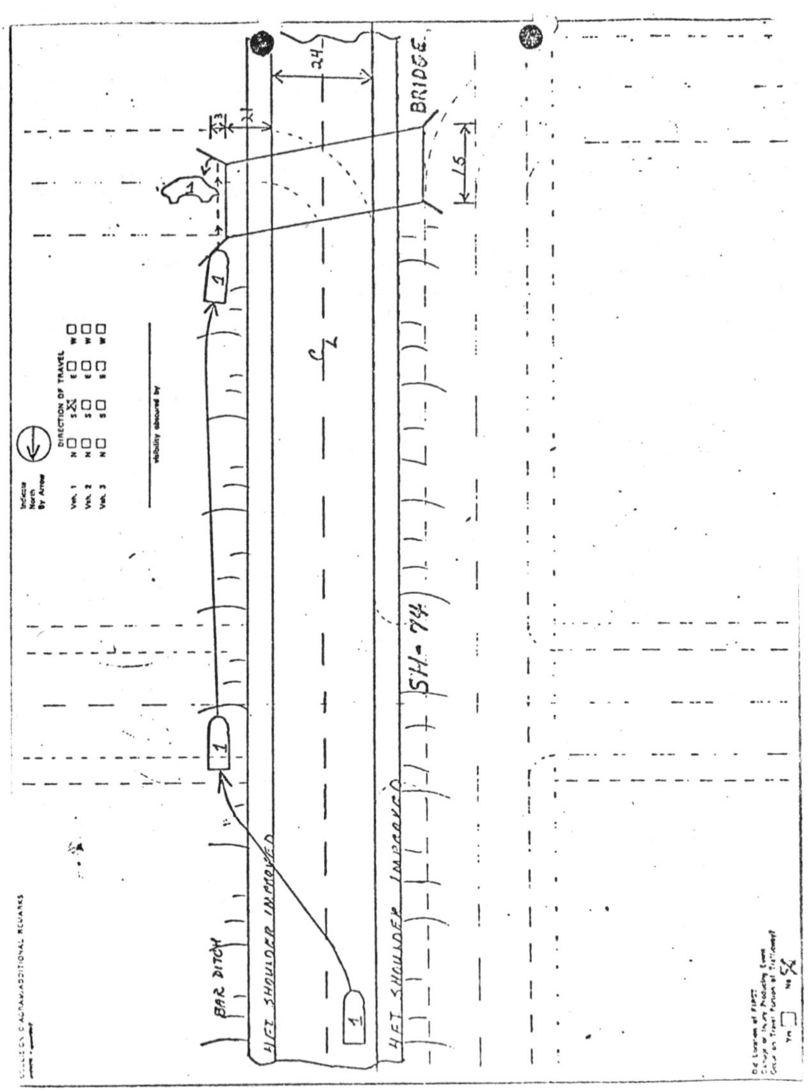
BAR DITCH
4 FT SHOULDER IMPROVED
SH-74
BRIDGE
24
15
DIRECTION OF TRAVEL

SUPPLEMENT to the OKLAHOMA OFFICIAL POLICE TRAFFIC COLLISION REPORT

Reporting Agency: OKLAHOMA HIGHWAY PATROL

Date: Nov. 13, 1974 Day of Week: Wednesday Hour: 7:30 PM County: LOGAN

Total Number Motor Veh. Involved: 1 No. Killed: 1

Distance From Nearest City Limits: 7.3 Miles S ☒

CRESCENT (Name of Nearest City)

ON: STATE HIGHWAY 74

AT/NOT AT INTERSECTION: 500 ft. south of County Road

REMARKS

Sign here: Tr. Rich Fagan #381 — Badge No. 381 — ONE — AR — Date of report: 11-15-74

OFFICIAL POLICE TRAFFIC COLLISION REPORT

OKLAHOMA — Reporting Agency: OKLAHOMA HIGHWAY PATROL

Date: November 13, 1974 — Day of Week: Wednesday — 7:30 PM — County: LOGAN

CRESCENT — Distance From Nearest City Limits: 7.3 Miles — S

No. Motor Vehicles Involved: 1 — No. Injured: 1 — No. Killed: —

On: STATE HIGHWAY 74

Not at intersection: 500 ft. South of COUNTY ROAD

County No. 42 — East 0090 — North 0169

Time Notified: Date: 11-13-74 Hour: 8:05 PM

Arrived At Scene: Date: 11-13-74 Hour: 8:15 PM

Unit 1

Driver: SILKWOOD, KAREN GAY — 848-6181

Address: 836 N.W. 48th Okla. City, Okla. 73118

License: 76 OKLA. 456780131 — Operator

Age: 28 — Race: W — Sex: F — Date of Birth: 2-19-46

Veh. ID No.: SBC1014541 — Driver Training: NO

Vehicle: 73 Honda Civic 2 Dr. C

License Plate: 74 OKLA. YF 8261

Owner's Name: Driver

Address: Same as Driver

55 MPH legal speed — 50-55 MPH before contact — 40-45 MPH at contact — $2000.00 estimated damage

Veh. removed to: CRESCENT — by: SEBRING WRECKER

Injured: SILKWOOD, KAREN GAY — 836 N.W. 48th OKC 848-6181 — 28 F W — Veh. 1

Injured Taken To: 1. LOGAN CO. HOSP., GUTHRIE, OK — By: GUTHRIE FIRE DEPT. AMBULANCE — Time left scene: 8:45 — Time arrived hospital: 9:00

Investigation made at scene: Yes — Investigation completed: Yes — Chemical Test: Blood

Tr. Rich. Fagan — Badge No. 381 — District: ONE — Date of report: 11-15-74

Sheet 1 of 2 Sheets FATALITY • yes

OKLAHOMA

OFFICIAL POLICE TRAFFIC COLLISION REPORT

Do not write in this space

Reporting Agency: Oklahoma Highway Patrol

No.

Date: 10-31-74 Day of Week: Thursday Hour: 1:15 AM PM County: Logan

Guthrie — Name of Nearest City (If outside city limits) — Distance From Nearest City Limits: 3.8 Miles — N S E W

Total Number Motor Veh. Involved: 1 — No. Injured: 1 — No. Killed: 0

LOCATION: IN (city or town) — ON SH 33 (name or number of street or highway) — AT (name of intersecting road or landmark)

NOT AT INTERSECTION: ft. North / South / East / West OF (show nearest intersecting street or highway) — OF: Control Location

STATE HIGHWAY CODES: Hwy Class — Control No. — Ins. I.D. — Location

County No. 42

County Section Line Grids Or City Street Codes: City No. — East 0 1 7 2 — North 0 2 0 0

Loc. Feat. — Collision Codes — Infor. Pos. Cl.

Time Notified: Date: 11-22-74 Hour: 4:35 PM

Arrived At Scene: Date: 11-23-74 Hour: 5:00 PM

Unit 1

Driver: Silkwood Karen Gay (last, first, middle, phone no.)

Address: 836 NW 48th Oklahoma City, Ok. (street or RFD, city and state, zip code)

License: 75 Okla. 456780131 (exp. yr., state, number) — [X] Operator — Chauffeur — other

Age: 28 Race: W Sex: F Date of Birth: 2 19 46 (Mo. Day Year) — Highschool Driver Training: Yes / No

Veh. ID No.: SBC101451

Vehicle: 73 Honda 2-dr. (year, make, model, style) (C) size — towed vehicle

License Plate: 73 Okla. YF-8261 (year, state, number) — Lap belt: Equipped Yes, In Use —; Shoulder belt —; Crash helmet —; Ejected: Yes

Owner's Name: Same as driver

Address: same as driver

Is Veh. Operable? Yes [X] No

55 MPH (legal speed) 40-45 MPH (before contact) estimated speed 10-15 MPH at contact — Burned? Yes — $ 200.00 estimated damage

Veh. removed to: — by: Martin Wrecker

Unit 2

Driver / Pedestrian / Animal / Train, etc. — (last, first, middle, phone no.) — Address — License — Age — Sex — Race — Date of Birth — Veh. ID No. — Vehicle — License Plate — Owner's Name — Address — Is Veh. Operable? Yes No — legal speed MPH, estimated speed MPH before contact, MPH at contact — Burned? Yes — $ estimated damage — Veh. removed to: — by:

INJURED OR WITNESSES

	Injured/Witness	Name (last, first, middle)	Address	Phone No.	Age	Sex	Race	Veh. #	Position in vehicle
1	Injured [X]	Silkwood, Karen Gay	836 NW 48th, Ok. City, Oklahoma		28	F	W	1	X
2									
3									
4									

Injury Type: Head — Trunk-External — Trunk-Internal — Arm Leg — Crash Helmet — Ejected? — Shoulder Belt: Equipped, In Use — Lap Belt: Equipped, In Use

Injured Taken To: 1. Treated By: Dr. Shields, Clarence

2. By: 11-1-74 5 p.m. &

3. By: 11-4-74

Time left scene: — Time arrived hospital:

Damage to property other than vehicles: $ — Owner — Address

Investigation made at scene? Yes [X] — Investigation completed? Yes [X] — Operator's report given to driver — Veh. 1 [X] — Photos taken? No [X] — CHEMICAL TEST: Driver #1 Yes X — Driver #2 — Pedestrian — INT & FIN: No X

Name: last, first, middle — Citation: — Citation No.

SIGN HERE: [signature] (Officer's rank & name) — Badge No. 92 — District or Division 1 — Reviewed by — Date of report: 11-24-74

WHAT VEHICLES WERE GOING TO DO (Unit 1: X — 1. Go ahead)

WHAT VEHICLES DID (Unit 1: X — 8. Entered other lane)

TYPE OF ROAD (X — 3. Two lanes)

TRAFFIC CONTROL (X — 9. No control)

ROAD CHARACTER (X — 3. Straight-downgrade)

CONDITION OF DRIVERS AND PEDESTRIANS (X — 7. Condition not known)

POINT OF FIRST CONTACT ON VEHICLES (X — 5. Rear-right)

LIGHT — WEATHER (X — 4. Raining)

ROAD CONDITION (X — 2. Wet)

ROAD SURFACE (X — 1. Concrete)

LOCALITY (X — other: Rural)

VEHICLE CONDITION (X — 1. Apparently normal)

COLLISION DIAGRAM

HILL CREST

61FT

FENCE POST

10 FT GRASS SHOULDER

NO PASSING ZONE

Did Location of FIRST Damage or Injury Producing Event Occur on Travel Portion of Trafficway? Yes ☐ No ☒

REMARKS: (Comments that will clarify report) (Refer to vehicles by number)

Vehicle one went out of control, backed off a 13-foot embankment, and struck a fence post appr. 40 feet north of the north edge of the roadway and appr. 110 feet northwest of control post number 42-06-10.00. (Due to the fact that this accident was not reported to this Department, all information was obtained 23 days after the crash and is at best an anachronistic report.)

UNSAFE, UNLAWFUL, OR OTHER ACTION (this section – primarily for general statistics and administrative purposes)

Unit 1	Unit 2		Describe	Unit 1	Unit 2		Describe
		1. Failed to Yield				10. Improper Overtaking	
		2. Followed too Closely	Wet Pavement			11. Improper Parking	
‹		3. Unsafe Speed				12. Inattention	
		4. Made Improper Turn				13. Wrong way on –	
		5. Changed Lanes Unsafely				14. Improper Start from –	
		6. Stopped in Traffic Lane				15. Other Improper Act or Movement	
		7. Failed to Stop				16. Not Known – or – No Improper Action	
		8. Unsafe Vehicle				17. Other Action – not directly related to collision	
		9. Left of Center				18. Pedestrian Action	

UNITED STATES
ATOMIC ENERGY COMMISSION

DIVISION OF INSPECTION
REPORT

44-2-339

By Kenneth H. Jackson   Dated December 19, 1974

Title: KERR McGEE, NUCLEAR DIVISION, CIMARRON NUCLEAR FACILITY, CRESCENT, OKLAHOMA - ALLEGED FALSIFICATION OF QUALITY ASSURANCE RECORDS

BRIEF OF FINDINGS

This investigation was predicated upon a request from Mr. John A. Erlewine, General Manager, Headquarters, to conduct appropriate investigation at the Cimarron Nuclear Facility of the Kerr-McGee Nuclear Division, Crescent, Oklahoma, in response to allegations by a representative of the Oil, Chemical, and Atomic Workers International Union that quality assurance records pertaining to the production of Fast-Flux-Test-Facility driver fuel pins were being falsified.

It was alleged by confidential informants to a member of the Oil, Chemical, and Atomic Workers International Union staff during personal and telephone interviews, both of which were tape recorded, that the falsification of quality assurance records and other related irregularities pertaining to the quality of fuel pins being manufactured by Kerr-McGee's Cimarron Nuclear Facility (KM) for the Westinghouse Hanford Company (WHC), prime contractor for construction of the Fast-Flux-Test-Facility at Richland, Washington, were occurring at the KM facility. A transcript of these tape recordings was provided, along with the identities of the persons voicing the complaints. Thereafter, the confidential informants were reinterviewed, and substantially reconfirmed their statements to the union official.

A composite of the total information obtained on this subject resulted in the following allegations:

1. Alleged falsification of photomicrograph negatives of weld test samples;

2. Alleged improper use of quality control sample analytical data by KM supervisors and employees;

3. Alleged irregularity in pellet inspections; and

4. Alleged falsification of computer input data.

Investigation with regard to item No. 1, supra, disclosed that the driver fuel pins are assembled in gloveboxes and that the final two steps in the assembly involves welding of top and bottom end caps to the fuel pin cladding. It was disclosed that the welding operation is a semiautomatic process with a maximum number of 60 fuel pins per fuel pin lot being assembled in a rack and the end caps welded separately; that the first and last weld (prior to and at the conclusion of welding the actual pins) are sample welds in which an end cap is welded to a short piece of cladding; and that the sample welds, consisting of two for the top end caps and two for the bottom end caps, are then sent to the KM Metallography Laboratory for destructive testing. It was determined that the sample welds are split longitudinally; that one-half of the weld is mounted into a bakelite-type substance; that it is then ground, polished, and subjected to a visual examination with a metallograph; that the laboratory analyst at that point visually inspects the weld surface at a range varying from 10X through 500X for evidence of porosities, inclusions, cracks, and intergranular attack; and that based upon his observations, the analyst either accepts or rejects the weld in question. It was stated that acceptance of the sample welds at this point in effect clears the fuel pin lot in question insofar as the welding of end caps-to-cladding is concerned, and that a rejection of a sample weld requires additional sample examinations, and/or in some cases, the actual destructive testing of a "hot" or loaded fuel pin in order to ascertain the integrity of the welds.

It was disclosed that after the visual inspection above described, the analyst is required to take a 4" x 5" black and white exposure of the throat area of the sample weld at 100X; that this exposure is developed in an adjoining darkroom by the analyst into a negative; and that the negative is then used to print a number of black and white 4 x 5 photographs, which are thereafter submitted as part of the required certification documentation when the fuel pins are shipped to the customer (WHC).

Investigation disclosed that one laboratory analyst during the period of approximately February through July 1974 used a felt-point pen to "touch-up" negatives prior to printing of the 4 x 5 black and white photographs, and in so doing, covered defects in the negative to prevent their showing up on the photographs. An examination of the negative files disclosed a total of 53 negatives possessing questionable characteristics, and that of this number, an estimated 40 possessed inkspots in varying degrees and quantities. (Due to the degree of technical expertise required to make a judgement as to the nature of the defects covered by these inkspots, the actual significance of the touching-up was not resolved during the course of this investigation; however, the negatives in question were secured for later examination and resolution.)

The analyst, who is no longer employed by KM, acknowledged touching up the negatives in question, and a signed statement was obtained from him to this effect. In this statement, he said, in substance, that he acted alone without either the consent or knowledge of management or supervision; that he made the alterations in an effort to conceal small artifacts occurring during the darkroom processing of the negative itself; that such artifacts, possibly the result of defective emulsion, static electricity, particles of lint, etc., were sufficient cause for his supervisors to reject any subsequent photograph from such a negative; and that since rejection of the photograph caused him to have to rerun the entire weld sample, he perfected this technique to avoid the necessity of performing the additional work. He also states that at no time did he endeavor to cover defects in the weld itself by the use of this technique, nor did he have any intent to defraud, or to falsify, records as such.

It was further disclosed that one other laboratory analyst had knowledge of the touching-up of negatives; that none of the other analysts had engaged in such practices; and that at no time prior to the investigation was management or supervision made aware of the actions of the analyst in question. An examination of records in the Metallography Laboratory, performed in conjunction with the investigation of this allegation, also disclosed some irregularities insofar as proper documentation of changes, eradications, etc., to records of weld test sample examinations and findings, although such discrepancies were not considered falsification in fact, nor substantive in nature.

Investigation with regard to item No. 2, supra, disclosed that quality control samples, provided by WHC, and accompanying analytical data for the individual samples are used to verify the continued adequacy of the KM processes and analytical techniques.

It was disclosed that on each occasion when the KM General Analytical Chemistry Laboratory performs analyses on samples from pellet lots, that a quality control sample is also analyzed; that after analysis of the quality control sample, the analyst is required to relate his findings to a designated company official who alone possesses the analytical data supplied by WHC for the samples in question; and that this official, after referring to the data, then advises the analyst whether his findings are "in" or "out" of the specified ranges.

It was alleged that such quality control sample analytical data was obtained possibly surreptitiously by one analyst and one supervisor; that, therefore, this analyst always "passed" the quality control analyses test; that as a result of his apparent success on such analyses, he was routinely assigned to perform the majority of such analyses; that he obtained excessive amounts of overtime work as a result; and that the integrity of the sample analysis program, in general, was questionable as a result of his possessing the "answers."

It was stated by three or four analysts that they felt sure the analyst in question possessed the quality control sample analytical data; that the KM official charged with responsibility for safekeeping of the sample analytical data and to whom analysts were required to go for verification of such analyses, kept the data in his desk drawer; that his office was located in the laboratory area; that a number of keys available in the area fit all desks, including this official's; that it would have been a simple matter for anyone to have, therefore, obtained surreptitious access to the desk and information; that they found entries in the analyst's personal notebook which appeared to be the analytical data; and that they reported him to supervision.

One analyst stated in a sworn statement that she personally saw such data in the possession of the analyst in question, and said also that the suspect informed her the data she saw was sample analytical data.

It was disclosed that continuing rumors to the effect this analyst possessed the data in question caused an internal inquiry to be conducted; that the results thereof were inconclusive; but that the analyst was subsequently transferred to another duty assignment.

It was further disclosed that this analyst had received considerably more overtime wages than his co-workers; that he had conducted an estimated 75-80 percent of the quality control sample analyses which was responsible for the overtime; and that, according to supervision, he had received these assignments because of his knowledge, experience, and general expertise in analytical chemistry.

In a signed statement, the analyst in question denied any access to quality control sample analytical data, although he acknowledged that on one occasion a supervisor had provided him the answer to a particular analysis.

It was disclosed that on some occasions, the supervisor of the General Analytical Chemistry Laboratory changed the slope of graph lines which interconnected points, as nearly as possible, on graphs representing analytical findings as to nitrogen, halides, etc., in samples. An examination of such records disclosed that the graphs are worksheets; that the data contained thereon reflects efforts to maintain close in-process checks and controls; that while the supervisor does not sign or date the changes he makes on such workpapers, they represent only a supervisor's judgment being imposed over that of a less knowledgeable analyst; and that the changes made were within the parameters of the specifications.

Investigation with regard to item No. 3, supra, disclosed that pellets are visually inspected for the presence of chips, cracks, discoloration, etc., by production operators while they are being assembled in a "V" trough immediately prior to being weighed, measured, and inserted into fuel pin cladding. It was disclosed that prior to approximately June 1974, the production employees assigned to this operation physically picked up and/or rotated each pellet 360° in the course of visually inspecting them; that the specifications for such inspections called for only one-half of the surface to be visually inspected; and that KM supervision elected to discontinue that 360° inspection effort which was over and above that required; and that the change in emphasis resulted in higher production by the operators at this duty station. There was no evidence to indicate unsatisfactory and/or defective pellets were passed as a result of this change. Operators advised that they are still encouraged to discard defective pellets and that they do so.

Investigation with regard to item No. 4, supra, disclosed that the confidential informant who made this allegation advised that he was employed as a statistical clerk; that he was responsible for calculating series of analytical data; that worksheets listing such data and calculations were prepared; and that on one occasion, a laboratory supervisor obtained such a worksheet from him, took it to the laboratory, thereafter returned it without making any changes to the entries; but that at a later time, the computer printout of such data was not factually identical to that supplied on the worksheet. The informant said that as he recalled, the data in question pertained to pellet lots Nos. 27 and 28, produced in July or early August 1974.

Investigation, including an examination of all worksheets and supporting data, for pellet lots Nos. 20 through 30 failed to disclose any evidence of misrepresentation, falsification, etc. It was stated by supervisors that with no more information to go on than that supplied by the informant, it would be impossible to identify the records in question. They advised, however, that constant analyses and checks are conducted on the in-process materials, and that in all probability, the worksheet obtained from the informant was used for such a check. They also advised that computer printouts, as such, are part of the certification data package sent to WHC and is subject to detailed review and verification by both KM and WHC.

BACKGROUND INFORMATION AND INTERVIEWS OF ADOLPHUS O. PIPKINS, JR.

TO File DATE November 19, 1974

FROM J. H. Reading SUBJECT Interview of Rick Fagan, State Patrolman

Attached is a copy of the accident report submitted by Patrolman Fagan of Karen Gay Silkwood. Preliminary interview with Fagan revealed the following:

There were no tire or skid marks on the hard surface which would indicate there had been any impact with any other vehicle. He further related that the tire marks in the grass adjacent to the highway did not indicate an attempt by the driver of the vehicle to regain control prior to hitting the bridge abutment.

He stated that upon gathering up the personal affects at the scene of the accident, he observed a partially filled capsule, one full capsule, and two hand-rolled cigarettes which he assumed to be marijuana. He further stated there was some white powder in the bottom of Miss Silkwood's purse. He advised the two cigarettes and the capsules were turned to Narcotics Division, State Crime Bureau. He further advised the doctor, name unknown at this time, Guthrie Hospital, inspected the capsules and advised him that they were of a barbiturate base. This can be verified after State Crime Bureau chemists analyze them. When Fagan was asked if he had submitted the purse to be analyzed for what the white powder might be, he advised that he had surrendered the purse to the parents of Karen Silkwood. He further advised that his original inspection of the interior of the vehicle revealed a red notebook and two bundles of paper, 8½ x 11, in the vehicle. His second inspection of the vehicle was with the AEC inspectors in Crescent, Oklahoma, where the vehicle had been stored after the accident. At this time, these papers and notebook was checked for contamination and proved to be negative. During this inspection, he noted the contents referred to the Kerr-McGee operations and labor negotiations at the Cimmaron facility. It should be noted that the patrolman assumed the people that inspected the car for radiation were AEC inspectors. It is possible they were AEC inspectors and some of our technicians from the Cimmaron facility.

He further states the car and its contents were released to Drew Stephens with the concent of Miss Silkwood's parents. For the information of the reader of this report, Drew Stephens was the current boyfriend of Miss Silkwood's. He was also

File
November 19, 1974
Page 2

processed for radiation after this recent episode. Miss Silkwood's identification indicated that she was still living at 836 NW 48th Street, which has been the residence of Stephens for some time.

Patrolman Fagan stated that he had observed the wrecker driver remove the car from the crash scene and that during the act of extracting the car from the area of the bridge abutment, the back of the car was damaged.

J. H. Reading

JHR:kkh

TO	File	DATE	November 21, 1974
FROM	J. H. Reading	SUBJECT	Pipkins

On November 21, 1974, call was received from Mr. J. Arnette, Pinkerton Security, and the following information was revealed from Mr. Nasworthy regarding Pipkins:

Pipkins is not licensed to operate in the State of Texas, even though he has an office in Dallas. One of Pinkerton's officers called Pipkins under the pretense of wanting him to investigate an accident and Pipkins advised he could not get involved in another case at this time because he was in trouble in Oklahoma City.

Mr. Nasworthy further advised that there is an Oklahoma City ordinance wherein a person like Pipkins cannot relate his findings within the City limits. He can relate them at Crescent for instance, but not in Oklahoma City. If he violates this ordinance, the person to contact would be Jim Nutter who is on the City Board.

Mr. Nasworthy further advised Dr. Tonn, Professor at Rice University, Houston, Texas, is the foremost authority on investigating automobile accidents of this nature. He suggested we may consider contacting him for his opinion on the cause of the accident.

TO **File** DATE November 21, 1974

FROM **J. H. Reading** SUBJECT Adolphus O. Pipkins, Jr.
5346 Mercedes Street,
telephone number 214-
826-2100 Dallas.

The nature of the following information is background on the above subject who was retained by the labor union currently negotiating a contract with Kerr-McGee at the Cimmaron Facility:

As related in the Daily Oklahoman newspaper dated November 19, 1974, Mr. Pipkins alleges that the automobile accident in which Karen Gay Silkwood, alias Karen Sue Meadows, was fatally injured was caused by another vehicle hitting the rear of Miss Silkwood's car causing her to loose control of it. The writer contacted Captain Wimberly, New Mexico State Police Intelligence, Santa Fe, New Mexico, and he advised he would investigate and advise the writer of any pertinent data about Mr. Pipkins.

On November 20, 1974, Captain Wimberly called the writer and advised the following:

Pipkins was employed by the Albuquerque Police Department from 1951 to 1954. He resigned in 1954 to go into business for him-self. He further advised that Pipkins had attended a short course in Albuquerque in 1955. This short course consisted of two weeks training and was an extension of the Northwest Traffic Institute. He further advised that Pipkins was unable to pass the required examinations to obtain a license to operate as a private investigator in New Mexico. Therefore, he moved his base of operations to Dallas, Texas, in 1963. He also advises that Pipkins had problems with the IRS in 1955. Captain Wimberly states he understands that Pipkins has attended college in the Dallas area. Captain Wimberly advised that an ex New Mexico State Policeman by the name of Joe Holland, telephone number 505-243-4537, Albuquerque, New Mexico, was more knowledgeable of Pipkins character and business operations. He stated that Holland had appeared in many civil law suits in which his testimony contridicted that of Pipkins. In the event of a law suit in this area, Joe Holland would probably make a good rebutal witness.

TO	J. H. Reading	DATE	November 22, 1974
FROM	W. C. Gentry	SUBJECT	Interview of ADOLPHUS O. PIPKINS, JR., reference his investigation of the KAREN GAY SILKWOOD death

On the afternoon of November 22, 1974, the writer telephonically contacted ADOLPHUS O. PIPKINS, JR., at Houston, Texas, at which time, PIPKINS furnished the following information regarding his activities in the investigation of the death of KAREN GAY SILKWOOD and the related automobile accident on State Highway 74, 7 3/10 miles south of Crescent, Oklahoma, on November 13, 1974:

Without stating any reasons, he said that he wanted to apologize for any criticisms he had caused the Kerr-McGee Corporation as a result of his making this investigation.

PIPKINS stated that on November 12, 1974, he was contacted regarding his availability to make instant automobile accident investigation by an attorney by the name of FREDERICK BARRON of Dallas, Texas, and that BARRON at that time, gave him the name of STEVE WODKA who was at that time in Washington, D. C. BARRON stated that WODKA would give him the background necessary for making the investigation. PIPKINS stated that he telephonically contacted WODKA immediately and that WODKA furnished him the background and requested that he make the investigation. PIPKINS stated that at that time, he realized he could be stepping into an embarrassing situation and told WODKA that there would be no hanky-panky, that he would make the investigation, and that he would strictly report the facts as he found them. He stated that after his conversation with WODKA, he went to Oklahoma City on November 16, 1974, and contacted DREW STEPHENS who represented himself to be the boyfriend of the deceased. He stated that STEPHENS took him to the damaged car which is somewhere in a garage in northwest Oklahoma City, where he examined the car and later proceeded to the scene of the accident at which place he made a series of photographs of treadmarks which he could identify as having been made by the automobile in question. He stated that the only thing that he found in the accident in his investigation that would adversely affect the position of Kerr-McGee was damage to the left rear of the automobile indicating that it could have been hit by another vehicle from the rear, causing the accident.

During the conversation with PIPKINS, it was pointed out to him that he had not seen the automobile involved until three days after the accident and that damage to the left rear

J. H. Reading
November 22, 1974
Page 2

of this automobile could have resulted from other causes during that interum. This he admitted.

During the conversation, the writer ascertained that PIPKINS was on the Albuquerque, New Mexico, Police Department, for a period of four years during the early 1950s and that he and the writer had numerous common acquaintances among FBI Agent personnel who worked in the Albuquerque area. After the conversation, PIPKINS stated that as soon as he finishes writing his report, he will personally see that a copy of it is brought to Jim Reading or the writer in Oklahoma City at the same time he submits it to his principals. PIPKINS stated that he was paid for this investigation by WODKA.

PIPKINS especially requested that members of the union not be advised that he was contacted by the writer regarding this investigation.

W. C. Gentry

WCG:kkh

TO	File	DATE	December 2, 1974
FROM	W. C. Gentry	SUBJECT	Accident Investigation

On November 26, 1974, HAROLD SMITH, an employee of TED SEBRING, Ted Sebring Ford, Crescent, Oklahoma, was contacted by the writer and ROY KING, Personnel Director, Cimmaron Facility, and he advised as follows:

He stated that on the night of November 13, 1974, SEBRING called him from somewhere near the scene of the accident and had him bring another wrecker to assist in removing the wreck from the scene where it was wedged under the edge of a concrete culvert. When asked whether or not the rear end of the vehicle was damaged, SMITH stated that they had probably damaged it all over as it was necessary to bang it around considerably in removing it from the wreck scene. He said he paid no attention to whether or not they damaged it because it was "already totalled".

SMITH did say that when he arrived on the scene the wrecked vehicle was laying with its rear end against the south wing of the concrete culvert but that it had been pulled around and moved about by SEBRING in his attempts to remove it from the ditch.

W. C. Gentry

WCG:kkh

TO **File** DATE **December 2, 1974**

FROM **W. C. Gentry** SUBJECT **Accident Investigation**

On Saturday, November 23, 1974, GEORGE MARTIN, who operates MARTIN's Conoco Station on old Highway 77, Guthrie, Oklahoma, was interviewed by the writer and J. H. READING, and furnished the following information:

He stated that on November 1, 1974, he was contacted at approximately 10:30 a.m. by KAREN SILKWOOD who stated that on the evening before while driving to Guthrie, Oklahoma, from the Cimmaron Facility, she dodged a stray cow on State Highway 33 approximately four miles west of Guthrie and that in doing so, her car ran off of the road on the left hand side as she was coming east and ran across the bar ditch and into the fence. She advised MARTIN that the car was still there and asked him to meet her at the scene of the accident at 1:30 p.m. that day, November 23, 1974. MARTIN stated that he went to the scene and KAREN SILKWOOD was there. He stated that the car, during the accident, had changed ends and backed across the bar ditch into a fence post. He said the only damage done to the car in this mishap was a broken right tail light lense which was broken when the car slid into the fence post. He recalled no further damage to the vehicle. He stated that after he pulled the car onto the highway, SILKWOOD drove it from the scene.

He stated that he was called with his wrecker to the scene of the accident in which SILKWOOD was killed on State Highway 74 and was on his way there. He says he has a radio receiver on Highway Patrol band in his wrecker and that he received orders over this radio to return to Guthrie at about five miles from the scene of the accident. He stated that he has never found out why they stopped him and sent him back. He said that he assumed that the Sebring wrecker from Crescent, Oklahoma, had arrived at the scene before he, MARTIN, got there and that he was not needed.

W. C. Gentry

WCG:kkh

TO **File** DATE **December 2, 1974**

FROM **W. C. Gentry** SUBJECT **Accident Investigation**

On November 26, 1974, TED SEBRING, owner of Ted Sebring Ford in Crescent, Oklahoma, was contacted by the writer and ROY KING, Personnel Director at the Cimmaron Facility, and he stated that at approximately 8:30 on the evening of November 13, 1974, he was notified of an accident approximately 7 1/2 miles south of Crescent, Oklahoma, on State Highway 74. He stated that when he reached the scene, several people including State Patrolman RICK FAGAN were at the scene and that he understood the car was at first on its side and top and it had been turned over in order to get the driver, KAREN SILKWOOD, from the car. He said that she was laying on a stretcher near the car when he arrived. He stated that he made a preliminary check of the car to see whether or not she had personal property that should be taken. He said he found her purse laying immediately in front of the wrecked vehicle. He further stated that while looking in the glove compartment he saw either an invoice or an estimate from Eskridge Honda - Oldsmobile Company for body work done on instant car to the rear end and he said the amount on this statement or estimate was somewhere in the neighborhood of $300.00. He said he had returned this statement to the glove compartment but as it was called to his attention, he made a brief examination of the rear end and found no body damage at that time. He stated that he hooked on to the rear axle of this wreck but was unable to do so. He stated that after he had worked for 30 minutes to one hour trying to remove this wreck from its position from the culvert where he had pulled it in attempting to remove it, he called his partner, HAROLD SMITH, to bring another wrecker. SMITH arrived at the scene and the two wreckers combined were able to remove it and he hauled it into his garage in Crescent, Oklahoma.

He stated the following morning when he arrived at the shop, he pulled the wreck into the yard as he usually does on wrecks. He stated on the afternoon of November 14, 1974, DREW STEPHENS appeared and wanted to take custody of the wrecked car. SEBRING said that he refused to let DREW STEPHENS take the car until he had called long distance to SILKWOOD's parents at Nederland, Texas. He stated that after this authorization, STEPHENS wrote him a check for his charges and removed the car from his premises.

TO	File	DATE	November 21, 1974
FROM	J.' H. Reading	SUBJECT	Mr. Nasworthy's report of Pipkins

On November 21, 1974, Mr. Nashworthy of Pinkerton Security advised as follows:

Pipkins returned the call to one of Pinkerton's men who had asked him to investigate an accident. Pipkins was extremely upset and advised Pinkerton's man that the labor union had gotten him in a lot of trouble over this incident in Oklahoma City. He advised that his specialty was not investigating car accidents but heavy truck accidents. Pipkins advised that he would send a copy of his report to the Oklahoma Department of Public Safety and this report will probably be available as a matter of public record on Monday. He advised that his rate of pay is $300.00 a day and $40.00 a hour for investigating accidents. Pipkins further advised that he would be an extremely reluctant witness.

Mr. Nasworthy advised that if he received any further information, he would immediately contact the writer.

J. H. Reading

JHR:kkh

Note: Information from another source revealed that Drew Stephens stated the cost for Pipkins investigation was $1,000.00, and that he was hired by legislative assistant Steve Wodka.

PINKERTON'S, INC.

Client	Character of Case
Kerr McGee	Investigation

Office of Origin	File No.	Status
Oklahoma City		Final
Reporting Office	**Report Made By**	**Date of Report**
Dallas	H. W. B.	Thursday, November 21, 1974

Received details and commenced investigation of A. E. Pipkin, later learned to be Adolphus O. Pipkin, 5346 Mercedes Street, Dallas, Texas.

Information Operator
Dallas, Texas

Revealed a listing for A. O. Pipkin and Accident Reconstruction Labs, Inc. as 741-2100. The business address was listed as 318 Pearl, Dallas, Texas. It was later learned this was an answering servic and the new answering service number is 826-2100. The home number is unlist

City Directory
1974 Metro Edition

Did not show a listing for Accident Reconstruction, but did list Adolphus O. Pipkin and wife Bonnie as homeowners at 5346 Mercedes Street, Dallas, Texas.

CRIMINAL RECORD

Informant #1

Revealed Adolphus O. Pipkin does not have a criminal record in Dallas, Texas.

COURT RECORDS

Federal District Court
Dallas, Texas

District Court
Dallas, Texas

County Court
Dallas, Texas

Commissioner's Court
Dallas, Texas

Search of all available records in denoted courts from 1950 to present, revealed only one record pertinent to Pipkin and is recorded as follows:

71-10590-DR2
Pipkin vs. Pipkin
Dated: 10/8/74

In this divorce suit, A. O. Pipkin brought suit against his wife, Bonnie J. on grounds

FORM 81 9,65 PTD. IN U.S.A.

J. P. Court Records
Dallas, Texas

of incompatibility. Bonnie waived any actic According to the file, the divorce was highl amicable. The suit revealed they had been married 9/16/67 in Columbus, Ohio. There were no children of the union. Th grounds for the divorce were incompatibility. The attorney for Pipkin was James C. Barber. The final decree for the divorce was 2/16/72. Mrs. Pipkir was allowed to retain her maiden name of Bonnie June Coody.

PUBLIC RECORDS

Secretary of State
Charters Division
Austin, Texas

Revealed neither Accident Reconstruction, In nor Accident Reconstruction Labs, Inc. is lis ed as operating in Texas or as being formed in Texas.

Texas State Board of P.I. and P.S.A.
Mr. McQuarter
Austin, Texas

Stated neither Pipkin nor his busine are licensed in the State of Texas. He stated he was familiar with Pipki work in New Mexico, primarily Albuqu que, but was not aware he was working in Texas. He stated the board usually did not license such operations. McQuarter stated Dr. Tonn, P.H.D. in physi at Rice University was the primary authority in the South, Southwest and Wes in accident reconstruction because his knowledge, technique and reputation were irrefutable. He said Dr. Tonn was quite active in this area and could testify primarily on physical evidence from photographs of accidents.

Professional Engineers of Texas
Registration Branch
Austin, Texas

Stated Adolphus O. Pipkin was not a registered engineer in the state of Texas.

NEWSPAPER FILES

Dallas Times Herald
Dallas, Texas

Dallas Morning News
Dallas, Texas

Search of clip files, art files and microfilm records from 1925 to present revealed no information concerning Pipkin or his firm.

BUSINESS REPUTATION

Of twenty-six private investigation firms contacted, only one was famili with Pipkin.

evidence". When asked about working from pictures, he stated the majority of his work was in this vein. He said rarely is he able to see an unchanged accident scene.

No further information was obtained, and the operation was discontinued ter preparing this report.

SUPERVISOR'S COMMENTS

It would appear another facet of Pipkin's creditility would be a similar investigation in the Albuquerque area. Further work in the Dallas area will be done upon request.

50/BR
11/22/74

Note: Report page 3 is deleted.

FBI INVESTIGATION

HAROLD SMITH was interviewed on Monday, December 9, 1974, by W. C. GENTRY and J. H. READING at the Sebring Ford Motor Company in Crescent, Oklahoma, at which time he furnished the following information:

He said that he is 52 years of age, resides at 202 West Jackson Street, Crescent, Oklahoma, telephone number 969-2916, and is employed as sales manager at Sebring Ford Motor Company in Crescent, Oklahoma. He said that this small company also operates a wrecker service and that at about 9 p.m., November 13, 1974, he received a telephone request through the local Police Department that his employer, TED SEBRING, needed assistance at the scene of a one car accident on State Highway 74; at a point approximately one miles south of the intersection of State Highway 33 and 74. This place is at a point of about 7½ miles south of the city of Crescent. SMITH said SEBRING had answered this call with their wrecker at about 8 p.m. that date.

SMITH advised that he proceeded to the scene to assist SEBRING. He said that at the point of the accident, Highway 74 runs north and south and that a large concrete culvert runs under the highway at this point. See Exhibit 3. This culvert is approximately 5½ feet high from the bottom to the top of the mouth and about 6 feet wide. SMITH said that when he arrived at the scene, he parked his pickup truck headed south in the bar ditch just north of the accident scene and left his lights on in order that their would be light to work by. He said that another car had parked south of the accident on the same side of the road with its lights on so that the scene was fairly well lighted.

Regarding the culvert, it has two retaining walls, one on each side, which flare out and slope down from each top corner of the culvert to the ground. SMITH stated that at the time he arrived the wrecked automobile, a badly damaged 1974 Honda Civic hatch-back, was sitting in a position with the rear to the south and slightly to the west and that the winch line was hooked to it under the rear end and had pulled this car up near to or against the retaining wall on the south side. He said that the truck had not used the power winch but had simply tried to pull the wreck up over the outer end of this sloping retaining wall and had been unable to do so. SMITH said he unhooked the truck and moved it over to the east towards the right of way fence. He said that he then raised the A-frame and that one KENNY VALLIQUETTE, who had been assisting TED SEBRING, then threaded the winch line through the snatch block at the top of the A-frame and rehooked it to a towing ring attached to the frame of the Honda at the rear.

SMITH said that this procedure caused the line to raise and pull on the Honda at the same time. He said that he stood on the east side of the Honda and waived to TED SEBRING who was operating the winch. He said SEBRING engaged the winch and the Honda slowly came out of the ditch, over the lower part of the retaining wall and out. (See Exhibit 1). SMITH said that while he could not actually see this car go over the retaining wall he knew from the position of the car and the pull that it must necessarily pass over the lower sloping end of the retaining wall. He said that he heard the grind and the rasp of the metal against the concrete wall as the Honda was pulled out. When the Honda was pulled to level ground, he unhooked the line from the rear, raised and hooked on to the front and SEBRING towed the wreck into Crescent, Oklahoma, and stored it in the Sebring Ford Motor Company garage.

SMITH stated that he did not examine the part of the wreck which came over the retaining wall inasmuch as the car was totalled when it crashed into the stream bed and the retaining wall at the time of the accident. He stated that during the operation of removing the wreck, other dents may have been put into the wreck also but he is sure that the left rear of the bumper and that part of the fender below the bumper was damaged during the removing operation.

On November 25, 1974, MIKE DAVIS and RICHARD HARRIS interviewed DENNIS FORD at which time he advised as follows:

He stated that he had talked to HAROLD SMITH, the wrecker service that handled the November 13, 1974, wreck. SMITH had an opinion based on the Oklahoma Highway Patrol statements that there were no brake skid marks and SILKWOOD was traveling at about 55 mph. HAROLD SMITH related to him that when he arrived at the scene of the accident, the Highway Patrol and other parties (he was not sure who they were) said that they had taken documents and drugs from the car before he got there. HAROLD SMITH was asked to witness the removal of a sample of a strong solution in a tomato juice can in the car which might have contained alcohol and also of various documents. J. DAVIS' wife, a nurse, supposedly knew of drugs being present. According to HAROLD SMITH, one of these documents was a letter from Ottawa, Canada, inviting KAREN to a drug party stating that the machine would roll and have everything ready when the invitee arrived. (Apparently this refers to the rolling of marijuana cigarettes). HAROLD SMITH further related to FORD that DREW STEPHENS came to the garage on Friday, November 15, 1974, and wanted to take the car. HAROLD SMITH did not like DREW's hippy-type look and would not have any business with him. Reportedly, DREW made three calls to KAREN's parents on Friday and received the car. The third call was apparently made from HAROLD SMITH's place of business. SMITH is the new car manager of Ted Sebring Ford in Crescent, Oklahoma.

FORD advised that he did not see a dent in KAREN's car before her death.

On November 25, 1974, RICHARD HARRIS and MIKE DAVIS interviewed HOWARD HAMMON at which time he advised as follows:

HAMMON stated that a few days after KAREN SILKWOOD's accident on October 31, 1974, he saw her after dinner at the General Lab. He advised they had discussed nothing in particular. SILKWOOD indicated to him that she felt bad and her back was bothering her because of the car accident a few nights before when she had to dodge a cow on Highway 33. She had told him that she banged up the rear of the car because she had backed into a ditch.

On November 27, 1974, MIKE DAVIS and RICHARD HARRIS interviewed JON HARRISON at which time he advised as follows:

HARRISON relates that he saw KAREN GAY SILKWOOD's automobile the day after October 31, 1974. He had been walking around during the lunch hour or perhaps after work. He saw some dents on the rear of the car (left rear one-quarter). He was familiar with the car, which was a Honda Civic and he did not think it was too dirty. There was some damage on the back and he perhaps looked because he had heard of her incident in which she ran off the highway between here and Edmond on Highway 74. He had heard that she had lost control while trying to miss a cow that was on the highway. He advised that she was okay but a bit shaken up. The car with pulled out on Tuesday.

On December 6, 1974, ANDY KISER, Manager of Clark's Hub Cafe at Crescent, Oklahoma, was contacted and interviewed by W. C. GENTRY regarding the union meeting which KAREN GAY SILKWOOD attended the evening before the fatal accident.

He stated that he had seen her at the meeting and that after the meeting had started in the rear room of the cafe, she came to the front and used the telephone for a few minutes and then returned to the meeting room. He stated that he paid no particular attention to her but noted nothing unusual in her talk or actions.

KENNY VALLIQUETTE, a contract pumper, age 41, was interviewed by W. C. GENTRY and J. H. READING at his home, 619 West Van Buren, on the afternoon of December 9, 1974. The reason for this being a brief interview was that VALLIQUETTE had to leave as he had what he termed an important engagement. He related that on the evening of November 13, 1974, KENNETH HART and TED SEBRING were guests at his house and that at about 8:00 p.m., SEBRING received a request to report to an accident scene approximately 7½ miles south of Guthrie on State Highway 74. He stated that he and KENNETH took SEBRING to the Sebring Ford Motor Company where SEBRING obtained a wrecker and drove south. VALLIOUETTE advised that after SEBRING drove the wrecker out, he closed the garage door for him and that he and HART decided to follow SEBRING to the wreck inasmuch as it had been described by SEBRING as being a very serious accident.

He advised that when he got to the wreck scene, which was in the stream bed on the east side of the road where a culvert goes under Highway 74, they ran on past the scene, made a U-turn, and came back and parked so that their lights were on the scene. He said SEBRING got in the car with them to warm up a little and that they then got out to remove the wrecked car from its location in the stream bed where a large concrete culvert runs under the highway. He said that an Oklahoma Highway Patrol Trooper was on the scene but that he did not notice very many people stopping and interferring with the work. He said that the wrecked Honda hatch-back was on its wheels in the stream bed headed in a southwesterly direction with the wrecked front end near the retaining wall which slopes from the top of the culvert in a southeasterly direction. He said this retaining wall slopes into the ground about 7 feet east or southeast out of the culvert. He said that he attached the winch line to the Honda after observing a towing ring which was welded on to the underneath rear of the Honda. He said that when the truck moved forward, it pulled the back end of the Honda around and its left door came open and caught on the retaining wall and acted as a pivot and the Honda swung around with the left rear end near or against the sloping retaining wall. He stated that as the Honda was against this wall, the wrecker truck was unable to move any further to the south. VALLIQUETTE said that at this time, HAROLD SMITH arrived on the scene and that he relocated the truck further east between the bar ditch and the right of way fence. VALLIQUETTE stated that he himself raised the A-frame and threaded the winch lines through the snatch block and rehooked it on the ring at the back of the Honda. He stated that

SEBRING then engaged the power winch and the Honda slowly came out of the ditch. He said that he heard the noise made by the car as the metal crunched and rasped against the concrete wall. VALLIQUETTE stated that he is sure that the left rear bumper and that part of the fender below the bumper were damaged at that time. He stated that he could not actually see the contact because he was not down low but he heard the grinding and crunching as the vehicle came in contact with the retaining wall and further advised that there was only soft mud in the remaining path of the car. He stated that after the Honda was pulled and on level ground, they unhooked from the rear end, raised the front end with the winch, and SEBRING towed the car into the Sebring Ford Motor Company at Crescent, Oklahoma.

The attached photographs were taken by J. H. READING of the accident scene on Highway 33 approximately four miles west of Guthrie, Oklahoma, of KAREN GAY SILKWOOD, on October 31, 1974. The car was removed from the accident scene by GEORGE MARTIN of Martin's Conoco Service Station, Guthrie, Oklahoma. He was contacted to remove the car from the ditch on November 1, 1974, and did so. It was very appar nt from Mr. MARTIN's description of the scene and the tire marks left at the scene what had happened. The embankment KAREN SILKWOOD had run off of is described as follows:

It is a fill across a creek approximately 15 feet high. The incline of the highway fill where her car left the roadway is approximately 40 degrees. Mr. MARTIN related that KAREN SILKWOOD had told him that she was traveling east on Highway 33, that it was raining and late at night, that she observed a cow in the middle of the highway, that she lost control of her car, and skidded backwards off of the highway or the embankment into the creek bottom on the north side of the highway. She advised that she had injured her neck and shoulders and was extremely frightened. Upon observing the tracks her tires made, it is apparent that the top of the embankment where she first went over that the car was airborne for approximately 8 feet. Upon the car coming into contact with the embankment, it hit some debris. The debris consisted of some rather large chunks of concrete which were semi-embedded in the soil plus some tree limbs, some of which were four inches in diameter. Both the tree limbs and the concrete had evidence of fresh scars or damage to them. After the initial impact, the car continued down the embankment backwards until striking the fence posts with its right rear (fence post is indicated in photographs). The path of the vehicle descending the embankment could be described as diagonally west to east.

From observation of the accident scene, it is understandable why KAREN SILKWOOD complained of neck injuries and so forth. She was treated for these injuries by Dr. MYHEW at the Guthrie Hospital. She described the damages to her car from this incident as the right rear lense of her taillight being broken. It is entirely possible that the steering mechanism and additional damage could have occurred to her vehicle from the debris encountered going over the embankment.

On November 27, 1974, FRED SULLIVAN, an employee of the Cimmaron Facility, was interviewed by W. C. Gentry and gave tha following information:

He said that on the evening of November 13, 1974, he was in his car alone following a car drived by LOLL CODWIN. He stated that when they neared the point where SILKWOOD's accident had occurred, there was only one person there who had just drived onto the shoulder and had gotten out of his pickup truck. SULLIVAN stated both he and CODWIN stopped and GODWIN proceeded to the end of the culvert. GODWIN yelled to him to call the Oklahoma Highway Patrol because of the accident. He stated that he left and went to the nearest farmhouse and called the patrol. He then returned to the accident scene. He said that when he looked down and saw the wrecked vehicle, it was laying in the ditch at the east end of the culvert. It was laying partially on the left side and top with the front end pointed southwest. He advised that all four wheels and the bottom of the vehicle were visable to him because of its position. He stated that some people who had stopped helped turn the car over so the victim could be removed. He stated that shortly after the victim was removed from the wreck and put on the stretcher, he and GODWIN left.

GODWIN was unavailable for interview and it was learned that he is to report for a three week stay at the Presbyterian Hospital where he will have surgery of some nature. As soon as possible, he will be interviewed regarding his impression of the accident and any information he may have regarding it.

On December 9, 1974, Oklahoma Highway Patrol Trooper RICK FAGAN, was contacted by W. C. GENTRY and J. H. READING in his patrol unit about 5 miles west of Guthrie, Oklahoma. He was advised that it was just learned from the wrecker operator who removed the car of KAREN GAY SILKWOOD from the accident scene south of Crescent on November 13, 1974, had pulled the automobile from the bottom of the ditch at the east end of the culvert about 7½ miles south of Crescent, Oklahoma, and that the car was pulled out by its rear end.

Without further prompting, FAGAN stated, "You don't have to tell me that. I stood above it and watched the wrecker pull the Honda out by its rear end over the lower end of the south retaining wall of the culvert and that is where the left lower part of the Honda was damaged." No further interview was made with FAGAN at that time.

KENNETH HART was interviewed on the evening of December 9, 1974, by W. C. GENTRY and J. H. READING at the Kerr-McGee Cimmaron Facility, at which time he furnished the following information:

HART advised that he is 35 years of age, resides at 617 West Van Buren, Crescent, Oklahoma, and is employed at the Cimmaron Facility. He said that on the afternoon of November 13, 1974, he was at KENNY VALLIQUETTE's home which is next door to his residence and that TED SEBRING was also there. TED SEBRING received a call at about 8:00 p.m. from the Crescent, Oklahoma, Police Department advising that he had received a call to furnish a wrecker at an accident about 7½ miles south of Crescent, Oklahoma, where one person was pinned in an automobile. SEBRING asked VALLIQUETTE to drive him to his garage so he could get his wrecker. HART advised that after driving SEBRING to his garage, he waited until after SEBRING had driven the wrecker out and closed the garage door for him. HART said that inasmuch as this appeared to be a serious accident, he and VALLIQUETTE returned home to obtain some flashlights and proceeded to the accident scene to assist SEBRING. They proceeded to the accident scene in VALLIQUETTE's automobile.

HART said they drove a ways past the scene, made a U-turn and came back near the scene where they parked off the highway with the lights on. HART said that when they arrived at the scene, the wrecker was parked 8 or 10 feet off the highway with the rear end towards the ditch where the Honda automobile was located. He said that when he observed the wreck, the Honda's front end was badly mangled and was sitting on its wheels headed in a southwesterly direction pointed towards the sloping retaining wall outside the culvert. He said the front end of the car appeared to be 2 or 3 feet from this retaining wall. He said that it was a very cool night and that when they arrived, SEBRING got into the car with them to warm up a little. He stated that he then got out of the car. It should be noted at this point, that the driver of the wrecked vehicle had been removed from the wrecked automobile and had been taken to the Guthrie Hospital. He said that at this time there was a Oklahoma Highway Patrol Trooper, RICK FAGAN, present, who declined to furnish them the name of the victim but did state that it was a fatal accident and that he could not release information regarding the identity of the victim until the next of kin had been notified. He stated then that he began preparations for pulling the car out of the ditch and that they decided to try to pull the wreckage over the south down sloping retaining wall. HART stated that VALLIQUETTE pulled the winch cable from the spool on the wrecker down to the rear end which would have been the north end of the Honda and that he heard him say "Here is a ring that I can hook the winch line on". HART said that this turned out to be apparently welded to the frame of the Honda underneath the rear for the purpose of towing. He said that SEBRING then tried to pull the car by setting the winch line and driving the truck forward. He said this slid the rear end of the Honda around and that as it neared the retaining wall, the door came upon and swung out striking the retaining wall. He said this door being open acted as a fulcrum and pulled the rear end of the Honda more sharply towards the retaining wall and ended up being very close to the retaining wall or even touching it. He said that at about that time, HAROLD SMITH arrived at the scene and that SMITH relocated the wrecker further east and near the right of way fence where he could have an easier line of pull. He stated that VALLIQUETTE continued to assist and strung the winch line through the snatch block after the A-frame had been raised. HART stated that they then used the power winch and pulled the car in a southeasterly direction out of the stream bed and onto level ground. HART said that the wrecker then was unhooked from the Honda and the Honda was picked up by the mangled front end and towed to the Sebring Ford Motor Company in Crescent, Oklahoma.

HART stated that at the time the wrecked Honda was pulled from its position in the ditch, rear end first, it is very possible that it could have come in contact with the concrete retaining wall but he could not say positively that it had.

On November 21, 1974, DERRILL CODY, JIM READING, and RICHARD HARRIS interviewed GERALD BREWER, aka JERRY BREWER, at which time he advised as follows:

BREWER stated that he saw KAREN just before she left the union meeting on the night of her death and that he did not observe any indication of alcoholic comsumption. He said that he observed nothing except for the fact that she was very tired and physically worn out. He said he and FRANK MURCH both offered to take her home but she refused. He stated that JOHN KING and ROY CARVER, along with the AEC Investigators delayed her at the plant prior to the union meeting and that KING and CARVER had complained to her about her being off work on Friday, November 8. He stated that the union meeting on November 13, started at 5:00 p.m. and that KAREN showed up 10 or 20 minutes after 5:00 with FRANK MURCH having arrived at about the same time. She left at about 7:00 p.m. having told BREWER that she was going to Oklahoma City to meet DREW STEPHENS and STEVE WODKA, legislative assistant on the payroll at OCAW. He stated that so far as he knew DREW was acting on his own rather than in a union capacity and so far as he knows DREW is not presently a member of the union. BREWER further advised that he did not visit the scene of the accident and did not know about it until 9:30 a.m. on November 14. KAREN did not discuss any documents with BREWER and although he did not know what was in her car, the only one he was aware of is a spiral notebook which she had. He did not state whether she had it at the time.

KAREN SILKWOOD'S DRUG USE

Karen		+	Sherri	
$ 300	dope		$ 75	dope
66	Rent		20	Myst
−66	Rent		55	
300			20	Bobby
130	Gumby		35	
170			20	Bob
115	Myst		15	
55				
20	J.B. + Drew			
35				
20	Louis + J.D		46.33	
15			21.33	
			25.00	

Rent + Phone

Karen		Sherri
87.50	Rent	87.50
21.33	Phone	

Numerous associates of KAREN SILKWOOD were interviewed and indicate that she uses marijuana and did have the reputation of a marijuana user. More tangible evidence of her association with drugs is as follows:

1. SHERRI ELLIS, roommate of KAREN SILKWOOD, states that KAREN SILKWOOD and DREW STEPHENS had smoked marijuana in her presence on several different occasions.

2. Marijuana and hypodermic kit found in KAREN SILKWOOD's and SHERRI ELLIS' apartment during decontamination.

3. Marijuana found in KAREN SILKWOOD's purse at the time of her death.

4. Records of the Gilliam Prescription Shop # 3, 6719 North May Avenue, Oklahoma City, Oklahoma, reflect that KAREN SILKWOOD had purchased methaqualone on script ssued by Dr. CLARENCE SHIELDS, 6719 North May Avenue, Oklahoma City, Oklahoma. Records reflect that purchases were made of 30 tablets each on the following dates: August 13, August 19, September 25, October 4, October 13, and November 4, all in 1974. The prescription calls for SILKWOOD to take one tablet upon retiring. It may be of interest to note that period covered was a total of 180 tablets purchased for a period of 92 days.

5. Statement of EVELYN EMERICH indicated that KAREN spent the night with her and her son, CARL THEMMONS, on the night of October 16, 1974. She stated it was obvious that KAREN was under the influence of drugs. She stated that the reason KAREN spent the night with her was that she had observed KAREN in a cafe in Crescent and was aware that she would be unable to drive and in concern for her safety, invited her to spend the night at her house. Miss EMERICH also advised that KAREN SILKWOOD told her that she had attempted suicide on two occasions: once in Duncan, Oklahoma, and once in Oklahoma City, Oklahoma. The suicide attempts were made by the means of an overdose of drugs. She advised that a fellow employee by the name of CONNIE EDWARDS had responded to KAREN SILKWOOD's call for help on the last occasion.

6. CONNIE EDWARDS was interviewed about KAREN SILKWOOD's attempt to commit suicide by overdosing and she advised that in September, 1973, KAREN SILKWOOD had called her at her home in

Orlando, Oklahoma, and advised her that she had tried to kill herself by taking an overdose of drugs. CONNIE stated she drove to Oklahoma City to KAREN's apartment which at the time was behind the Lady Classen Cafeteria on May Avenue and upon entering the apartment, found KAREN in a stoop on the sofa. She advised that she got her to her feet and wanted to take her to the hospital and KAREN resisted. CONNIE advised that she induced KAREN to vomit repeatedly by placing her finger in her throat and after a short time KAREN seemed to respond. She advised she took her back to Orlando to her home and that KAREN spent the night with her.

7. Medical report from CHAPMAN indicates that autopsy discloses excessive amount of methaqualone in vital organs along with a slight amount of alcohol in the blood system.

8. Handwritten note retrieved by decontamination in SHERRI ELLIS' and KAREN SILKWOOD's apartment which indicates a purchase of $300.00 for drugs by KAREN SILKWOOD and a purchase of $75.00 for drugs by SHERRI ELLIS, a copy of which is attached.

Work schedule for the above mentioned period.

Records indicate she worked seven days during this period for a total of 51½ hours. Records indicate that she had five days off during this period. Records further indicate that she was off November 8 and 9, 1974, AWOL. November 10, 1974, at 10:00 a.m., she left for Los Alamos, New Mexico, to be tested for radiation. She was in Los Alàmos on November 11, 1974, and most of the day of November 12, 1974, returning by commercial aircraft to Oklahoma City, Oklahoma, arriving around 10:30 p.m.

She was accompanied by SHERRI ELLIS and DREW STEPHENS to Los Alamos, New Mexico. Upon their return to Oklahoma City, at approximately 10:30 to 11:00 p.m. on November 12, 1974, they went to DREW's home at 836 NW 48th, Oklahoma City, Oklahoma, where they spent the night.

SHERRI ELLIS stated that upon arrival at DREW's home they were met by one of her girlfriend's, DEBE, and also an ex-Kerr-McGee employee by the name of BILL STILLWELL. She further stated that they had bought some 190 proof alcohol in Albuquerque and other assorted liquors: these being described as survival kits which contain four, four oz. minature bottles of assorted liquors. SHERRI states she and her girlfriend stayed with KAREN, DREW, and BILL drinking Bloody Mary's made with the 190 proof alcohol until about 2:00 a.m. at which time, she and DEBE retired and she had no knowledge when KAREN or the others went to bed. When asked how much alcohol KAREN SILKWOOD had consumed, she stated she had no idea as they were mixing the drinks without the benefit of a shot glass. In other words, she stated they would pour the tomato juice in a glass and then would pour in enough alcohol which was measured by what she termed "eye balling" it. SHERRI stated she got KAREN up at 7·00 a.m. on the morning of November 13, 1974, after which KAREN left for the Cimmaron Facility to attend union negotiations.

To continue KAREN's activities after leaving DREW's house, the next contact was with FRANK MURCH. FRANK MURCH relates the following:

MURCH, CHAPMAN, and JACK TICE were to meet KAREN SILKWOOD at the Hub Cafe, Crescent, Oklahoma, at approximately 7:00 a.m., November 13, 1974, to discuss their labor negotiations that were scheduled for 9:00 a.m. at the Cimmaron Facility. MURCH said KAREN SILKWOOD did not show up and while they were en route from the cafe to the plant, she passed them en route to the cafe. He states she recognized them and made a U-turn and followed them to the plant. MURCH states they negotiated until lunch time and them that he, SILKWOOD, TICE, and CHAPMAN went to Crescent to eat after which they returned to negotiations at the plant. The negotiations broke up at approximately 3:00 p.m. at which time AEC inspectors wanted to interview KAREN SILKWOOD and MURCH further advised that they continued their interview until approximately 5:00 p.m. At this time, she broke up the interview by advising AEC inspectors that she had to go to the Crescent Cafe to a union meeting to advise the union membership of their negotiations that date. Prior to leaving the plant, she was escorted to the bathroom by Kerr-McGee Health Physicist for a controlled sample after which she drove to the Hub Cafe accompanied by FRANK MURCH. They discussed their negotiations with the union membership which numbered approximately 10 to 15 people and at approximately 7:00 p.m., she advised that she had to go to Oklahoma City, Oklahoma, to meet somebody. Subsequent information revealed this to be WODKA and BURNHAM: STEVE WODKA being a union representative from Washington, D. C., and DAVID BURNHAM being a New York Times reporter. MURCH states that he and TICE were concerned about SILKWOOD's ability to drive. He advised that she exhibited the following characteristics: She was nervous, appeared to be shook up, was very upset, and emotionally unstable. MURCH stated he was concerned about SILKWOOD's ability to drive and offered to drive her to Oklahoma City.

AEC inspector BILL FISHER was interviewed by public service and he related that during the above mentioned interview SILKWOOD appeared to be emotionally disturbed at times. He stated that she cried and advised the interviewing team that she was worried about her health. FISHER further advised upon checking the car for contamination on the night of the accident, they observed a whiskey flask in the wreckage and that he and WAYNE NORWOOD, Health Physicist, took samples of the contents of the flask for contamination tests. He stated the contents appeared to be tomato juice. FISHER was asked if the sample he had taken was tested for alcohol and he advised that it had been sent to Idaho Falls. A call from FISHER later that day in which he was interviewed about this, revealed that they had not checked for alcohol and that the sample could not be retrieved.

MIKE WILDING, Kerr-McGee Cimmaron Facility employee, advised that he had received information through another source that the sample submitted to Idaho Falls was a Bloody Mary and one would assume that a Bloody Mary contained alcohol.

WAYNE NORWOOD was asked if the sample he had in his possession could be tested for alcohol. He advised that he had tested it for contamination and that no contamination was indicated and that he threw the sample away.

Chapter 14

KERNENERGIE

For a day and a half, Dr. H. J. Stoecker, director of the Energy Research Program for the Federal Republic of Germany, sat very patiently at the Energy Technology Conference in Washington, listening to testimony of what he would later tell colleagues must surely be a "catastrophic" energy situation.

The jovial German made notes, shook his head, and conferred with a team of advisers. And, like advisers everywhere, they would smile and say: *"Ja, das ist richtig, Herr Stoecker, das ist richtig, ja, ja."* Everyone, it seems, has their share of "yes" men. . . .

After listening to many learned speakers eager to relate their country's resources and expertise, the message, Stoecker concluded, needed no further translation: life is tied to the domestic resources known as oil, gas and uranium. And there are those who have it and those who simply don't. In that regard, the German realized with a sigh, his beloved Deutschland would never win a beauty contest.

Worldly-wise, Dr. Stoecker would later rise to the podium, go through the preliminaries, test the wind, and then throw out his hands and wit to the audience of nearly a thousand.

Outside the sprawling Shoreham Hotel, azaleas splashed colors as the traffic whizzed along Connecticut Avenue while taxis darted in and out like so many worker ants. Inside, Stoecker's honest declaration, spiced with thick German humor, was preparing to bring the house down and, in some sections, even produce a spontaneous standing ovation.

"As a contributor," he began, looking like Charlie Brown clutching at a security blanket, "I feel a little guilty, because we simply have no oil, we have no gas, and we have no uranium. . . ."

There was one loud afterthought boomed a tall, grayheaded economist-planner from Jackson, Mississippi: "Hell, man, you-all in a heap of trouble this day and age."

Funny, yes, but there is also a message here, and those sitting at my table agreed that Dr. Stoecker's comment stressed anew our international dependence on each other. We have grown from a backwoods-hollow mentality to one of world consciousness. We have grown from a people jealous and defensive of our turf, to a people willing to share and, yes, learn from other cultures. Or have we? Is this the real or the ideal?

Certainly, the advent of nuclear energy, which translates from a gentle German word, *kernenergie,* requires some rapid large-scale thinking along the line of priorities. The world will not say *ja* or *nyet* or *kan* or *oui* just because the United States may decide to declare a moratorium on nuclear power. The world will go on testing and developing its resources in this vital area. And that is exactly what it is, a *vital* area.

Energy research and development in foreign countries tends to be aimed at alleviating near-term energy problems. While research associated with oil and gas exploration, recovery and fuel conversion technology continues, the primary program emphasis is directed at electrical generation with nuclear power.

The foreign effort is concentrated in six countries, each with budgets that total more than $100 million. These include:

France . . . . . . The current (1975-76) estimated research budget is $700 million with 60 percent devoted to current and future nuclear technology. Also under development are coal mining technology and oil and gas exploration, storage and transport.

West Germany . . . . A current (1975-76) budget of $350 million with lion with 65 percent geared to nuclear and 25 percent to coal extraction and utilization.

United
Kingdom . . . . A current (1975-76) budget of $350 million with 60 percent devoted to nuclear power.

Canada . . . . . . The current (1975-76) budget is $220 million with 50 percent alloted to derived fuels mainly in conversion of coal and tar sands. Some 30 percent is devoted to nuclear technology.

Japan . . . . Here the current figure is $200 million with 80 percent devoted to the nuclear option.

The primary focus for U.S. multilateral cooperation is the International Energy Agency (IEA). Through the IEA an extensive program of cooperation is underway in nine technological areas such as coal technology, radioactive waste management, and hydrogen. Cooperative programs in nuclear energy have been conducted for several years with Euratom and the International Atomic Energy Agency.

For example, the United States is not alone in its selection of the Liquid Metal Fast Breeder Reactor (see Chapter 11) for highest priority in its development program. Other major nations—France, Great Britain, Germany, Japan, Italy, Spain and the Soviet Union—have also made significant commitments to this program. France, Great Britain and the USSR have already completed demonstration plants, and Russia has an even larger facility under construction.

There is even a computer-based international communications network, through which nuclear power plant operators or planners can input operations data directly into a central computer and obtain the same data via systems-supplied data terminals. So, the atom has indeed gone global.

India was the first of the developing countries to exploit nuclear energy as a primary power source with the operation of the Tarapur Power Station, which has two boiling water reactors. Initial refueling operations utilized underwater television equipment and other special tools to detect and repair defective components inside the reactor. Even initial fueling was handled entirely by Indian engineers and technicians, with faster deliveries of replacement fuel attained by having the fuel fabricated under license in India.

On the other side of the fence, or perhaps "wall" is a better term, the German Democratic Republic in December, 1973, opened the Nord Atomic Power Station, the second atomic power reactor in the GDR built to Soviet design. Others are under construction.

Nuclear power stations are also either under construction or operational in other socialist countries, such as Bulgaria, Hungary, and Czechoslovakia.

Nor have the Soviets limited the loan of their nuclear expertise to their friends and neighbors. At the First International Conference on the Peaceful Uses of Atomic Energy, held in Geneva back in 1955, delegates reviewed reports from throughout the world. One particular item of interest was the operation of the first atomic power station in Obninsk, Russia. At that time, it had been operational for one year, and Soviet delegates—as they have at subsequent meetings—explained technical characteristics of a reactor's operating principles to a very interested assembly. Since that date, other countries have taken the ball, or rather atom, and run with it toward a goal of infinite energy.

Robert Doyen, the American-trained director of a nuclear plant in the Meuse River Valley in Sweden, has been quoted as saying that initially he had the most difficulty with local anglers who feared that the water intake would kill fish. He settled the controversy by promising to erect, if necessary, an "electric fence," a weak underwater current, near the intake pipe to drive off fish. It hasn't proved necessary, yet.

"Now, the fishermen like the plant," Doyen has said. "They gather around the outlet pipe. The water is warmer there and it's where the fish collect."

And, by mid-1975, the government of Sweden responded to nuclear power with a decision to proceed with eleven plants that it is planning to start during the late seventies. Ten, with a firm commitment on two, are being eyed for the 1980s.

Germans, in general, have not complained a great deal about nuclear plans, but the West German government has come under fire in other countries because of its plans to sell nuclear material and technology to Brazil. Chief objections have focused on the fear that this might enable Brazil to make nuclear weapons and possibly destroy her neighbors.

Overall, it seems that the experience of German utilities in the construction and operation of some eleven nuclear power plants has been noticeably similar to the United States. Both countries agree on the importance of standardization and technical improvements as a means of shortening the planning, construction and licensing phases.

An official pamphlet distributed in Belgium by electric power firms says that nuclear power is the cleanest way to make electricity:

> Unlike fossil fuels, fissionable material does not make contact with the atmosphere. A nuclear reactor does not use oxygen, and does not emit carbon dioxide, or oxides of sulphur or nitrogen. All the reactor products remain confined within the uranium or are held in its sheathing. A very light diffusion of some of them, volatile or in the form of gas, may take place, but these products are absorbed by the water or gas that cools the reactor. That gas or water is suitably filtered, so that the radioactivity reaching the outside is zero or extremely small.

Nuclear potential in Israel and among the Arab states was the topic of a 1975 study by Drs. Robert J. Pranger and Dale R. Tahtinen, who head the Foreign and Defense Section of the American Enterprise Institute, a nonpartisan, research-oriented organization. The two researchers have been publicly quoted as saying that it is now certain Israel possesses nuclear weapons. Their study cited "mystery" surrounding operations at Dimona, one of two Israeli nuclear reactor complexes. There, the report noted, the Israelis can produce enough plutonium to duplicate the bomb dropped by the United States on Nagasaki at the end of World War II. In addition, Drs. Pranger and Tahtinen found that Israeli scientists have developed sophisticated laser beam techniques which allow for the production of uranium useable in atomic weapons.

By mid-1975, the only Arab states known to possess reactors—both built by the Soviet Union—are Iraq and Egypt. While these facilities are much smaller than those in Israel, the institute study predicted that Egypt could probably begin producing atomic weapons in six to ten years.

One of the major world powers, however, is *not* putting a top priority on the construction of nuclear power plants. Chang Chih-Hsiang, an official of the Liaison Office of the People's Republic of China, told me in Washington that oil production —more than nuclear power—is being stressed by the Chinese. Crude oil production in China went up by another 20 percent in 1974, to reach a level more than six times that of 1965 (the year before the Cultural Revolution), he said, noting that in 1974 more oil-bearing strata were located than in any previous year.

A major quantity of oil now flows through the 1,152-kilometre (715-mile) long pipeline which links the Taching Oil Field in the Northeastern China province of Heilungkiang with the North China port of Chinhuangtao. Started in the winter of 1970, the pipeline was completed some three years later after the shifting (often by hand) of some

14 million cubic metres of earth and rock. Pumping stations, spaced 60 to 70 kilometres apart, provide the pressure and temperature required to keep a steady flow of crude oil moving.

Quality assurance and the reliability of reactors is of worldwide interest. The French have advocated a definite move toward the use of international standards with the International Atomic Energy Agency in Vienna (IAEA) as the obvious mechanism for its early development. The emphasis on this question of quality assurance can be ascertained from a sampling of these national attitudes:

- The Federal Republic of Germany requires that specifications meet the German codes and be approved by a German AEC. Preoperational pressure tabs and ultrasonic "fingerprint" inspection serve as basic modes for future in-service inspection.
- The United Kingdom Atomic Energy Authority has a grading system for a contractor's quality assurance capability and requires chemical analyses of reactor steel.
- Spain and India also stress a degree of quality assurance by following the standards of the specific country supplying their nuclear hardware.
- The United States has evolved stringent regulatory requirements, as evidenced by the "Code of Federal Regulations," Title 10, Part 50. This program assigns the prime responsibilities and related authorities to plant owners. It also figures in quality assurance at the earliest phase of design, in addition to providing for independent design reviews and a system for audit.

In many cases, U.S. specifications and regulations are being followed by countries where no domestic rules have yet been developed. This was indicated during a symposium on operating and fueling of nuclear power plants which was held by the International Atomic Energy Agency in Vienna. There were 188 representatives from thirty-five countries and four international organizations. Table 13 illustrates the wide area of expertise that was available to participants.

Nuclear power is an old story in the United Kingdom. The first reactor started to produce in 1962, and Britain now has thirty-four major installations—more than any other country except the United States. It gets 11 percent of its power from the atom.

Leonard H. Leighton, who heads the Department of Energy in London, cited what he termed two predominant features that govern his country's planning, funding and execution of public sector energy research and development: (1) A basic philosophy—supported by the U.K. Government—of control and management of civil-applied research; and (2) Numerous public statutory bodies operating in the field of energy.

The term "public statutory bodies" was defined by Leighton as referring to nationalized energy supply industries, the U.K. Atomic Energy Authority and the research councils.

The U.K.'s Department of Energy is not so much a new, distinct agency as it is a reincarnation of the old Ministry of Fuel and Power which was started some thirty years ago to coordinate and develop energy resources and promote efficiency in their use.

Leighton, an Oxford graduate, described the general energy situation in the United Kingdom as quite good. There are, he said, ample coal reserves in the ground plus oil reserves off the Continental Shelf which could lead toward self-sufficiency by the mid-1980s and even stretch through the last decade of this century. What happens after that is anybody's guess. . . . Providing, I suggested, the United Kingdom isn't invaded by some energy-hungry neighbor like the United States, Russia, China, Italy, France, India, Bangladesh, or even Dr. Stoecker's West Germany. . . .

Twitching one eyebrow slightly, Leighton looked at me for one full minute, apparently wondering if I knew something that he didn't. Satisfied that there were no immediate plans for the para-military invasion of Britain, Leighton chose to ignore my war-mongering comment and concluded, very graciously, with this remark: "The civil nuclear program remains by far the largest single program that we now have. And, with that, the greatest concentration of effort is on the Sodium-Cooled Fast Breeder Reactor."

When oil prices started to soar, the Japanese government soon announced a major nuclear program that would raise Japan's current 4 percent usage to 25 percent by 1985. As a result, the "Sunshine Project" was one of several ambitious technological programs that were launched by Japan in 1974. Its completion date is scheduled for the year 2000. According to Dr. Shogo Sakakura, the project's senior officer, "Sunshine" is aimed at alleviating the energy crisis resulting from a dwindling supply of petroleum resources. Until recently, Japan had what was thought to be an abundant supply of readily accessible oil, but this is no longer the case, and the Japanese are now busy tapping other sources.

Dr. Sakakura, a brisk but pleasantly professional individual, told me quite rapidly—and with frequent smiles—that the energy consumption in his country increased from 100 million kilolitres in oil equivalent (primary energy supply) in 1960 to 340 million kilolitres in 1971. "This represents an annual increase of 11 percent," he said, "which means our growth rate is twice as fast as the world's average annual increase in energy consumption during the same period."

As a result, Japan's energy consumption grew by 1970 to the second largest volume in the non-Communist world—second next to the United States. At present, said Dr. Sakakura, "we fill 73.5 percent of our primary energy requirements with petroleum, for which we are almost totally dependent on imports."

During 1974, approximately 2.7 billion yen ($8,667,000) was appropriated for "Sunshine Project" and during 1975 some 4.8 billion yen ($15,408,000) were put to work. Energy sources regarded as priorities include:

(1) Solar
(2) Geothermal
(3) Coal Gasification and Liquefaction
(4) Hydrogen
(5) Supporting Research (a kind of general catchall)

The Japanese energy problem is promoted on a national scale and geared toward total cooperation from national research institute organizations, universities and private enterprises.

Canada, that friendly neighbor to the north of the United States, has a research and development program that seems geared toward two options. According to Charles H. Smith, assistant deputy minister and director of the Canadian Government Energy R&D Task Force, these areas are: (1) reducing the energy demand, while (2) making better use of domestic resources.

Smith is a tall, scholarly type who looks like he might enjoy an occasional bloody game of rugby in-between scanning energy flow charts and rock-collecting in the lower Amazon Basin. He is a graduate of Yale with a strong interest in geology, for which he devoted two years of intensive study in South America. He is also a highly recognized professional who has published thirty-seven different scientific papers.

Government figures show that during a 1973–74 period, primary energy usage in Canada was at 7,400 trillion BTUs, or a gasoline equivalent of forty-three gallons per week per person. Oil accounted for 44 percent, gas for 21 percent, and coal, 8 percent.

Canada's accessibility to water sites, Smith said, makes hydroelectric power its most renewable resource and accounts for 75 percent of its power source. However, nuclear reactors, totaling some 12,000 megawatts, are either in operation, under construction, or already committed under the research development program. Radioactive wastes are currently being stored in massive geological formations which, Smith said, appear most practical for the present technology.

Down south of the border, Mexico's situation is a progressive one, according to Juan Eibenschutz, executive secretary of the Commission on Energy. Approximately ten years ago, he said, Mexico realized the need for improving and strengthening its technological status, and three research and development institutions evolved: (1) Institute of Petroleum, (2) Electric Industry Research Institute, and (3) National Commission on Nuclear Energy, which was recently changed to the National Institute of Nuclear Energy. Since oil and gas provide 92 percent of Mexico's energy needs, the Institute of Petroleum has "developed into our most important research institution in the energy field," Eibenschutz explained.

For a specific area of research, whether national or international, it has been necessary to initiate a study on a broad analysis basis. In 1971, for example, Mexico established its National Council of Science and Technology. Other countries, both large and small, show a similar pattern. Initially, nations make a technical evaluation of techniques that are either available or being investigated. Next, the economic aspects of such use, together with social implications and the consequences to public health and environment are probed. From such rough evaluations (compare them to mini-environmental impact statements) emerge a main problem and a host of secondary ones. In short, it's something like opening a Pandora's Box that has a whole bunch of black, swirling goodies inside. Still, of such is the oft-untraveled way of progress.

Even in the case of research related to a secondary problem, repeated evaluation remains essential. While it is necessary in the first place for determining the state of knowledge on a subject and for gap-filling in technology, it is equally vital for supervising and guiding the overall progress of R&D on an international scale.

One main aspect is feedback—also known as input and output—

between the results obtained and those hoped-for. Again, in such repeated evaluations, it is necessary to consider the primary areas of (1) technical feasibility, (2) economic merits and (3) social implications.

W. van Gool, chairman of the National Energy Research Steering Group (LSEO) for the Netherlands, outlined the following phases applied by the Dutch to energy research:

- Evaluation, analysis, assessment
- Basic research
- Exploratory theoretical and/or experimental research
- Applied theoretical and/or experimental research
- Applied experimental research on a larger scale
- Research on a semi-industrial scale (pilot plant)
- Prototypes of commercial installations

It would seem realistic, van Gool noted, "to suggest that the LSEO's involvement with the research should end when positive results are obtained in the phase involving the operation of semi-industrial installations. Further commercial development will have to be left to the industry concerned, with NEOM playing an important part in certain cases." (NEOM stands for the Netherlands Energy Development Corporation and generally corresponds to the Energy Research and Development Administration in the United States.)

In the Netherlands, van Gool continued, industrial experts are brought in during the later stages of evaluation. "After all," he reasoned, "it would be a waste of funds provided for research and research capacity to enter into a costly semi-industrial phase, if industry were not prepared to apply the system in the future, even if, in the technical sense, the research had been successfully concluded."

In the United States, one of the most current and viable examples of this division of labor between government, utility and private industries working toward a common goal is the Clinch River Nuclear Island near Oak Ridge. (Table 9 in Chapter 11 provides an excellent rundown on this operation and how it is applied.)

The Netherlands' current expenditure for nuclear research and development totals Fls. 200 million a year ($74,100,000). This amount includes Fls. 5 million ($1,852,500) for pure research. In 1974, the total budget for the Netherlands Organization for the Advancement of Pure Research (ZWO) was Fls. 107 million ($39,643,500).

Nuclear fission efforts have been directed primarily toward developmental work and the construction of large pilot plants. Roughly half the money has been spent in the form of contributions to international programs devoted to the SNR-300 and research on Fast Breeder Reactors. A large amount of the funds made available to the RCN (Netherlands Nuclear Research Centre) is spent on research into environmental and safety aspects of nuclear energy. Van Gool also noted that the Dutch are intensifying research into the storage of radioactive wastes.

Due to the limited nature of the Netherland's research potential, much of its R&D work is enhanced by closer looks at what other members of the international community are studying. As a direct result of the energy crisis, a number of countries—including the Dutch—met in Washington in mid-1975 and participated in a consultation that resulted in the formation of an international agency on energy. This agency has been specifically charged with regulating such fields as energy conservation, development of alternative energy sources, and R&D related to new technologies and the specific aspects of uranium enrichment.

Mention should also be made of the proposals of the Commission of the European Communities, dated July 17, 1974, which are contained in a document called "Energy for Europe: Research and Development." Since its initial session, the committee has already drafted program proposals in the fields of systems analysis, geothermal energy, solar energy and hydrogen economy. Work is also proceeding on a new Five-Year Plan for nuclear fusion research covering the 1976 to 1980 period.

Extensive research into and the development of magnetohydrodynamic (MHD) installations have led many countries to realize that this form of energy conversion—converting heat directly into electrical energy—shows potential. An MHD generator shows the "possibility" of: (1) high conversion yield, (2) limitation of polluting processes, and (3) competitive installation and production costs per KW or per KWH.

It should be stressed, however, that MHD is still very much in the experimental stage just like solar or geothermal. There is a 25 MW prototype power station in the vicinity of Moscow which is currently generating power at the rate of 6 MW. Natural gas is used to fuel this generator. One clear advantage of such a station is that MHD can reach an efficiency of up to 65 percent as compared to 40–42 percent for traditional thermal stations.

A main problem encountered in such research is difficulty in coupling MHD plasma generators to nuclear reactors. For this reason, several Western European countries, like France, Britain, Germany and Italy, have decided to limit development of MHD generators until time proves them more practical. Researchers in the United States and Russia view MHD in combination with the combustion of natural gas and coal. As a result, most installations are located, as often as possible, near mines.

In the Netherlands, such research is located at the University of Technology in Eindhoven. It has resulted in the generation of 250 KW with high capacity densities of 80 NW/M^3 for short periods. In order to interest industrial development, the scale of this project was to be enlarged in 1976—an effort requiring international cooperation.

In 1974, the Smithsonian Science Information Exchange, Inc., an agency associated with the Smithsonian Institution in Washington, D.C., officially began a study on current energy research. Its data collection specifically involves the United States, Canada, the Federal Republic of Germany, the United Kingdom, Italy, France and the Netherlands. The Smithsonian is expected to publish its findings sometime in 1976.

The purpose of European research into nuclear fusion is to clarify the fundamental physical and technological aspects of the various thermonuclear reactor systems that have been proposed. This is especially true regarding the magnetic occlusion systems of the Tokamak variety. Results here warrant expectations that upscaling (going to larger models) will make it possible to enter an area of plasma parameters where the required thermonuclear regimen is closely attained.

As a result, European fusion laboratories are projecting a preliminary study that would create a "Joint European Tokamak" to combine efforts and conserve costs. Independently, large programs are underway at the Lebedev Institute in the USSR, and at the French Atomic Energy Laboratory at Limeil. Study is also being conducted at Garching in West Germany, and at Osaka and Nagoya in Japan.

The Soviet Union took a major lead in the fusion sweepstakes with the announcement in mid-1975 of the successful operation of the world's first fusion power station based on the time-tested Tokamak system. It is now chugging away near Moscow, some five years ahead of the original schedule.

Such controlled thermonuclear fusion represents the ultimate source of inexhaustible energy. In the long run, it is almost entirely pollution-free and is something the world's greatest minds have been trying to perfect since the end of World War II. The Russians, with their not-so-sophisticated machinery and "quaint little labs," made a few important turns and proved that a fusion reactor can go operational. Fusion, unlike fission, cannot blow up—only *out,* and that spells energy.

Confinement continues to be a difficult problem to cope with. Once plasma (ionized gas) is heated, it is critical to hold it steady for longer than the briefest fraction of a fraction of a second. You've got to contain it without allowing it to touch anything solid and keep it (nuclear plasma) burning at a not very cool 100,000,000° to 200,000,000° Centigrade.

I remember the look of horror on my neighbor's face when I came bouncing through her front door, clutching the morning edition of the *Tennessean* and pointing to a story on page two about the successful operation of the Russian Tokamak system in Moscow. As far as I was concerned, it should have been headlined over the masthead.

"They've done it. They've got a thermonuclear fusion plant going. Read this," I screamed, shoving the crumpled mass at her.

Poor Glenda, God love her. No one could want a better neighbor. She took one look at my red, bleary eyes (from writing a book on nuclear power) and asked if I needed a scotch or brandy!

"Make it on the rocks," I muttered, fusing inside. Nobody understands . . . Nobody cares. . . .

So, you're right on target if you get the impression that I'm excited about the prospects of fusion. It appears as the most viable and safest energy source for the future. Again, however, the problem is how to get there from here.

While we are on this subject of Russia and fusion, a chapter follows which details how the USSR views nuclear power and some of its major installations. It's all true, courtesy of the Embassy of Soviet Socialist Republics, Washington, D.C., and their press agency, Novosti, in Moscow.

Table 13: Operating Experiences—Agenda for the IAEA Symposium on Experience from Operating and Fueling of Nuclear Power Plants.

It illustrates the wide area of expertise that is available through the international nuclear community. This was a symposium on experience from operating and fueling of nuclear power plants which was held at the International Atomic Energy Agency in Vienna. Some 188 participants from 35 countries and 4 international agencies were presented.

Paper No. (SM-178)	Title	Author	Country
SESSION I Review of Experience from Nuclear Power Programs—1			
101	Performance and Reliability of Operating USA Nuclear Power Plants	D. J. Shovholt	United States
48	Summary Report on Performance of Nuclear Power Stations in the Federal Republic of Germany	W. H. F. Huenlich	Germany
SESSION II Review of Experience from Nuclear Power Programs—2			
102	Experience with Canadian Nuclear-Electric Generating Stations	E. K. Keane	Canada
103	Experience from Operating and Fueling a Nuclear Power Plant in an Industrially Developing Country (invited paper)	M. Dayal	India
55	Measuring Performance of Nuclear Power Plants	D. B. A. Chase and J. Iljas	IAEA
18	Analysis of Generation Availability at the CEGB Magnox Stations over the First 10 Years of the Magnox Program	R. S. Gow	United Kingdom
46	Experience Gained in the Operation of Spanish Nuclear Power Plants	L. Alvarez de Buergo	Spain
SESSION III General Experience from Operation—1			
24	Nuclear Component Reliability and Performance in Westinghouse PWRs	R. G. Hobson and R. B. Cambien	Belgium
49	Operating Experience with San Onofre Nuclear Generating Station	H. L. Ottoson and W. R. Gould	United States
8	Experience from Operation of Stade Nuclear Power Plant During First Fuel Cycle and First Refueling	A. Gotz	Germany
14	Experience from Operation of Oskarshamn Unit 1	O. G. Gimstedt and S. O. Hilding	Sweden

Continued

Table 13—Continued

7	Oskarshamn 1, Experiences from Load Rejection with the ASEA-ATOM BWR	T. I. Sterner	Sweden
11	Operating Experience of the Ardennes Power Plant	A. Marchal	Belgium
33	Analysis of the Performance of the Systems at the Santa Maria de Garona Nuclear Power Plant After Two Years of Commercial Operation	E. Velasco Garcia	Spain
SESSION IVa General Experience from Operation—2			
32	Highlights of Experience During the Commissioning and Initial Operation of Karachi Nuclear Power Plant (KANUPP)	S. M. N. Zaidi	Pakistan
45	In-Core Instrumentation and Reactor Feedwater Control Systems: Operating Experience at Tarapur	K. V. Chalam, B. Mazumder, P. G. S. Mony, and D. P. Paranjpe	India
22	Significance of Present and Future Operation of the Kahl Experimental Nuclear Power Station (VAK) for Construction and Operation of Economically Efficient Nuclear Power Stations	M. Ellmer and M. Wolff	Germany
SESSION IVb Experience from Reactor Systems—1: Major Components			
25	Experience in Operation and Maintenance of Westinghouse Steam Generators	L. Marique and O. A. Willson	Belgium
52	Maintenance and In-Service Experience at Large Nuclear Power Plants	G. B. Lloyd, D. G. Bridenbaugh, and R. L. Turner	United States
16	Operation Performance of Steam Generators and Circulating Pumps in the Primary Circuit of KWO	A. Mayr	Germany
17	Problems of Sodium Generators Expanded by the Example of a Leakage in a Steam Generator of the Sodium-Cooled KNK Nuclear Power Station	W. Marth and K. Dumm	Germany
5	Repairing the Primary Heat Exchangers and a Circulation Pump at KWL	O. Deublein	Germany
21	Correlations of Neutron Embrittlement with Composition and Damage Fluence Observed from Power-Reactor Pressure-Vessel Surveillance Programs	C. Z. Serpan, Jr. and J. R. Hawthorne	United States

Continued

Table 13—Continued

SESSION V Experience from Reactor Systems—2: Waste-Management Systems			
28	Waste Management Experience of Westinghouse PWRs	J. L. Gallagher, R. Billingham, J. A. Battaglia, and R. T. Marchese	Belgium
19	Fission-Product Release After Reactor Shutdown	N. Eickelpasch and R. Hock	Germany
31	Activity Due to Corrosion Products in the Primary Circuit of the Chooz Power Plant	G. Frejaville, A. Marchal, P. Beslu, and A. Lalet	France
20	Transport of Irradiated Fuel and Control Rods from a Decommissioned Power Reactor	W. A. Pryor	United States
SESSION VI Experience from Fueling—1			
50	Nuclear Fuel Experience in Westinghouse PWRs	W. J. Dollard, F. W. Kramer, and E. Passig	United States
	Experience from Fuel Performance at KWO	H. Schenk	Germany
6	Six Years' Experience with Core Performance at KRB Power Station Gundremmingen	N. Eickelpasch	Germany
29	Experience with Fuel and Core Performance at Tsuruga Nuclear Power Station	Y. Kuge, R. Yamasaki, and W. Shinoda	Japan
9	EL-4 Fuel: Technological Development and Experience of On-Load Refueling	F. Decool, Y. Gouchen, G. Orsier, and C. Ringot	France
13	The Physics of CANDU On-Power Fueling: From Design Objective to Pickering Demonstration	A. A. Pasanen	Canada
SESSION VII Experience from Fueling—2			
26	Westinghouse PWR Reactor Refueling Experience and Improvements	R. Evenepoel, M. Fabris, and L. Katz	Belgium
43	Core Performance and Fuel-Handling Operations at Tarapur	T. Tirupataiah, K. S. N. Murthy, and R. Rajaram	India

Continued

Table 13—Continued

47	Reactivity Behavior of the Dodewaard 50-MW BWR Core Before and After the Use of Gadolinium as Burnable Poison	J. C. Bruggink	Netherlands
12	Verification of BWR Physics Characteristics from Wurgassen Startup and Related Experiments	P. Killian, D. Pleuger, and J. Schulze	Germany
4	Core Management, In-Core Instrumentation, and Use of the On-Line Computer on the Windscale Advanced Gas-Cooled Reactor	J. H. Leng, J. McGhee and I. G. Smith	United Kingdom
2	The Nuclear Assurance Corporation Performance Program	J. H. Nail and J. J. Stobbs	United States, Switzerland
SESSION VIII Operational, Preoperational, and Startup Testing, and Personnel Training			
27	Standard Westinghouse Nuclear Steam Supply System, Startup Program	H. M. Couez and M. F. Carteus	Belgium
3	Safety-Related Experience in the Startup Phase of the BWR Nuclear Power Plant at Wurgassen	H. A. Ritter and S. Wiesner	Germany
35	Initial Operating Experience with the Vandellos Nuclear Power Plant	J. Rierola Puigserinanell	Spain
44	Experiences from Operational Testing Program Including Engineering Safeguards at Tarapur	K. S. N. Murthy, R. Rajaram, and C. G. G. Rao	India
34	Reactor Operator Training Programs Utilizing Nuclear Power-Plant Simulators	P. F. Collins	United States
10	Training of Personnel for the Power Plants of Electricite de France	M. Feger and M. Brosson	France

Panel Discussion: Quality Assurance Programme for Nuclear Power-Plant Operation: Its Meaning and Importance

Chairman: O. G. Gimstedt, Sweden

Panel Members

E. K. Keane, Canada
F. Decool, France
H. J. Schenk, Germany
M. Dyal, India
D. J. Skovholt, United States
R. Skjoldebrand, IAEA

Chapter 15

MY FRIEND, THE RUSSIAN

A solid, black-tipped fence surrounds the grey-stone building, while outside, pacing like a well-fed Bengal tiger, stands a very pleasant on-duty Washington police officer. Usually, there are two blue-uniformed officers: one to watch the rear alley and another to check the front entrance.

Rumor also goes—which I later found out was true—that there are FBI agents manning elaborate cameras and monitoring devices located somewhere in the numerous buildings that stare blankly toward 1125 Sixteenth Street N.W.

The Embassy of the Union of Soviet Socialist Republics, some three brisk blocks or so from the White House, is a frequent target of bomb threats, curiosity-seekers hoping to see a real-live Communist, and unnamed others who sometimes—as a form of the ultimate sacrifice for one protest or another—chain themselves to its worn but sturdy fence. The majority of Americans, I am told, prefer to simply stay away. It's less complicated.

As a journalist often in Washington on various assignments, I have always been intrigued by the embassy. Until recently, I had no assignment that really justified my going up, knocking, and simply saying "hello." After all, one has to be professional in these matters which smack of the forbidden.

Flying back from the West Coast toward a series of interviews in the capital, I realized that my time had finally come. Nuclear power was the key and certainly a legitimate one. It was at Los Alamos and Oak Ridge that I had learned just how much the world owes the Russian scientists for their assistance and willing cooperation. Shared knowledge is not a one-way street, with the U.S. giving all and taking none, and we apparently do need each other much more than the average John Doe Citizen realizes or appreciates. At least, that's what this native-born Southerner, who grew up on fried chicken and "Masters of Deceit," discovered with an uncomfortable and unforgettable jolt.

It was Dr. John F. Clarke, director of the Thermonuclear Division at the Oak Ridge labs, who helped reverse my thinking which still has a tendency to view the United States as the "firstest with the mostest." I guess you might term it home town pride, except that being realistic, we really aren't the only grain of sand on the beach. So, as the beer commercial goes, "We are all in this together." It does, however, take a little bit of rethinking.

At any rate, there was Dr. Clarke, commenting on Ormak, which is Oak Ridge's contender in the fusion sweepstakes.

"Russians *know* about this?" I whispered, glancing at the complicated equipment with its wires and strange gadgetry.

"Yeah. They come through all the time . . . one group or the other," he replied, not yet tuning in to what I was thinking.

"You let *them* in here?" I sputtered. "They could steal our ideas, photograph everything, even wipe us out."

The whole conversation was beginning to take on the dimensions of an updated version of "Dr. Strangelove." And Dr. Clarke, beginning to be amused by my surprised expression, proceeded to very neatly slice my red, white and blue balloon.

"Ormak," he smiled, "is patterned after the Russian Tokamak system. We got this idea from 'them.' They've been advising us on what kinks to avoid."

So, as I said, it does take a little rethinking on a global scale. Despite what dissident writer Solzhenitsyn has implied, the thawing caused by détente is not something new in scientific circles. In fact, very little ice has ever formed between the learned men and women of Russia and America. A cloistered mentality in academic circles merely serves to create a polarization that blocks progress on all sides. It serves no interest whatsoever, as scientists are well aware.

One example that goes back some twenty years involved the operational principles of the first atomic power station in Obninsk, Russia. The year was 1954, and one year later, at the first International Conference on the Peaceful Uses of Atomic Energy (Geneva), there were the Russians relating both failures and successes to a very interested assembly—many of whom were still struggling with reactor design.

Today, there is a regular exchange program involving our scientists and theirs at places like Los Alamos, Oak Ridge, Argonne National Laboratory and Stanford.

You might remember that it was the Russians who put up the first rocket and the first men and only got seconded when we got the "smarts" and applied our sophisticated technology. Reactors are going the same way. Prior to 1968, Princeton was tinkering in fusion with a device called a Stellarator, Oak Ridge and California's Livermore were into mirrors, and Los Alamos was playing with Theta pinches. Finally, after a lot of failures and very few successes, a team of British physicists elected to go to Moscow to find out what was going on and why the Free World was getting so confused over this entire plasma mess. Taking the cue, Oak Ridge "smashed" its mirrors and went into Ormak (a play on the first letters of Oak and Ridge); Princeton dismantled its Stellarator and in nine months gave birth to an illegitimate but made-in-America, Russian Tokamak.

As one scientist noted, "The Russians stick to those fine old-fashioned things, keep it simple and push it a little further. They still hang chandeliers in their labs for light."

Another colleague added: "One does not want to muddy the fact that the Russians did the work. We don't want to say that it's our idea. It's not. We picked it up and carried it with them, but without their help, it wouldn't have been possible. . . ."

For example, Drs. N. G. Basov and O. N. Krokhin (authors of English translation, *Soviet Physics,* 1964) in the USSR in 1962, proved analytically that lasers could produce plasmas of the required temperature to start fusion reactions. Based on this evidence, they then initiated a program which has produced the most advanced experimental results to date.

The release of thermonuclear energy in a controlled manner, with its promise of a relatively clear and inexhaustible supply of energy, continues to receive an increasing amount of attention in the Soviet

Union and elsewhere, especially since a Tokamak machine went operational in mid-1975.

There is one additional thought voiced by Dr. Clarke which made a lot of sense. It would have been easy, he observed, for the Soviets to have simply kept quiet and not told anybody anything. Then they could have startled the world by coming out in 1980 or 1990 with a working fusion machine. "They just didn't do that," he said. So, this is the thinking that prompted me to ask for Soviet help and advice on my research involving the international aspects of nuclear energy.

Thinking it would probably sound insane over the telephone and preferring eyeball-to-eyeball contact, I decided to go directly to the embassy from National Airport, rather than first arranging a definite interview. Washington traffic is something else, and after parking several good country blocks away, I finally arrived, literally breathless, at the gate.

It had apparently been a long uneventful day and the officer eyed me and my attaché case curiously. . . .

The cameras must be going wild, I muttered to myself, suddenly apprehensive as I realized exactly how the situation must look: here I was, just off the plane from ultra-secret Los Alamos, and the first place I head is the Soviet Embassy. They probably think I'm here to report to my Russian masters. Since the cloak-and-dagger stuff is just not my bag, I could feel the beginnings of a red rash forming on the upper part of my neck and a small cluster of hives grouping around the left ear lobe.

A few weeks later, during an informal question-and-answer session called by a curious FBI, I would throw up my hands and stare incredulous at the two pictures they produced that showed me first entering, and then leaving the embassy with a Russian official—we were both smiling and it looked just like we were on our way to the White House for dinner.

A slight discrepancy in the two photographs caused a further problem that required an official explanation. One shot showed me wearing a dress, and in the other I wore a sporty pants suit. The FBI wanted to know why I had changed clothes inside the embassy. . . .

"It's just a little baffling and not a usual occurrence," one agent explained kindly, adding, "you don't have to answer if you would rather not."

Silence. I was beginning to feel like the "Big Bird" of Sesame Street being interviewed by Timothy Frog. . . .

"Simple," I replied, "I went there on two separate occasions. Your camera caught me on a Tuesday and then a Wednesday."

There was an "oh" and a flurry of note-taking on blue-lined, white legal-size pads. . . .

"My fellow Americans," I grinned. "Surely, you don't think that if I were a real spy that I would be so dumb as to go to the front door of the Soviet Embassy to make contact with my control."

"Yeah," the other agent replied in an official manner, indicating that all fun and games were over. "It's so ridiculous that it's a good possibility. It's the kind of direct approach that they just might try. Now, *what did you give them from Los Alamos?"*

Funny, how unfunny some situations can become. Eventually, the matter was resolved to everyone's satisfaction, but I learned first-hand how serious it is for an average and naive citizen to dabble a toe in cold international waters. Sharks abound and many of them have red, white and blue stripes. . . .

However, back to the embassy and where it all started.

Glancing at the officer and feeling like a country girl on her first outing in the big city, I said, "Hi. How do you get in?"

"You push the gate open," he replied dryly.

"You're right. Absolutely right," I blushed.

He grinned and shook his head.

After making the rounds of major nuclear installations in the United States, I had grown accustomed to being searched, finger-printed, and photographed with my permission. In fact, my brief visit to Los Alamos was nothing short of a trip into 1984—the one George Orwell wrote about in such vivid and unglowing terms.

So, as I approached the heavy, wooden doors and waited for the opening buzzer, I fully expected the "third degree." If the Americans are hyper, I reasoned, then imagine what an interview with the Communists will be like. Sodium Pentothal, the whole works, I thought grimly. Still, the experience was worth it.

The doors creaked open and I stepped into a small alcove that was quiet and cool. Directly in front stood a row of tall glass doors that enabled one to see out but not in. I stared quietly at my own pale complexion, wondering if someone was watching on the other side.

"Yes?" A heavy, masculine voice startled my thoughts and beckoned me inside. I moved toward a small parlor and a smiling Russian (he was blonde-headed and about my age) who sat squarely behind a small bullet-proof glass partition. There were office chairs and a small bookcase with works on Marxism-Leninism and current Soviet magazines.

Clutching my attaché case, I tried to look cool despite a thumping heart and explained that I was in Washington as a journalist, researching a book on nuclear power. "Peacetime uses," I grinned.

The deskman looked at me for a moment, probably wondering if I was for real, and then said, very kindly: "But, of course, just one moment while I contact our scientific officer. Please make yourself comfortable."

Feeling very Tennessee and extremely tired, I sank back into a large leather chair and listened as my presence was explained to someone on an upstairs extension.

"I don't speak Russian," I apologized to the young man.

"No problem," he said, struggling with the English words. He then stared at me and I stared back, trying to think of something normal to say under the circumstances. Nothing came, so we waited in silence.

In a few minutes another kindly-faced man emerged who identified himself as an official of the embassy and a nuclear physicist. His name was Doctor Sergey Zaitsev, of Moscow.

Smiling, he crossed the room and sat down in a leather chair as though he were prepared to spend the rest of a busy day just listening to what brought me to his door.

"Now, Jacque," he began, pronouncing my name correctly (that was fifty points already), "what is your problem and how can we help you?"

I tried to show him my press credentials, but he merely shrugged them off, saying quietly, "I believe you. There is no need for these."

Trust was established and on that note, we became instant friends.

During the days, weeks, and months ahead, this gentle Russian and I would talk or correspond on many things—nuclear power, governments and man's striving for peace. He spoke to me as a father, as a citizen proud of the Soviet Union and its accomplishments, and, perhaps most important, as a human being concerned about the state of the world. And I listened and replied as a writer, a mother, and a citizen equally proud of my own country. Neither of us compromised our position.

On that first visit, however, Dr. Zaitsev left me with one thought that still lingers. "Jacque," he asked softly. "Have you been to us before?"

"No," I replied.

"Then you are a brave journalist. Most Americans are afraid to come and see us . . . They are afraid."

There was a pause then and heaviness in the air. I understood quite well what Dr. Zaitsev was not saying aloud. Finally, I replied wearily: "And that, my friend, is why we have wars and misunderstandings, why we hate: because people are afraid of each other. They don't take time, or perhaps the courage, to talk and to listen."

The Russian nodded.

At this point, I want to make one very quick observation based on experience. There will be some who will read this chapter and instantly think that here is a gal who was conned by the Communists on a grand scale. "The Russian knew a good thing—press-wise—when he saw it. Durn Reds are smart," or so the reasoning might go. An even smaller group might possibly speculate that here was a very polished KGB officer who was eyeing a potential recruit for the Soviet cause. That's the reaction: got to be a motive somewhere, other than simple human trust.

One fact is clear and it covers everyone. If we are to survive as a people, then we've got to break down the barriers imposed by prejudice and outright fear. Vance Packard wrote that we are a nation of strangers. He was right. In fact, we are a *world* of strangers—tiny, poisonous piranhas, girded to the teeth and ready to slash and cut anyone who comes too near. We've got to meet and trust as human beings rather than enemies espousing conflicting ideologies. The key is trust and not preconceived notions. The pressure to turn the latch comes from global re-thinking, and you have to ask "just what have we got to lose or win?" Another war? It is well to remember that our sticks and stones are much sharper now and deadlier. The human heart has far more power than the strongest laser. Unfortunately, in too many cases, it is not operating at its total capacity. But the potential is definitely there, and it may well be the international scientific community, through joint nuclear research, may help clear some of the clouds that have gotten in our way. . . .

In this capacity, Dr. Zaitsev has proven to be a valuable resource; the material contained in this chapter was provided by him and the

Novosti Press Agency in Moscow. As to motive, perhaps it is reflected in a letter he sent shortly after my first visit: "Your friend," he closed. And that's how I read it. . . .

Fuel and power resources in the Soviet Union—specifically combustible shales, coal, oil and natural gas—represent an enormous amount of energy. The figure, released from the Council for Problems of Energetics, USSR Academy of Sciences, is about 6.6 trillion metric tons of standard fuel (note: one kilogram of standard fuel produces 7,000 kilocalories). Of these total reserves, more than 1.8 trillion metric tons are already available for extraction and use under present technical and economic conditions. As to the quantities of surveyed resources, coal comes first, then oil, gas, and combustible shales. While the reserves of oil and gas are smaller, they are nevertheless sufficient to guarantee Russia's unimpeded economic development for a long time to come.

Now, a figure like 1.8 trillion is impressive but also abstract. Simplified, it means that the utilization of fuel and power resources in the Soviet Union, as in certain other industrially developed countries, has been doubling every ten years. This is a fact of life. Therefore, the available mineral reserves are estimated to last just about 200 years and these comprise only a portion of Russia's total resources. There is also disposal hydropower and atomic energy. As to the first, power stations can be built on rivers with a total annual capacity of more than one trillion kilowatt-hours. In addition, the fuel mineral reserves not yet available for extraction come to about 4.4 trillion metric tons.

The Soviets have also learned that the total capacity of a thermal power station should be as large as possible. There are now more than thirty of these stations in the country, each with a capacity of a million or more kilowatts. During the USSR's current Five-Year Plan, nineteen thermal plants, with a capacity of 2.4 million to 4 million killowatts each, will be built.

Unified power transmission systems make possible additional economies, including conservation of fuel. There are more than 100 district power systems in the USSR, grouped into eleven zonal grids. Eight of these zones are combined to form a high-capacity unified power system for the European part of the country. The system will shortly incorporate the remaining three grids, and a single power grid for the country will thus take shape, involving an area of some 8.5 million square miles.

Distribution of fuel resources is a major problem faced by the Soviet Union. The European part of the country—with most of the population and basic industry—accounts for only about 10 percent of the overall fuel reserves, while the Asian part, east of the Urals, has the other 90 percent. For example, the KanskAchinsk coal basin in Siberia, which has reserves of more than 1.2 trillion metric tons of fuel, is a far 2,000 miles from Moscow. The northern oil and gas deposits are also a considerable distance from the country's main consumption centers.

An analogy might be drawn in the United States where there are tremendous reserves of oil shale located in areas where there is insufficient natural water supplies to process it. Mother Nature is like that sometimes. . . .

The Soviet Union is also conducting research to determine new energy sources. Interestingly enough, an experimental tidal electric station was recently built in the country's northern European section, and preparations are now underway for the construction of another, much larger facility. The Russians view ocean tides as tremendous sources of potential energy, and, unlike other countries, they have the large areas of land mass needed to accommodate the technology involved in utilizing this distinctive form of energy.

In addition to MHD research, which was detailed in a previous chapter, there is one major geothermal power station now operational in the Soviet Union. Considerable research is also being done on the generation of electricity by a controlled thermonuclear reaction involving the fusion of hydrogen isotopes. This method would make a sizeable contribution toward solving the world's fuel resources problem for hundreds of years because it utilizes ordinary sea water. One drawback has been that extremely high temperatures of from 70 million to 100 million degrees (at least) are required for nuclear fusion. An experimental station is now operating near Moscow.

Present-day nuclear electric stations use uranium fission, and several of these have been built in the USSR. They are located in areas where the reserves of conventional fuels are low, such as the Central European part of the country, and in remote sections like the far North.

In 1974, Russia generated some 975 billion kilowatt-hours of electricity. By the end of 1975, power production is expected to reach 1 trillion kilowatt-hours. The greater part of this power will be produced by thermal power stations, followed closely by hydroelectric power stations. At present, according to Dr. Pyotr Neporozhny, USSR Minister of Power and Electrification, the contribution of nuclear power is fairly modest—third place—but this ratio is expected to change as the nuclear industry's share of power increases from year to year.

In 1954 Obninsk became the first atomic power station to begin operation. It first went operational with a mere 5,000 kilowatts, and today, twenty years later, a single unit of the Leningrad Nuclear Power Plant turns out a capacity of 1 million kilowatts. The plant has carved itself a permanent place in the history of technology alongside the first steam engine, the first automobile and the first airplane. One of the incidental but still very important results of this station is that it has been instrumental in training designers of future plants—and on an international level.

In 1964, the first unit of the Novo-Voronezh Atomic Power Station—one of the largest in Europe—became operational, and in 1973, a new power reactor with a capacity of 440,000 kilowatts was added to this plant. Performance testing of the first atomic reactor with a capacity of one million kilowatts has gotten well underway at the previously mentioned Leningrad station. Plants are now under construction at Kursk and Smolensk, as are reactors in the Caucasus.

Construction of a reactor is also underway in a northernmost station in a settlement called Bilibino, Magadan Region, in the permafrost area. Its first turbine generator produced electricity for industrial purposes in early 1974. The station has the unique function (unique by American standards) of supplying power mainly to the local gold-mining industry.

The main trend in atomic power engineering throughout the world is the use of water-moderated, water-cooled reactors like the ones in use at the Novo-Voronezh station, but the Soviet Union is also researching other reactor types. A multi-channel one is unique to the Leningrad station and, when operational, this facility should have a capacity of 2 million kilowatts, according to Dr. Neporozhny's figures.

Construction in the USSR has followed two stages: the first is characterized by atomic stations with thermal-neutron reactors, the second by plants with fast-neutron reactors. The latter appear more promising, as indicated by U.S. research on the LMFBR, whose main attraction is the ability to produce more fuel than is consumed. That translates "economy" in any language, and in this specific area, the

Soviet Union is conducting tests at the Shevchenko Nuclear Power Station on the Mangyshlak Peninsula. This station has a fast-neutron reactor with a capacity of 350,000 kilowatts—the largest of its kind in the world.

It has been the experience of the Russians that, even in these early stages, one kilowatt hour of electric power produced by the atomic power stations in the European part of their country is cheaper than one produced by larger thermal power stations. They further predict that this ratio will drop to half in the very near future.

As with their colleagues elsewhere, Soviet scientists are acutely aware of the increasing need for energy in the face of depleted resources. In an interview, Academician Lev Artsimovich, research supervisor, and his co-workers, Vitaly Shafranov (doctorate in physics and math) and Vyacheslav Strelkov (masters in physics and math), echoed a familiar litany:

"World requirements in power are increasing so quickly that the resources of all the existing types of non-nuclear fuel may prove to be quite insufficient in the coming decades.

"The even faster rate of organic fuel combustion pollutes the atmosphere. It also leads to irreversible changes in the biological and energetic equilibrium of nature. Moreover, oil, coal and gas are a valuable *chemical* source material, and it is economically inexpedient to burn it up in thermal power plants. *Future generations will look upon that as barbarity.*"

Nor do the stocks of nuclear fuel for reactors—based on the fission process—pose as a lasting answer (mainly due to radioactive wastes disposal) for the Russians. That's why fusion has nctted a fair share of research. . . .

Artsimovich then noted (through a translator) that reactors based on the synthesis of light nuclei do not produce durable radioactive substances and harmful waste. In addition, the radiation level stays quite low. "Still more attractive," he explained, "is the fuel for synthesis reactors—heavy hydrogen which is widespread in nature. If we succeed in taming thermonuclear reactions, mankind will draw on the world's oceans for electric power."

Scientists in the Soviet Union, the U.S. and Great Britain officially began tackling the fusion problem some twenty-five years ago, right after the first thermonuclear reaction produced after the explosion of a hydrogen bomb.

"At that time," the three Soviet scientists recalled, "it seemed that everything would proceed very quickly. The idea of magnetic thermal insulation enchanted the scientists by its attractive shape, and it looked as if nothing but engineering problems remained to be solved. However, the door to our thermonuclear El Dorado was not to be opened so quickly."

Nature, it seemed, operating within its own laws, posed difficulties that caused hopes to evaporate just as quickly as the walls of the chambers in which the unsubdued plasma rages.

Note: As stated in a previous chapter, in order to produce a controlled thermonuclear reaction, an exotic substance is needed, i.e., a greatly rarefied gas heated to the temperature of hundreds of millions of degrees which exists but for a few seconds. And therein lies the problem—confinement of a plasma that reacts to confinement like a glob of jello wedged between two rubber bands.

Russia's Tokamak installations have offered hope and many systems elsewhere in the world have converted to this method. An installation of the Tokamak family consists of a toroid-shaped discharge chamber resembling a doughnut placed upon the iron core of a transformer. The chamber is filled with hydrogen or deuterium under a pressure that is a million times lower than that of the atmosphere. A current is excited in the gas, heating it to the temperature of about ten million degrees. But the soft pinch of the plasma coil proves to be unstable, that is, any minor changes in its shape lead to complete annihilation within a very short time. This is why physicists, in order to inhibit such instabilities, have to create additional great magnetic fields in the chamber, building, as it were, a framework upon which the plasma coil rests for a few hundredths of a second. An installation of this type is called *closed,* because the magnetic field is fully confined in the chamber.

Research into creating a magnetic thermonuclear reactor had been started in the Soviet Union in 1951 at the I. V. Kurchatov Institute of Atomic Energy. A number of fundamental ideas were suggested to underlie all subsequent experimental and theoretical research. Also, a new science, known as the physics of high temperature plasma, was originating. Initially, there was practically no plasma theory; scientists in the Soviet Union and elsewhere knew of some effects, but this knowledge was extremely simplified and there were no contact points between theory and experiment. Difficulties were further aggravated by the fact that the new science called for new techniques—those of high pulse currents, magnetic fields and high vacuum. Conventional methods for measuring matter proved to be useless. And this was when a new sphere of science, plasma diagnostics, appeared.

Initial experiments in the Soviet Union brought out the first regularities which encouraged theoretical research. The late fifties saw the development of the hydrodynamic theory of equilibrium and stability of plasma pinch, which served as a foundation for subsequent experimental research. The major results of this research were reported at the Second International Conference for the Peaceful Uses of Atomic Energy in Geneva in 1958. At that time the installation was officially christened "Tokamak," the Russian abbreviation for Toroidal Chamber with Magnetic Coil. In 1964, experiments were being conducted on four such facilities. Further research developments exerted a great influence on the creation of a neoclassic theory of energy and particle transfer in toroid plasma systems. This new theory resulted in a sharp reduction of the gap between theory and experiment. By 1968, marked successes were already scored and even pessimists realized the prospects of Tokamaks.

Scientists at the Kurchatov Institute of Atomic Energy—jointly with British scientists who by this time had well-developed methods for measuring plasma parameters with the aid of laser beam diffusion—carried out research which fully confirmed earlier results. At an international conference in Dubna (USSR) in 1969, the Tokamaks completely won over any sceptics. This was probably the instant in which the triumphant march of these installations through the laboratories of the world began.

In addition, Soviet scientists are continuing intensive research into other traditional lines of thermonuclear activity: open traps with magnetic mirrors, and heating of the plasma while compressing it by a quickly intensifying magnetic field. It is likewise essential to mention an entirely new concept in plasma physics which appeared in recent years—the interaction of a powerful laser beam with a solid, heating it to thermonuclear temperatures.

Current scientific papers from Soviet experimenters are reporting the first fusion reactions actually induced by converging electron beams on fuel pellets. Such an achievement adds another candidate to the competing approaches for extraction of nuclear energy by fusing small atoms into bigger ones. Dr. Gerold Yonasc, director of an American program in electron beam fusion at Sandia Laboratories near Albuquerque, termed the Soviet runs "a significant step." He was contacted shortly after telephoning to congratulate his Soviet counterpart, Dr. Leonid I. Rudakov, at the Kurchatov Institute in Moscow.

Russian scientists report that their converging electron beams have crushed pellets to 100 times their original density, resulting in temperatures of almost 20 million degrees Fahrenheit. The fuel inside the pellets was deuterium—a form of heavy hydrogen with basically two particles in its nucleus instead of the one found in most hydrogen.

In the Soviet experiments, enough of these nuclei fused to release more than a million neutrons. The fuel for American fusion experiments is normally a mixture of deuterium and tritium, an even heavier form of hydrogen with three nuclear particles.

Regardless of initial successes, the Soviet neutron production is reportedly far below what would be required for a practical energy source.

An interesting conclusion to what might be termed the philosophy of nuclear power from a Soviet vantage point, is reflected in several comments made to me by Dr. Zaitsev at the embassy. I was attempting to contrast the number and location of nuclear power reactors in the United States with those in the USSR. Together we scanned a U.S. map (public source material from ERDA) which detailed reactor locations and their kilowatt capacity.

After a short while, it became obvious that something was wrong. The Russian scientist was frowning as he looked at the drawing which listed some 200 reactors—and the figure has increased—scattered throughout the country, but mainly in high population areas along the East Coast.

Finally, he sighed. "So many," he began slowly, trying to piece together the right English words. "The safety must be the first plan. The human should not be sacrificed just as payment of reactors. Our breeder reactors are being built where density of population is very low. For instance, in Siberia, in the northern part of the Soviet Union."

It was obvious that Dr. Zaitsev was not being critical. Rather, he was making an observation. Of course, we may reason, Russia has the land mass and mineral resources to easily accommodate scattered reactors in such low density areas. This is very true. Still. . . .

"You must be very careful," he cautioned in a fatherly manner. The words trailed off, broken by the noise of Washington traffic, and I wearily nodded agreement.

Sitting there in the embassy of that foreign country, sipping tea and watching the late afternoon sun push through the thick curtains, so many thoughts raced and blared through my mind. The Russians have been right so many times regarding this particular technology. . . .

I glanced across at Dr. Zaitsev, who was watching me intently. These people, I thought, the Communists—our so-called enemies—working quietly and patiently like so many snails bunched together.

Meanwhile, the Free World shoots off fireworks and generally finds that such added frills don't really produce the goods.

It is difficult sometimes, many times, I think, to be an adult, because there are no easy answers to questions like "kernenergie." It was easier "way back when," I suppose, but we who are children of this twentieth century must be careful, because the human, as my friend has said, must not be sacrificed. It cannot be. . . .

There was a slight rustle then as Dr. Zaitsev, sensing the pensive mood, offered a silver bowl covered with Russian chocolate.

"For your children," he smiled. "That they might enjoy."

For all our children, the children of the future, careful thought must be given. There are hazards inherent in nuclear power, but what greater hazards are there in a highly technologicalized world faced with fuel famine? Leaving behind our seventeenth century mentality, we have the means to safely progress into the twenty-first century and beyond.

As Einstein prophesied years ago, "To the village square we must carry the facts of atomic energy and from there must come the voice of America." Is the market place then ready to awaken, or must the Grand Silence descend? The decision rests clearly in our hands.

Soviet Academician Dr. Andrei Budker and American scientist, Dr. Glenn Siborg, of the Institute of Nuclear Physics share equations during a work session.

A Soviet physicist examines a cryogen component, an installation for producing low, freezing temperatures, at the Russian Institute of Thermophysics.

Critical Mass

One aspect of the Soviet power industry is this 440,000 KW nuclear reactor belonging to the Novovoronezhskaya power system.

below: This Soviet MHD unit—converting heat directly into electrical energy—located near Moscow is under major study as an alternative power system.

EPILOGUE

". . . that everything we know is almost nothing compared with what remains to be discovered . . ."

Descartes

Humanity has ever been a race of dreamers. Self-propelled flight, healing the incurably ill, touching the man in the moon—all of these feats of science were once dreams of an ambitious Neanderthal man who even then thought to look up at the sky and began to wonder how to capture some of its power of life. Many thousands of years later we are still considering the matter. Energy is life, particularly to the race of dreamers who have created this world of machines and internal exploration. And just as the body atrophies when it is restricted to a motionless state, so society stagnates when it attempts to restrict its own natural growth—when it tries to ignore its own need for energy.

As I write this, the epilogue, we Americans find ourselves in a most perilous position regarding our fuel and energy sources. By the simple means of reading the daily newspapers one can easily see that this condition is a common one we share with the entire Western industrial world, and the question that leaps to my mind is why are so many powerful consumer groups, politicians and, yes, even the President of the United States so adamantly opposed to the development of nuclear energy? Admittedly there are dangers, but as I have indicated in this book, few accidents directly related to the operation of a nuclear plant have occurred. In point of fact, the number of deaths in coal mine disasters, off-shore oil rigs and chemical plants far exceeds the estimate of many nuclear experts who try to predict accident probabilities in the operation of nuclear facilities. Daily, huge trucks loaded with combustible fuels and lethal chemicals roll through the most populous portions of our cities—with nary a cry as to their possible danger. The possibility that a terrorist would be able to construct a bomb from hi jacked plutonium is analogous to giving a two-ton block of steel, along with a hammer and chisel, to an auto mechanic and then telling him go ahead and make a Rolls-Royce. Yet many critics would have us believe that every second-year chemical engineer or physics student has the capability to build a nuclear bomb.

Question: Do we Americans harbor an overwhelming guilt regarding nuclear energy because of the use of atomic bombs against Hiroshima and Nagasaki during the Second World War? If so, it is time to put the needs and compulsion of wartime decisions behind us and face the compulsion of our own cultural survival.

Question: Are the power and oil companies attempting to grab off all sources of energy, fossil and nuclear, by creating artificial shortages? Are they possibly backing some of the anti-nuclear groups until such time as they have gained control? Ridiculous? An example: In 1972 the construction of the Alaskan Pipeline was halted by Congressional action, accompanied by wild huzzahs from conservationists and environmentalists groups. The oil companies with their fantastic economic resources and their powerful Washington lobbies raised not a murmur. Then in 1973, the boom was lowered. Without warning we experienced an alleged shortage of gasoline and fuel oil, the so-called OPEC nations were hastily organized, prices of crude oil were increased 100, 200 percent, the cost of gasoline and heat fuel soared. Long lines formed at gas stations, while our economy was literally turned upside down—industries closed, workers thrown out of jobs, a spiraling round of inflation and unemployment. Our Western allies, particularly Italy and England, were almost bankrupted in order to pay the increased cost of overseas fuel. A new term, "double-digit inflation," entered our vocabulary of least-liked words. This, dear reader, was a classic example of economic and energy blackmail applied by "foreign powers," and if one were to project this to its reasonable conclusion a scenario would emerge in which we, the Western world, would be under the control of those whose hand is on the oil spigot. This may seem somewhat heavily loaded in an effort to promote the cause of nuclear energy, but that is not the case. I have merely tried to chronologize a series of events that has affected the economic life of the United States and its Western allies.

There are several alternatives to nuclear energy. Solar energy is most definitely a promising possibility, but still some fifteen or twenty years in the future. There is MHD, magnetohydrodynamics, which certainly can serve some of our energy needs; development of thermal energy is again another very good possibility but some years in the future. Which leaves nuclear energy as the most immediate source that can get us from this shaky present into that brave new future.

Another point we should consider is that a great many manufactured products depend upon a petroleum base. Do we have the right to burn up this most valuable resource in the form of fuel? We are told that our present coal resources will last from 400 to 500 years, but that is at the *present* level of consumption. If we turn to the use of coal to replace oil, then this estimate will be drastically reduced. In addition, coal is also the raw material for many other synthetic products that the Western world uses.

It is extremely interesting to note that whenever the proposition to curtail the use or construction of nuclear facilities was put to the voters, in all instances the public voted overwhelmingly against restrictions. A classic example was the fight waged in California, where both the pro- and anti-nuclear forces mounted huge campaigns—whose cost to both sides ran into millions of dollars—and the outcome was a clear victory for nuclear energy.

Nuclear energy is a reality in almost every industrial country in the world including the USSR and mainland China. The growth and expansion of this power source in these countries is assiduously pursued, and we Americans owe it to ourselves not to curtail or restrict our own technological and industrial growth because of artificially created fears. The very plain fact is that the United States, in order to *survive,* needs energy. And this need will increase. We cannot just turn down the lights and return to some fanciful Garden of Eden or pastoral paradise that so many environmentalists and conservationists envision. Attempting to curtail American industrial and technical power would be analogous to suddenly stopping a train hurtling down a track at 100 miles an hour—sheer disaster. We Americans have the technology, the intelligence and the ability to solve whatever safety or other problems that might arise in the use of nuclear energy.

It is incredible that this technology that has opened up the way to the stars, that has given us a level of life that exceeds almost every country in the world, this technology that produces enough food to feed our country plus a good part of the world's population, this technology that holds the promise of finally defeating the four horsemen of the Apocalypse, should now be held in such low esteem. There is only one thing that technology cannot do, and that is lift our faces from the mud and force us to look at the stars. This we ourselves must do.

In spite of the many problems that seem to beset the world, I believe that we are living in the best of all possible times, the beginning of a new era for all of mankind.

ENERGY INFORMATION RESOURCES

Project Independence Report. Federal Energy Administration, 1974. Available at $8.35, from Superintendent of Documents, U.S. Government Printing Office, Washington, D.C. 20402.

Exploring Energy Choices. A preliminary report of the Ford Foundation's Energy Policy Project. 1974, 81 pp., $.75 prepaid, from Ford Foundation, P.O. Box 1919, New York, New York 10001.

A Time To Choose: America's Energy Future. Final report by the Energy Policy Project of the Ford Foundation. 1974, 511 pp., $10.95 ($3.95 paper). Ballinger Publishing Co., 17 Dunster Street, Cambridge, Mass. 02138.

The Nation's Energy Future. A Report to the President submitted December 1, 1973, by Dr. Dixy Lee Ray, Chairman, U.S. Atomic Energy Commission. Wash-1281, 171 pp., $1.95, from Superintendent of Documents, U.S. Government Printing Office, Washington, D.C. 20402.

U.S. Energy Prospects: An Engineering Viewpoint. Task Force on Energy, National Academy of Engineering, 1974, 117 pp. Available at $5.25 from Printing and Publishing Office, NAS, 2102 Constitution Ave., N.W., Washington, D.C. 20418.

Energy: Use, Conservation and Supply. Philip H. Abelson, Ed., 1975. Available from American Association for the Advancement of Science 1515 Massachusetts Ave., N.W., Washington, D.C. 20005. $12.95 ($4.95 paper).

Energy and the Future. Allen L. Hammond, William D. Metz, and Thomas H. Maugh II. 184 pp., 1973. Available at $9.95 ($4.95 paper), from American Association for the Advancement of Science, 1515 Massachusetts Ave., N.W., Washington, D.C. 20005.

Energy: A Glossary. AAAS Publication Number 73-17. Available on request from American Association for the Advancement of Science, 1515 Massachusetts Ave., N.W., Washington, D.C. 20005.

Toward A National Energy Policy. 13 pp., available from Mobil Oil Corporation, 150 East 42nd Street, New York, New York 10017.

The Nuclear Debate: A Call to Reason. A position paper June 19, 1974, Boston, Mass., Dr. Ian A. Forbes, Marc W. Goldsmith, Dr. Joseph P. Kearney, Dr. Andrew C. Kadak, Dr. Joseph C. Turnage, and Dr. Gilbert W. Brown. 43 pp. Available from Department of Nuclear Engineering, Lowell Technological Institute, Lowell, Massachusetts.

Nuclear Power—Our Best Alternative. Winter 1974, issue of Tennessee Valley Perspective, published by the Tennessee Valley Authority (Vol. 5, No. 2). Available from Office of Information, Tennessee Valley Authority, Knoxville, Tennessee 37902.

Nuclear Power and the Environment. Questions and answers, 64 pp. Available from the American Nuclear Society, 244 East Ogden Avenue, Hinsdale, Illinois 60521.

Reactor Safety Study: An Assessment of Accident Risks in U.S. Commercial Nuclear Power Plants. WASH-1400. Norman C. Rasmussen. Summary, 29 pp., available free on request from Technical Information Center, Energy Research and Development Administration, P.O. Box 62, Oak Ridge, Tennessee 37830. Main report available at $7.60 and appendices at $1.74 from National Technical Information Service, Springfield, Virginia 22151.

Nuclear Safety. Bimonthly Technical Progress Review prepared by the Nuclear Safety and Information Center, Oak Ridge National Laboratory, for the Energy Research and Development Administration. $10.80 per year from Superintendent of Documents, U.S. Government Printing Office, Washington, D.C. 20402.

Safety-Related Occurrences In Nuclear Power Plants As Reported in 1973. An annotated bibliography covering 1053 reports of safety-related occurrences during 1973. ORNL-NSIC 114, 378 pp. Available at $15 from National Technical Information Service, Springfield, Virginia 22161.

The Energy Crisis, Radioactive Waste, Nuclear Power Plants—How Safe Are They?, Heated Water From Power Plants, The Breeder Reactor, and *Fusion: Energy Source for the Future.* Pamphlets available free of charge from Office of Information Services, Energy Research and Development Administration, Washington, D.C. 20545.

The Nuclear Controversy. Analyzed by Dr. Ralph E. Lapp, nuclear physicist and journalist. 93 pp. Available at $7.95, from Fact Systems, 537 Steamboat Rd., Greenwich, Conn. 06830.

Breeder Backgrounder and other materials on the Clinch River Breeder Reactor Project. Available from: Project Management Corporation, 617 Walnut Street, Knoxville, Tennessee 37902.

Everything You Always Wanted to Know About Shipping High-Level Nuclear Wastes. 49 pp. WASH-1264. For sale by Superintendent of Documents, U.S. Government Printing Office, Washington, D.C. 20402.

Photo Sources

United States Navy, Chief of Information, Washington, D.C.
United States Air Force, Systems Command, Arnold Engineering & Development Center, Tullahoma, Tennessee
Tennessee Valley Authority, Knoxville, Tennessee
Oak Ridge National Laboratory, Tennessee
Nuclear Regulatory Commission, Washington, D.C.
Energy Research & Development Administration, Washington, D.C.
Los Alamos Scientific Laboratory, New Mexico
San Onofre Nuclear Plant, California
Embassy of the Union of Soviet Socialist Republics, Washington, D.C.
Novosti Press Agency, Moscow, USSR
Naval Research Laboratory (Lloyd Carter), Washington, D.C.
People's Republic of China Liaison Office, Washington, D.C.
Federal Energy Administration, Washington, D.C.
Federal Bureau of Investigation, Department of Justice, Washington, D.C.

Index

Advanced Energy Systems, 36
Air Pollution Control, 37
Alaska, 31
Alaskan North Slope, 31
Allied Chemical Corporation, 48
American Enterprise Institute, 365
American Health Physics Society, 46
American Nuclear Society, 54
American Physical Society, 28
Annular Core Pulse Reactor, 238
Appalachian, 214
Arab, 36, 365
Argonne National Laboratory, 381
Artsimovich, Lev, 227, 389
Associated Press, 279
Atlantic Offshore, 31
Atomic Energy Act of 1954, 181
Atomic Energy Commission, 16, 21, 42, 53, 261-262, 266, 274, 279, 288-289
Atomics International, 54, 211, 214

Bartlett, Senator Dewey, 268
Basov, Dr. N. G., 381
Behnke, W. B., 211
Bell-System, 237
Bethlehem Steel, 38
Bibb, Dr. William R., 21-27, 153-154, 226
Big Rock Point Nuclear Plant, 178
Bistowish, Dr. Joseph, 49
Blackmarket, 276-279
Bogart, Larry, 15, 16, 48-49
Boiling Water Reactor (BWR), 178
Breeder, 87, 203-216
Brookhaven Report, 115
Browns Ferry Nuclear Plant, 52-53, 103-106, 152
Bulgaria, 364
Business & Professional Women's Club, 270

Calvert Cliffs Decision, 100-101
Canada, 24, 77, 363, 368-369, 372
Carbide, Union, 226
Cerenkov Reaction, 154
Chih-hsiang, Chang, 365
China, People's Republic of, 106, 132, 171-172, 278, 365
China Syndrone Theory, 15, 115-116
Christophers, 29
Church, Senator Frank, 290
CIA, 45
Clarke, Dr. John F., 226-227, 380, 382
Clean Air Act, 62, 69
Clinch River, 16, 42, 43, 153, 204-205, 211-212, 214
Coal, 33-34, 80, 82
Coalition of Labor Union Women, 290
Cohen, Dr. Bernard L., 28, 133
Colorado Plateau, 34-35
Comey, David, 46, 104
Commission of the European Communities, 371
Commonwealth Edison, 179, 212
Confinement, Inertia, 235
Connecticut Yankee Atomic Power Plant, 179
Council, Citizens Energy, 15, 48
Cronkite, Walter, 103
Culler, Floyd L., 101-114, 158
Czechoslovakia, 364

Dallas Morning News, 286
Dallas Times Herald, 286
Darling, James P., 43
Davidson, Dr. Robert H., 53-54
de Lorenzo, Dominic, 18
Deep, Farris, 79
Defense-in-Depth, 101
Deuterium, 222-223, 230, 235, 237, 390, 392
Dimona, 365

Dingell, Rep. John, 278
Dolphin, Dr. G. W., 136
Dominican Republic, 169
Doublet, 234-235
Dresden Nuclear Power Plant, 179
Dubna, 15
Dunlop, John T., 30

Eibenschutz, Juan, 369
Einstein, Albert, 225, 229, 393
Egypt, 365
Electromagnetic radiation, 55
Emergency Core Cooling System, (ECCS), 47, 102, 104, 112, 216
Energy Analysis, Institute for, 133
Energy Research & Development Administration (ERDA), 21-22, 29, 33, 42, 52, 61-62, 73, 76, 81, 90, 150, 206, 211, 213, 221, 226, 232-235, 237, 274, 392
Energy Resources Council, 61
Energy Technology Conference, 33, 361
Environmental Impact Statement, 43, 55, 213
Environmental, National, Policy Act, (1969), (NEPA), 42, 215
Environmental Protection Agency, 49, 73, 79
Environmental Quality Analysts Inc., 152
Extremely Low Frequency Communication, (ELF), 55
Exxon, 71

Fast Flux Test Facility (FFTF), 206
Federal Bureau of Investigation (FBI), 267-268, 290, 292, 379, 382-383
Federal Energy Administration, 61, 65-66, 73, 75
Fenstermacher, Dr. Charles, 223, 236, 239
Fermi, Dr. Enrico, 205
Fission, 14, 41, 110, 152-155, 225, 227, 370-373
Fossil fuels, 28
France, 68, 142, 153, 362, 372
French Atomic Energy Laboratory, 372
Fundy, Bay of, 76
Fusion, 14, 32, 152, 221-239, 372-373, 380, 390, 392

Garching, 372
Gas, 82
General Electric, 38, 130, 178, 211
Geneva, 391
Geothermal, 32, 73-74
Germany, Federal Republic of, 361-363, 366, 372
Geysers, 73-74
God, 98, 103, 116
Goldberg, Rube, 46
van Gool, W., 370
Government Institutes, Inc., 44
Government Transportation Permit, 158
Gravel, Senator Mike, 79
Green, Jim, 104
Green River Formation, 34-35
Gulf, 71
Gulf General Atomic, 178

Hanford, 33, 156, 208, 274
Hanford Engineering Development Laboratory, 206
Hanford Special Projects Shipment, 274
Hartsville Nuclear Plant, 30, 43
Health Physics Society, 49
Herwig, Dr. Lloyd O., 81
High Temperature Gas-Cooled Reactor (HTGR), 179
Hirsch, Dr. Robert L., 221, 223, 228, 233, 237
Holloway, Admiral James L., 180-181
Honickers, 50-51
Hope, Bob, 261
House-Senate Joint Atomic Energy Committee, 30
House Task Force on Energy, 32
Hughes, Nat B., 77
Humboldt Bay Power Plant, 178
Hungary, 364
Hydrogenation, HTG, 34

Implosion rocket effect, 235
Indian Point, 179
International Atomic Energy Panels, 102

International Energy Agency, 363, 366
Iodine, 152
Iraq, 365
Israel, 278, 365
Italy, 372

Japan, 363, 367-368, 372-373
Jesus, 169
Jewish, 169
Johnson Space Center, 88
Justice Department, 292

Kansas, Lyons, 150
Kaplan Fund, Inc., of New York, 47
Kelley, Clarence, 290
Kelp beds, 176
Kerr, Senator Robert S., 262
Kerr-McGee, 261
KGB, 385
Knuth, Dr. Donald F., 105
Krokhin, Dr. O. N., 381
Kruger, Paul, 76
Krypton, 152
Kurchatov Institute of Atomic Energy, 390, 392

Lapp, Dr. Ralph, 47
Larson, Dr. Ron, 83
Laser, 230, 391
Lawson's Criteria, 235
Lebedev Institute, 372
Leighton, Leonard H., 366-367
Leningrad Nuclear Power Plant, 388
Light Water Reactor (LWR), 207
Liquid Metal Fast Breeder Reactor (LMFBR), 203-216, 388-389
Lithium, 237
Livermore, 232-233, 238
Los Alamos Scientific Laboratory (LASL), 94, 130, 136, 223, 231, 232, 233, 236, 380-381, 383
Loss of Criteria, 236
Lynn, Jeri, 104

McBride, Charles, 168
McCormack, Congressman Mike, 32-34, 65-66, 89, 149, 203
McGee-Kerr, 261-293
McGee, Dean, 262

MHD, 90A, 90B, 371, 387
Magnetic Mirror, 230, 391
Manchester Guardian, 267
Manhattan Project, 210, 236
Manhattan Study, 137, 142
Marine Biological Consultants, Inc., 152
Masters of Deceit, 380
Mazzocchi, Tony (Natural Resources Defense Council), 284-288
Meltdown, 103
Metcalf, Senator Lee, 290
Methaqualone, 267, 273, 283-284
Mexico, 369
Middle East, 36, 86
Mobil, 71-72
Musgrave, Story (Astronaut), 88

Nader, Ralph, 46-50, 290
Nagasaki-Hiroshima, 43
Nashville! magazine, 15, 16
National Aeronautics & Space Administration (NASA), 87-89
National Airport, 382
National Council on Radiation Protection & Measurements, 109
National Council of Science & Technology, 369
National Energy Research Steering Group, 370
National Reactor Testing Center, 114, 156
National Safety Council, 42
National Science Foundation, 76, 87
National Security Council, 279
Natural Resources Defense Council, 140
Nautilus, 180
Navy, 54, 95, 104, 180-181, 234
Nemzek, Thomas A., 41, 211
Neporozhny, Dr. Pyotr, 388
Netherlands Energy Development, 370
Netherlands Nuclear Research Center, 371
New Mexico, 150, 262
New Times, 267
Newton's Third Law, 236
Nord Atomic Power Station, 363
Novosti, 373, 386
Nuclear Fuel Services, 278
Nuclear Regulatory Commission (NRC), 2, 52-53, 73, 97, 100,

104, 105, 109, 117, 151-152, 205, 275-277, 279
Nuclear Safety Information Center, 53

Oak Ridge, 21, 27, 36, 94, 97, 111, 130, 136, 150, 204, 208, 232, 380-381
Obninsk, 381, 388
Observer, The, 267
Ohkawa, Dr. Tihiro, 234
Oklahoma, 261
Oklahoma Highway Patrol, 267-272, 283, 287
Olos, Fred, 221
Olson, Lawrence (Special Agent), 268
OPEC (Organization of Petroleum Exporting Countries), 64, 67
Ormak, 227, 233, 380
Orwell, George, 383

Packard, Vance, 385
Paducah, Kentucky, 208
Palomares, 134
Paris Match, 267
Peach Bottom Atomic Power Station, 178
Pendleton, Camp, 167
Pentagon, 54, 171
Photosynthesis, 239
Pinch, Theta, 230-231, 381
Pipkin, A. O., 284-287
Plankton, 175
Plasma, 222, 227-228
Plutonium, 93, 109, 127-143, 203, 206, 228, 289, 264
Pollution, 82
Pollution Engineering magazine, 44
Postma, Dr. Herman, 27, 47-48
Power Engineering magazine, 44, 221
Pranger, Dr. Robert J., 365
Pressurized Water Reactor (PWR), 179
Princeton, 97, 233

Radiation, 46
Ramey, James T., 22
Range, Wayne, 168
Rasmussen, Norman C., 99
Raytheon Company, 88
Reactor, 39, 46, 91-119, 129, 203, 209-216, 264, 367, 392
Re-cycling, 37
Reagan, Governor Ronald, 170
Research/Development magazine, 44
Ribe, Dr. Fred, 231
Richmond, Dr. Chester R., 136-143
Rickover, Vice-Admiral Hyman G., 180-181
Rolling Stone, 127, 267
Rosenberg, 290
Rosenthal, Dr. Murray W., 36
Rudakov, Dr. Leonid I., 392
Russians, 132, 142, 171-172, 209, 222, 227, 228, 363, 371-373, 379-393

Sakakura, Dr. Shogo, 367-368
Salt beds, 149
San Clemente, 174
Sandia, 236, 238, 391
San Diego Gas and Power, 167
San Onofre Nuclear Plant, 117, 152, 167-181
Scaife Nuclear Laboratories, 28
Schofield, Dr. G. B., 136
Science magazine, 141
Scientific American, 45
Scyllac, 231
Seaborg, Dr. G. T., 138
Seafarer Project, 54-56
Seamans, Robert C. Jr., 90
Shafranov, Dr. Vitaly, 389
Shale oil, 34, 35
Shell, 71
Shevchenko Nuclear Power Station, 389
Silkwood, Karen, 261-293
Simon, William, 61
Skylab, 82
Smith, Charles H., 368-369
Smith, Harold, 288-289
Smog, 38
Sodium, 207-208, 211, 367
Solar, 32, 46, 81
Solid waste plants, 79-80
Sol-R-Tech, 83
Southern California Edison (SCE), 38, 74-75, 77, 167, 173
Southerners for Safe Power, 50
Southwest Experimental Fast Oxide Reactor, 211
Soviet Physics, 381

Stellarator, 381
Sternglass, Dr. E. J., 49
Stevens, Turney, 16
Stockton, Peter, 292-293
Stoecker, Dr. H. J., 361-362
Strachan, Alex Ronald, 173-176
Strelkov, Dr. Vyacheslav, 389
Sunshine Project, 367

Tahtinen, Dr. Daler, 365
Tamplin, Arthur, 46
Taylor, T. B., 134
Tennessean, The (newspaper), 45, 50, 103, 290-291, 293, 373
Tennessee Valley Authority (TVA), 16, 43, 52, 77-78, 103, 211-212
Terrorism, 45, 93-94, 135, 158, 209, 224, 265
Thermal, 81, 152, 174
Tides, 76, 387
Tokamak, 222, 226, 230, 232, 372, 380, 390
Tomorrow's Weapons, 131
Total environmental action, 82
Transuranium Registry, 132
Transchel, Milton H., 168-170, 181
Tritium, 152, 222-223, 230, 235, 237
Truman, Harry, 73
Turner, E. Winslow, 290, 293

Udall, Congressman Morris, 71
Union, The Oil, Chemical, and Atomic Workers', 266-268
United Auto Workers, 290
United Kingdom, 134, 136, 152, 363, 366, 367
United Press International, 290
Uranium, 82, 91, 95, 109-110, 152, 206, 228, 262-265, 361-362

Vallecitos Nuclear Center, 130
Vanguard I, 82
Vermont Public Service Corporation System, 79
Vietnam, 169

Wagner, Aubrey, 52, 157
WASH-740, 113-114
Wastes, 45, 80, 147-159, 228
Watson, J. E., 211, 213
WCBS, 103
Western Electric, 237
Westinghouse Electric, 180, 214-215
Wind power, 79
Wodka, Steve, 283
World War II, 37
WSM, 103
Wyoming, 262

Xenon, 152

Yonasc, Dr. Gerold, 391

Zaitsev, Dr. Sergey, 384-386, 392-393
Zarb, Frank, 61-70
Zen, 21
Zooplankton, 175
Zumwalt, Admiral Elmo R., 180